Protein Function Prediction for Omics Era

Daisuke Kihara
Editor

Protein Function Prediction for Omics Era

Springer

Editor
Daisuke Kihara
Department of Biological Sciences/
Computer Science
Purdue University
Hockmyer Hall
249 S. Martin Jischke Drive
47907-2107 West Lafayette
IN, USA
dkihara@purdue.edu

ISBN 978-94-007-0880-8 e-ISBN 978-94-007-0881-5
DOI 10.1007/978-94-007-0881-5
Springer Dordrecht Heidelberg London New York

Library of Congress Control Number: 2011925680

Printed on acid-free paper

Springer is part of Springer Science+Business Media (www.springer.com)

Preface

Elucidation of protein function has been a central question in molecular biology, genetics, and biochemistry. The importance of computational function prediction is increasing because more and more genome sequences are being determined by genome sequencing projects. Recent advancement of sequencing technologies further achieves surprisingly fast speed for sequencing complete genomes. It is clear that genome sequencing will become a routine in biological and medical studies in very near future. In addition, it is noteworthy that structural genomics projects have been launched for over few years, which are producing an increasing number of protein structures of unknown function. Besides the flood of protein sequences and structures, other types of large scale biological data, including protein–protein interaction data, gene expression data, are awaiting biological interpretation. Thus, the post-genomics era has entered to the second phase, the omics era, when various types of large-scale biological data are generated and referred to each other toward systems level understanding of organisms and life. Obviously function prediction is indispensable for capitalizing the rich sources of the omics data.

It has been 20 years since FASTA and BLAST, the most commonly used homology search tools, were developed. As exemplified by the fact that the first complete genome was finished 6 years after the two homology search tools were developed, the circumstance of biological research has dramatically changed since then. The appearance of omics data has brought different needs and sources for function predictions. Conventional use of homology search methods is not necessarily most suitable for analyzing large scale data. For analyzing data which have many genes included, large coverage in function annotation is essential. For biological interpretation of large-scale data, detailed biochemical function assignment to genes is not always necessary. A broad class of function, or low-resolution function, is still helpful to understand functional unit of genes and speculate biological background of coordinated behaviour of genes. Omics data is not only the targets for analyses, but also provide additional sources for elucidating functional relationships between genes. Thus, in recent years we observe emerging development of computational function prediction methods, which use various sources and techniques to address the needs of biology of this century.

In this book, we provide a snapshot of this emerging field by providing reviews of notable computational methods and resources. In Chapter 1, we state the current situations of protein function prediction and overview computational frameworks. Chapters 2, 3, 4, and 5 address sequence-based function prediction methods. In Chapter 2, Chitale and myself review two methods we have developed, which exploit function information from PSI-BLAST searches more thoroughly than conventional usage. In Chapter 3, Kim and his colleagues discuss the use of conserved gene clusters for genome annotation. Chapter 4 by Uchiyama discusses issues in the ortholog classification and introduces an algorithm for ortholog group construction and a database for comparative genomics for microbial genomes. In Chapter 5 Livesay et al. present a sequence-based functional site prediction method, which identifies a local region as functional site whose mutation pattern is restricted by phylogenetic constraints.

The next five chapters, Chapters 6, 7, 8, 9, and 10, address structure-based function prediction. In Chapter 6 Orengo and her colleagues analyze structural conservation in protein superfamilies and describe an approach for assigning functional subfamilies based on global structure comparisons between inter and intra superfamilies. In the subsequent chapter, the Liang group describes global and local structure alignment methods which align structures in sequence-independent manner. The local alignment method is applied to identify conserved atoms in functional pockets of a family of protein structures (Chapter 7). Chapter 8 by Chikhi, Sael, and myself describes pocket shape representation and comparison methods which use two dimensional and three dimensional moments. The methods are applied for predicting binding ligand molecules for a pocket. Chapter 9 by Ahmad overviews computational methods for DNA binding sites prediction ranging from available datasets, computational techniques, to properties of proteins that can be used for input for prediction. In the subsequent chapter, Ondrechen and her colleagues describe a method for predicting functionally important residues in proteins by computing theoretical titration curves for ionisable residues (Chapter 10).

Finally, we move on to omics data driven approaches and omcis data resources in Chapters 11, 12, 13, 14, and 15. In the first chapter in this section, Chapter 11, Kinoshita and Obayashi discuss the use of protein tertiary structure, particularly protein surface shape, to predict molecular function and to use protein–protein interaction and expression data for predicting cellular function of proteins. Chapter 12 by Tian et al. overview types of omics data as well as computational approaches for integrating various omics data for function prediction. In Chapter 13, Wong and his colleagues discuss the use of indirect interactions in addition to direct interactions in protein–protein interaction networks for function prediction. The idea was also appplied for protein complex prediction and cleansing interaction data. Chapter 14 by the Kanehisa group overviews KEGG and GenomeNet resources, which contain genomic, chemical, and systems (e.g. pathways) information of organisms. In particular they discuss their recent developments including databases of plant secondary metabolites, crude drug molecules, and prediction tools for metabolic pathways and enzymatic reactions. In the last chapter, Chapter 15, Mori, Wanner and their colleagues describe GenoBase, which contains high-throughput experimental data for *E. coli*, including the single-gene deletion library, phenotype screening, genetic interactions, and protein–protein interactions.

There are many other existing methods and databases and new approaches are being published month by month in this active research field. Nevertheless, chapters in this book cover almost all the types of function prediction approaches. Thus, I believe this book successfully provides comprehensive overview of this exciting and important field. I believe this book is informative for those who are interested in developing new approaches and also for biologists who are looking for tools and resources for elucidating protein function. In closing, I would like to thank all the authors of chapters in this book. It is very fortunate to have leading experts of the field as the authors. I am also thankful to the editors in Springer, Dr. Meran Owen and Ms. Tanja van Gaans, for their patience and professional work. At last, I would like to share with the readers the happiness and the excitement to observe dramatic changes of biology in the omics era, which are made possible by brilliant ideas and dedicated efforts by researchers across the world.

West Lafayette, Indiana Daisuke Kihara

Contents

Contributors

Shandar Ahmad National Institute of Biomedical Innovation, Ibaraki, Osaka, Japan, shandar@nibio.go.jp

Dukka Bahadur KC Department of Bioinformatics and Genomics, University of North Carolina at Charlotte, Charlotte, NC, USA, dukkakc@gmail.com

Rayan Chikhi École Normale Supérieure de Cachan, Bruz, Brittany, France, rayan.chikhi@ens-cachan.org

Meghana Chitale Department of Computer Science, College of Science, Purdue University, West Lafayette, IN, USA, mchitale@purdue.edu

Hon Nian Chua Institute for Infocomm Research, Singapore, hnchua@i2r.a-star.edu.sg

Alison Cuff Department of Structural and Molecular Biology, University College London, London, UK, cuff@biochem.ucl.ac.uk

Bhaskar DasGupta Department of Computer Science, University of Illinois Chicago, Chicago, IL, USA, dasgupta@cs.uic.edu

Benoit Dessailly Department of Structural and Molecular Biology, University College London, London, UK, benoit@biochem.ucl.ac.uk

Xinran Dong School of Life Sciences, Fudan University, Shanghai, China, xrdong@fudan.edu.cn

Joe Dundas Bioinformatics Program, Department of Bioengineering, University of Illinois Chicago, Chicago, IL, USA, jdunda1@uic.edu

Susumu Goto Bioinformatics Center, Institute for Chemical Research, Kyoto University, Uji, Japan, goto@kuicr.kyoto-u.ac.jp

Minoru Kanehisa Bioinformatics Center, Institute for Chemical Research, Kyoto University, Uji, Japan; Human Genome Center, Institute of Medical Science, University of Tokyo, Minato-ku, Tokyo, Japan, kanehisa@kuicr.kyoto-u.ac.jp

Daisuke Kihara Department of Biological Sciences; Department of Computer Science; Markey Center for Structural Biology, College of Science, Purdue University, West Lafayette, IN, USA, dkihara@purdue.edu

Sun Kim School of Informatics and Computing, Indiana University, Bloomington, IN, USA, sunkim2@indiana.edu

Kengo Kinoshita Graduate School of Information Science, Tohoku University, Sendai, Miyagi, Japan; Institute for Bioinformatics Research and Development, Japan Science and Technology Agency, Chiyoda-ku, Tokyo, Japan, kengo@ecei.tohoku.ac.jp

Masaaki Kotera Bioinformatics Center, Institute for Chemical Research, Kyoto University, Uji, Japan, kot@kuicr.kyoto-u.ac.jp

David La Department of Biological Sciences, Purdue University, West Lafayette, IN, USA, davidla@purdue.edu

Heewook Lee School of Informatics and Computing, Indiana University, Bloomington, IN, USA, heewlee@indiana.edu

Joslynn S. Lee Department of Chemistry and Chemical Biology, Northeastern University, Boston, MA, USA, lee.jos@husky.neu.edu

Jie Liang Bioinformatics Program, Department of Bioengineering, University of Illinois at Chicago, Chicago, IL 60607, USA, jliang@uic.edu

Guimei Liu School of Computing, National University of Singapore, Singapore, liugm@comp.nus.edu.sg

Dennis R. Livesay Department of Bioinformatics and Genomics, University of North Carolina at Charlotte, Charlotte, NC, USA, drlivesa@uncc.edu

Hirotada Mori Graduate School of Biological Sciences, Nara Institute of Science and Technology, Ikoma, Nara, Japan; Institute for Advanced Biosciences, Keio University, Tsuruoka, Yamagata, Japan, hmori@gtc.naist.jp

Toru Nakayashiki Graduate School of Biological Sciences, Nara Institute of Science and Technology, Ikoma, Nara, Japan, nakayashiki@bs.naist.jp

Takeshi Obayashi Graduate School of Information Science, Tohoku University, Sendai, Miyagi, Japan; Institute for Bioinformatics Research and Development, Japan Science and Technology Agency, Chiyoda-ku, Tokyo, Japan, obayashi@ecei.tohoku.ac.jp

Mary Jo Ondrechen Department of Chemistry and Chemical Biology, Northeastern University, Boston, MA, USA, M.Ondrechen@neu.edu

Christine Orengo Department of Structural and Molecular Biology, University College London, London, UK, orengo@biochem.ucl.ac.uk

Vikas Rao Pejaver School of Informatics and Computing, Indiana University, Bloomington, IN, USA, vpejaver@indiana.edu

Oliver Redfern Department of Structural and Molecular Biology, University College London, London, UK, ollie@biochem.ucl.ac.uk

Ren Ren School of Life Sciences, Fudan University, Shanghai, China, 06300720116@fudan.edu.cn

Lee Sael Department of Computer Science, College of Science, Purdue University, West Lafayette, IN, USA, lee399@purdue.edu

Rikiya Takeuchi Graduate School of Biological Sciences, Nara Institute of Science and Technology, Ikoma, Nara, Japan, r-takeuchi@bs.naist.jp

Weidong Tian School of Life Sciences, Fudan University, Shanghai, China; Institute of Biostatistics, Fudan University, Shanghai, China, weidong.tian@fudan.edu.cn

Yukako Tohsato Department of Bioscience and Bioinformatics, Ritsumeikan University, Kusatsu, Shiga, Japan, yukako@sk.ritsumei.ac.jp

Toshiaki Tokimatsu Bioinformatics Center, Institute for Chemical Research, Kyoto University, Uji, Japan, tokimatu@kuicr.kyoto-u.ac.jp

Ikuo Uchiyama Laboratory of Genome Informatics, National Institutes of Natural Sciences, National Institute for Basic Biology, Nishigonaka 38, Myodaiji, Okazaki, Aichi 444-8585, Japan, uchiyama@nibb.ac.jp

Barry L. Wanner Department of Biological Science, PurdueUniversity, West Lafayette, IN, USA; Institute for Advanced Biosciences, Keio University, Tsuruoka, Yamagata, Japan, blwanner@purdue.edu

Limsoon Wong School of Computing, National University of Singapore, Singapore, wongls@comp.nus.edu.sg

Natsuko Yamamoto Graduate School of Biological Sciences, Nara Institute of Science and Technology, Ikoma, Nara, Japan, nyamamot@lif.kyoto-u.ac.jp

Yuanpeng Zhou School of Life Sciences, Fudan University, Shanghai, China, 07300700055@fudan.edu.cn

Computational Protein Function Prediction: Framework and Challenges

Meghana Chitale and Daisuke Kihara

Abstract Large scale genome sequencing technologies are increasing the abundance of experimental data which requires functional characterization. There is a continually widening gap between the mounting numbers of available genomes and completeness of their annotations, which makes it impractical to manually curate the genomes for function information. To handle this growing challenge we need computational techniques that can accurately predict functions for these newly sequenced genomes. In this chapter we focus on the framework required for computational function annotation and the challenges involved. Controlled vocabularies of functional terms, e.g. Gene Ontology, MIPS functional catalogues, Enzyme commission numbers, form the basis of prediction methods by capturing the available biological knowledge in the form, suitable for computational processing. We review functional vocabularies in detail along with the methods developed for quantitatively gauging the functional similarity between the vocabulary terms. We also discuss challenges in this area, first pertaining to the erroneous annotations floating in the sequence database and second regarding the limitations of the functional term vocabulary used for protein annotations. Lastly, we introduce community efforts to objectively assess the accuracy of function prediction.

Introduction

With the advances in technology, whole genome sequencing for new organisms is no longer an enormous project. Numbers of genomes are being sequenced every year adding the tremendous amount of data available for computational investigators. As shown in Fig. 1, the number of entries of genomes in KEGG database [1] have almost doubled form year 2007 (~ 600 genomes) to year 2010 (~1,200 genomes).

D. Kihara (✉)
Department of Biological Sciences; Department of Computer Science; Markey Center for Structural Biology, College of Science, Purdue University, West Lafayette, IN 47907, USA
e-mail: dkihara@purdue.edu

D. Kihara (ed.), *Protein Function Prediction for Omics Era*,
DOI 10.1007/978-94-007-0881-5_1,

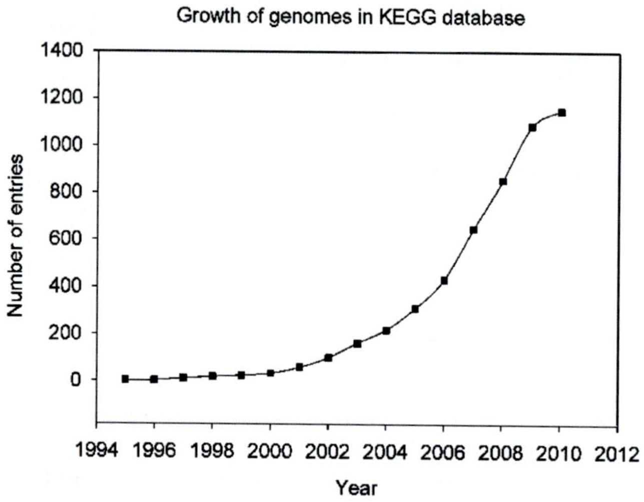

Fig. 1 Growth of genomes in KEGG database from year 1995 till 2010. Yearly release information of KEGG data was obtained from GenomeNet (http://www.genome.jp/en/db_growth.html)

The pace of accumulating sequence data will only increase, in fact, the new generation technology can sequence microbial genome within a couple of days [2, 3].

However, it is still a daunting task to correctly assign functional annotations to these newly sequenced genomes based on their sequence information. It is not feasible to conduct conventional experimental procedures on this entire stockpile of sequences for recovering the functional information, and this has triggered the need for methods that can consistently assign functions to unknown proteins [4–8]. Conventionally in this scenario researchers have focused on using homology or sequence similarity to transfer annotations to newly sequenced proteins using popular homology search algorithms such as BLAST [9] and FASTA [10, 11]. Although considering homology is a genuine way of inferring function in the light of evolution, practically, it is not always trivial to extract correct function information from a sequence database search result. Another weakness of the conventional homology searches is that a considerable portion of genes in a genome are left as unannotated. In Fig. 2, we have analyzed the number of annotated genes in the genome sequences taken from the KEGG database [1] (version March 2010). We have examined the genomes to separate the number of genes that have unknown annotations characterized by keywords mentioned in the caption for the figure. This gives us a crude idea about the percentage of unknown genes in each genome. It can be seen from Fig. 2 that for around 50% of genomes in the database we know functional characteristics of less than 60% of genes in there. Even for well studied model organisms such as *Saccharomyces cerevisiae* (82.4% annotated), *Escherichia coli K-12 MG1655* (64.9% annotated), *Arabidopsis thaliana* (66.3% annotated), a significant number of genes have no annotation. Therefore, new methods in this area are required to improve the function prediction accuracy as well as the genome annotation coverage.

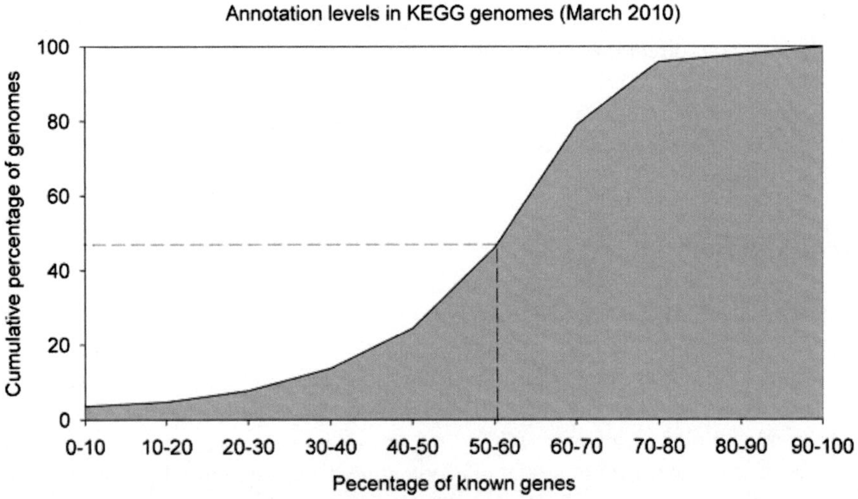

Fig. 2 Annotation levels of genomes in KEGG database. 1,172 genomes in KEGG database were analyzed to separate the number of annotated genes from unknown genes (entries in the database annotated with terms "hypothetical", "putative", "unknown", "uncharacterized", "predicted", "no hits", "codon recognized", "expressed protein", and "conserved protein"). The figure shows cumulative percentage of genomes having specified percentage of annotated genes

As the first chapter in this book, we explain the fundamental information, which lays the framework of computational protein function prediction. We first summarize controlled functional vocabularies and evaluation measures for accuracy of protein function prediction. Along with this, we would like to draw readers' attention to challenges in this area, first pertaining to the erroneous annotations floating in the sequence database and second regarding the limitations of the functional term vocabulary used for protein annotations. Lastly, we introduce community efforts to objectively assess the accuracy of function prediction.

Controlled Functional Vocabularies

For managing computational function prediction we need to transform the descriptive biological knowledge into qualitative and quantitative models, which requires robust and accessible biological information system. Protein functions or annotations have long been described with vocabularies that are conventionally used within each research community or research group. Thus, there have been cases that essentially same annotations are described with different terms across different species and research communities. However, such situations hinder computational handling of functional information, including extraction of function information of genes from databases and summarizing such information to predict function. A practical solution for this is to unify the functional terms used for functional annotation of genes. In recent years controlled sets of functional vocabularies have been developed along this direction. Below we describe several ontologies, including Gene Ontology (GO) [12], Enzyme Commission (EC) number, [13], MIPS functional catalogue [14] (FunCat), Transporter Classification System [15], KEGG orthology [16], and the other efforts of constructing ontologies.

Gene Ontology

The Gene Ontology (GO) Consortium [17] of collaborating databases has developed a structured controlled vocabulary to describe gene function. GO vocabulary terms are arranged in a hierarchical fashion using a Directed Acyclic Graph (DAG) and are separated into three categories: Biological Process (BP), Molecular Function (MF), and Cellular Component (CC). One or more terms from each category can be used to describe a protein. Cellular component indicates to which anatomical part of the cell the protein belongs to, for example, *ribosome (GO:0005840)* or *nucleus (GO:0005634)*. Biological process terms indicate assemblies of molecular functions which achieve a well defined task through a series of cellular events. Examples of biological processes are *carbohydrate metabolism (GO:0003677), regulation of transcription (GO:0045449)* etc. Molecular functions represent activities carried out at molecular level by proteins or complexes, for example, *catalytic activity (GO:0003824)* or *DNA binding (GO:0003677)* etc. Thus each GO term will have a category and an identifier in the format GO:xxxxxxx associated with it, along with a term definition to explain the meaning of the term. For example, term *protein binding* is referred using identifier *GO:0005515* and its definition says following *Interacting selectively and non-covalently with any protein or protein complex*. The vocabulary is arranged as a DAG where each term can have one or more parents. Figure 3 represents the tree structure obtained for the term *hemoglobin binding* showing all its parents till the root term *all*. As you go deeper in the hierarchy the terms become more specific.

All terms in GO other than the root term have either *is-a, is_part_of, positively regulates, or negatively regulates* relationship with some other more general term. For example as shown in Fig. 4 the term *glucose transport (GO:0015758)* is_a

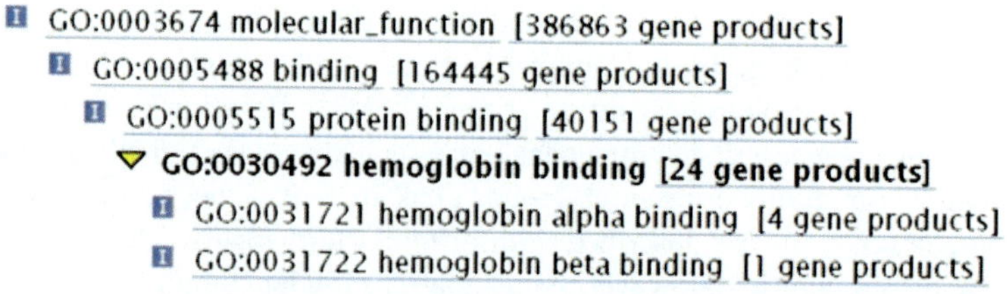

Fig. 3 Structure of Gene Ontology for term *hemoglobin binding* displayed using AmiGO browser (http://amigo.geneontology.org/cgi-bin/amigo/go.cgi) for GO terms. Against each term the number of gene products that are annotated with the given term in the GO database, is displayed

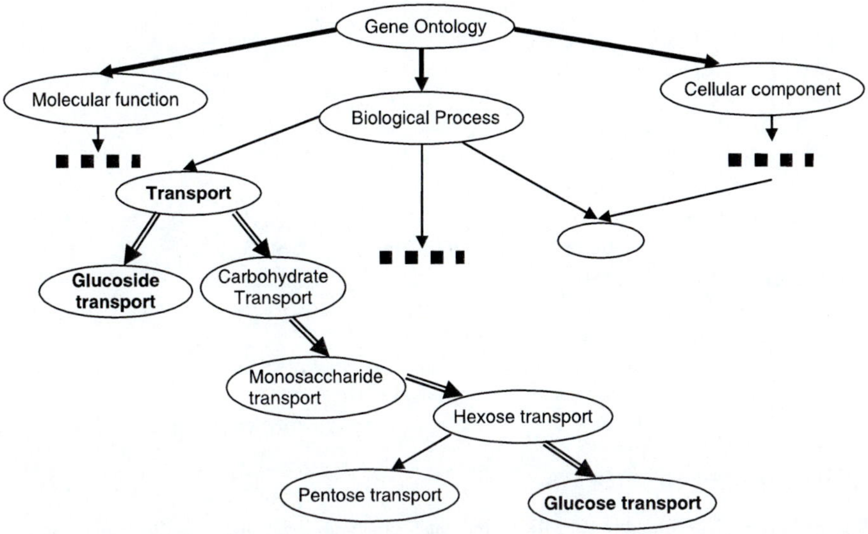

Fig. 4 Partial Gene Ontology hierarchy describing the ancestors of terms *Glucoside transport* and *Glucose transport*. Double lined arrows show the path to the Lowest Common Ancestor (LCA) of the two terms

Hexose transport (GO:0008645), which ultimately is_a *transport (GO:0006810).* Due to this relationship when a protein is annotated by term *X* then it is automatically annotated by all ancestor terms of *X* which are basically less specific descriptions of *X*. Similarly, some more relationships have been defined in GO, e.g. *B* is part_of *A*, which implies that when *B* exists it is part of *A*. For example, *mitochondrial membrane (GO:0031966)* is part of *mitochondrial envelope (GO:0005740)*. *Regulates* relationship is used in GO to capture the fact that one process can directly affect the manifestation of another process; this relationship has two sub-relations *positively regulates* and *negatively regulates* to capture the specific forms of regulation.

Association between a gene product and its GO annotation is generally based on one or more supporting evidences. GO has defined the evidence codes that help capture information about the source from which this association is obtained (http://www.geneontology.org/GO.evidence.shtml). Inferred from Electronic Annotation (IEA) is the only evidence code that is not reviewed by a curator indicating that assignment of annotation to the gene product is automatic. All curator-assigned evidence codes fall into one of the four categories; (1) experimental (e.g. Inferred from Direct Assay (IDA), Inferred from Genetic Interaction (IGI) etc), (2) computational analysis (e.g. Inferred from Sequence or structural Similarity (ISS), Inferred from Genomic Context (IGC)), (3) author statement (Traceable Author Statement (TAS), Non-traceable Author Statement (NAS)), and (4) curatorial statement (Inferred by Curator (IC) and No biological Data available (ND)). It should be noted that evidence codes do not indicate quality of annotation but only provide information about the source of annotation.

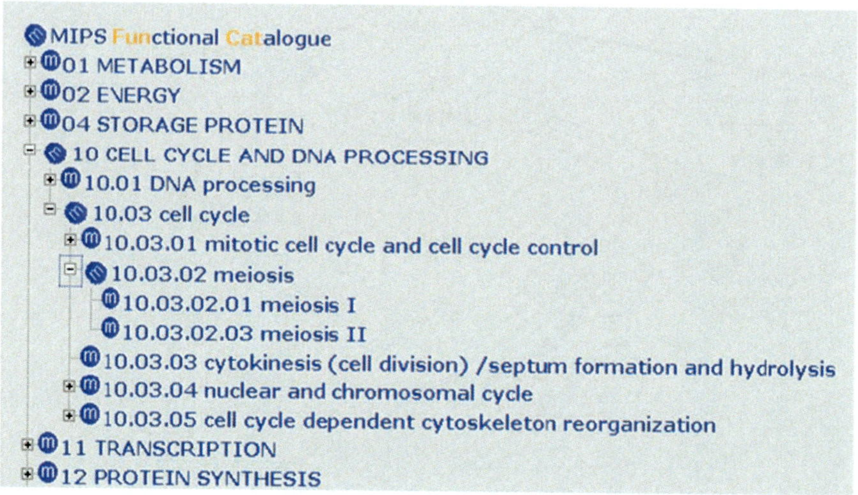

Fig. 5 Hierarchical structure of MIPS functional catalogue displayed partially using FunCat Database tool (http://mips.helmholtz-muenchen.de/proj/funcatDB/search_main_frame.html)

MIPS Functional Catalogue

Similar to Gene Ontology, MIPS Functional Catalogue (FunCat) [14] is a hierarchically organized species independent vocabulary (Fig. 5). FunCat is organized as a tree rather than a DAG. In FunCat there are 28 main catalogues, each of which is organized in a hierarchical tree structure. These main branches or catalogues cover features like *localization*, *transport*, *metabolism,* etc. FunCat currently contains 1,307 categories each of which is assigned a two digit number. FunCat identifier is represented as a series of category numbers separated by a dot based on the level in the hierarchy, for example *metabolism* is *01* and locates at first level, while *01.01.03.02.01* is *biosynthesis of glutamate* which belongs to most specific level.

Enzyme Commission Numbers

The Enzyme Commission (EC) numbers [13] are another functional classifiers that are used to classify enzymes based on reactions they catalyze. Thus as compared to the GO vocabulary, the EC numbers are reaction oriented and describe only the biochemical activity of proteins. In the enzyme nomenclature, each EC number consists of four numbers, i.e. EC x.x.x.x, each describing the enzyme at different levels of detail. There are six top levels of EC numbers from 1 to 6 which represent *oxidoreductases, transferases, hydrolases, lyases, isomerases*, and *ligases*, respectively. The next level of depth contains more details about the reaction, for example, EC number 2.1 indicates *transferase* (2 at the top level) involved in transferring

```
2. -. -.-  Transferases.
2. 1. -.-   Transferring one-carbon groups.
2. 1. 1.-    Methyltransferases.
2. 1. 2.-    Hydroxymethyl-, formyl- and related transferases.
2. 1. 3.-    Carboxyl- and carbamoyltransferases.
2. 1. 4.-    Amidinotransferases.
2. 2. -.-   Transferring aldehyde or ketone residues.
2. 2. 1.-    Transketolases and transaldolases.
2. 3. -.-   Acyltransferases.
2. 3. 1.-    Transferring groups other than amino-acyl groups.
2. 3. 2.-    Aminoacyltransferases.
2. 3. 3.-    Acyl groups converted into alkyl on transfer.
2. 4. -.-   Glycosyltransferases.
2. 4. 1.-    Hexosyltransferases.
2. 4. 2.-    Pentosyltransferases.
2. 4.99.-    Transferring other glycosyl groups.
```

Fig. 6 EC number hierarchy displayed partially as shown by ExPASy Proteomics Server (http://ca.expasy.org/enzyme/enzyme-byclass.html)

one carbon groups (1 at the second level) as shown in Fig. 6. The KEGG pathway database [1, 18] uses the EC numbers to indicate enzymes involved in metabolic pathways.

Transport Classification (TC) System

Almost all transmembrane transport processes are mediated by integral membrane proteins which are classified using Transporter Classification System [15] (http://tcdb.ucsd.edu/tcdb/). As compared to EC numbers which are focused only on function, TC classification is based on both function and phylogeny. According to this system, the transporters are classified based on five criteria and each of these provides one component of TC number for a protein. A TC number has usually five components, A, B, C, D, and E, where A corresponds to the transporter class, B corresponds to the transporter subclass, C corresponds to the family (or superfamily), D corresponds to subfamily, and E specifies the substrate transported as well as polarity of transport (in or out).

KEGG Orthology (KO)

The KEGG database includes the KEGG Orthology (KO) [16] database as one of its components [1, 18]. The primary purpose of KO is to provide the list of orthologous genes in genomes. KO is structured as a DAG hierarchy that can be effectively used for the definition of the function of ortholog groups. It has four levels with the first one consisting of five classes; *metabolism, genetic information processing, environmental information processing, cellular processes, and human diseases*, as shown in Fig. 7. The second level consists of finer functional sub-categories, third

Fig. 7 KEGG orthology displayed partially (http://www.genome.jp/kegg/ko.html)

- ▸ **01100 Metabolism**
- ▾ **01120 Genetic Information Processing**
  - ▸ **01121 Transcription**
  - ▸ **01122 Translation**
  - ▸ **01123 Folding, Sorting and Degradation**
  - ▸ **01124 Replication and Repair**
- ▸ **01130 Environmental Information Processing**
- ▾ **01140 Cellular Processes**
  - ▸ **01151 Transport and Catabolism**
  - ▸ **01141 Cell Motility**
  - ▸ **01142 Cell Growth and Death**
  - ▸ **01143 Cell Communication**
- ▸ **01150 Organismal Systems**
- ▸ **01160 Human Diseases**

level consists of KEGG pathways and fourth one corresponds to functional terms. The unique feature of the KO is that each entry has links to pathways and reactions as well as orthologous genes and hence it is convenient to annotate a set of genes with KO function terms and identify pathways where the genes belong to [16].

Other Biological Ontologies

Along with the aforementioned vocabularies for protein function, there are some other interesting ontologies that provide annotations to proteins in different domains specifically for particular species or research communities. Smith et al. [19] have developed Open Biological and Biomedical Ontologies (OBO) Foundry which consists of a collaborative effort to merge ontologies, where we can find a wide variety of open biological ontologies listed on their project website (http://www.obofoundry.org/). The ontologies include Protein Ontology developed by Protein Information Resources (PIR, http://pir.georgetown.edu/pro/), which encompasses evolution and multiple protein forms of a gene, Chemical Entities of Biological Interest (CHEBI) developed by the European Bioinformatics Institute, which classifies structures of biologically relevant chemical compounds, and ontologies for phenotype and anatomy of individual organisms. Such efforts are helping standardize the representation of domain knowledge across research communities and increase its application. By combining different ontologies, function prediction methods which output GO terms could be expanded to predict other types of ontology terms, such as phenotype.

Definition of Functional Similarity

Definition of functional similarity for protein pairs is important when comparing predictions with actual annotations of proteins to compute the prediction accuracy. A quantitative functional similarity score is also used as the target function to be optimized in the course of developing a function prediction method. In this section we overview several metrics proposed for quantifying functional similarity using the function ontology. We use the GO here since the proposed metrics are developed for the GO. However, application of the metrics to the other ontologies should be straightforward. For a review on this topic, refer to Sheehan et al. [20].

The simplest technique that can be used to compare annotations is head to head comparisons [21, 22] where we check for exact matches. Its key disadvantage is that the information embedded in the vocabulary structure is not used. Vocabulary structure relates terms to each other and with head to head comparisons we will be penalizing inexact predictions that are close to the actual ones on the GO DAG. Set based similarity measures have been developed based on head to head comparisons to match the two objects described using a set of features. Tversky et al. [23] use Eq. (1) to describe similarity between two objects a and b which have feature sets A and B respectively, as some function F of features that are common, that only belong to A and that only belong to B.

$$sim(a,b) = F(A \cap B, A - B, B - A) \tag{1}$$

Another technique [21, 24, 25] that is commonly used for GO annotations is to base the similarity on the minimum path length between a pair of terms on DAG or on the fact that ancestors are less specific representation of the same term in DAG hierarchy. This technique can suffer from drawback that not all parts of GO are developed equally and not all terms at the same depth in the structure represent same biological details.

Some techniques describe a protein as a binary vector with 1's and 0's specifying presence and absence of terms in the annotation set of a protein. The similarity between two such vectors can be defined as a cosine distance (Eq. (2)), where p_i and p_j are vectors describing annotations of two proteins. Instead of binary values, the terms can also be represented as weights based on their frequency of occurrence in the database reflecting how specific they are [26, 27].

$$sim(p_i, p_j) = \frac{p_i \cdot p_j}{|p_i|\,|p_j|} = \frac{p_i \cdot p_j}{\sqrt{p_i \cdot p_i} \cdot \sqrt{p_j \cdot p_j}} \tag{2}$$

In the function prediction category in CASP7 [21], the assessors designed a score based on the depth of common ancestor between predicted and actual GO terms as shown in Eq. (3). Each annotation is compared to its closest target prediction which forms a "computable pair", and the total score is given by the sum of depths of common ancestor of all computable pairs normalized by the maximum possible value of score. Along with this they have also used the head to head comparison of GO term predictions for comparing different methods.

$$\text{GOscore} = \frac{\text{sum of common ancestor depths of computable pairs}}{\text{sum of annotated terms depth}} \tag{3}$$

Resnik [28] has defined the Information Content (IC) of a term c based on the frequency of the occurrence of that term in the database as explained in the Eqs. (4), (5), and (6), where each term's frequency depends on its children node in the vocabulary structure because of the is_a relationships in the GO.

$$\text{freq}(c) = \text{annot}(c) + \sum_{h \in \text{children}(c)} \text{freq}(h) \tag{4}$$

$$p(c) = \text{freq}(c)/\text{freq}(\text{root}) \tag{5}$$

$$IC(c) = -\log(p(c)) \tag{6}$$

He has developed a graphical method to compute similarity between two terms (say *c1* and *c2*) in the taxonomy, by using the IC of their Lowest Common Ancestor (LCA) term (Eq. (7)). Figure 4 illustrates the concept of LCA by showing that the LCA of terms *Glucoside transport* and *Glucose transport* in the GO hierarchy is the term *transport* which is common ancestor for both terms and is located at the maximum depth in the DAG. Lin [29] further extended this semantic similarity measure to include information content of both terms being compared along with the information content of the ancestor term (Eq. (8)). Lord et al. [30] have first applied this IC based semantic similarity technique from Eq. (7) to Gene Ontology vocabulary to compute functional similarity based on protein annotations.

$$Sim_{Lin}(c1, c2) = \max_{c \in \{\text{common ancestors of } c1 \text{ and } c2\}} (-\log(p(c))) \tag{7}$$

$$Sim_{Lin}(c1, c2) = \max_{c \in \{\text{common ancestors of } c1 \text{ and } c2\}} \left(\frac{2 \log p(c)}{\log p(c1) + \log p(c2)} \right) \tag{8}$$

These term based similarity scores were extended to develop a pair-wise protein similarity score by Schlicker et al. [31]. They combined Resnik's and Lin's scores to compute a semantic similarity score for a pair of GO terms as shown in Eq. (9). To compute the semantic similarity between pair of proteins A and B they used the pair wise similarity values between the GO annotations GO_A and GO_B of both proteins respectively. Then scores from two different GO categories were combined to finally compute the overall similarity between the given two proteins (Eq. (12)). As shown in Eq. (10), semantic similarity matrix S_{ij} holds the pair wise similarity scores for all pairs of annotations from GO_A and GO_B where set GO_A has N annotations and GO_B has M annotations. For these two sets the overall similarity score referred as *GOscore* is computed by finding best matched hits for annotations in one of the directions using either row wise or column wise average of maximums (Eq. (11)). Further as shown in Eq. (12) *BPscore* and *MFscore* values computed using annotation sets from each of these categories are combined to yield the final *funsim* score that represents semantic similarity between pair of proteins under consideration.

$$Sim_{\mathrm{Rel}}(c1, c2) = \max_{c \in \{\text{common ancestors of } c1 \text{ and } c2\}} \left(\frac{2 \log p(c) \cdot (1 - p(c))}{\log p(c1) + \log p(c2)} \right) \tag{9}$$

$$Sij = sim(GO_A^i, GO_B^j), \forall i \in \{1...N\} \; and \; \forall j \in \{1...M\} \tag{10}$$

$$GOscore = \max \left\{ \left(\frac{1}{N} \sum_{i=1}^{N} \max_{1 \le j \le M} Sij \right), \left(\frac{1}{M} \sum_{j=1}^{M} \max_{1 \le i \le N} Sij \right) \right\} \tag{11}$$

$$funsim = \frac{1}{2} \cdot \left[\left(\frac{BPscore}{\max(BPscore)} \right)^2 + \left(\frac{MFscore}{\max(MFscore)} \right)^2 \right] \tag{12}$$

Methods developed in the last few years have mainly focused on pair-wise protein similarity, but with the development of high throughput techniques we are frequently required to functionally interpret a computationally or experimentally determined set of proteins and check if they are functionally homogeneous [27, 32–37]. Earlier coherence of set of proteins was based mostly on the enrichment of annotations in the set [38, 39], but it has been shown that average number of enriched GO annotations in random groups is more than the number in coherent groups of proteins [37]. This has put forth the need to further develop better protein group coherence detection methods that can segregate groups of biologically relevant proteins from random ones.

Chagoyen et al. [27] use Eq. (2) for computing pair wise similarity between proteins in the set under consideration. Later they aggregate the scores across all pairs of proteins in the set S to obtain coherence score for the set as shown in Eq. (13). Statistical significance of this coherence score is computed in the context of reference set using hypergeometric distribution.

$$\mathrm{score}(S) = \frac{\sum_{i=1}^{|S|} \sum_{j=i+1}^{|S|} sim(p_i, p_j)}{|S|(|S| - 1)/2} \tag{13}$$

Pandey et al. [36] performed similar aggregation basing their pair wise protein similarity score on the information content of minimum common ancestor set to the sets of terms annotating two proteins. For annotations of proteins p_i and p_j they compute minimum common ancestor term set and find the number of proteins annotated by all of those terms, which is given by $| G_A(p_i, p_j) |$. Further the pair wise protein functional similarity score is given by Eq. (14) where G_r is set of all proteins. The pair wise scores for all pairs of proteins in a set S are averaged in Eq. (15) to obtain the coherence score for S.

$$\rho_I(\mathrm{p_i}, \mathrm{p_j}) = -\log_2\left(\frac{|G_{\Lambda(pi,pj)}|}{|G_r|}\right) \tag{14}$$

$$\sigma_A(S) = \frac{\sum_{i=1}^{|S|}\sum_{j=i+1}^{|S|} \rho_I(p_i, p_j)}{|S|(|S|-1)/2} \tag{15}$$

Zheng et al. [37] use probabilistic model to extract biologically relevant topics from GO annotation corpus and classify each word from MEDLINE document abstracts into these topics. A document is semantically represented as count of the number of words belonging to each of the topics. A bipartite graph called ProtSemNet is constructed by joining topics obtained from each document with the proteins associated with that document, where edge weights in the graph are based on the count of words for the topic. For evaluating functional coherence of group of proteins, they construct Steiner tree from ProtSemNet for the given group of proteins where the number of edges and total distance of the tree are used as two metrics for computing protein group coherence.

Aforementioned techniques offer an interesting new avenue in the domain of functional similarity by complimenting high throughput techniques which require formal analysis of groups of proteins.

Limitations of Homology Based Function Transfer and Erroneous Database Annotations

As an increasing number of genomes are being sequenced, more and more genes are annotated computationally mainly by using homology search tools, i.e. BLAST [9] or PSI-BLAST [40], and assigned annotations will be eventually stored in the public sequence databases [41, 42]. Once these annotations are included in the databases, they will be used as a source of function information in the annotation of new genomes. Computational annotations based on homology, however, are not always trivial [43–45]. There are numerous cases where proteins with high sequence identity have different functions [46]. Galperin and Koonin discussed major causes of questionable function assignments. These include taking into account only the annotation of the best scoring database hit, insufficient masking of low complexity regions, ignoring multi-domain organization of the query proteins or the database hits, and non-orthologous gene displacement [47]. It should be also reminded that proteins which have multiple seemingly unrelated functions in a single region (moonlighting proteins) further add complications to description of protein function [48].

Indeed several studies report potential wrong annotations to genes in genomes. Brenner compared annotations by three groups to the *Mycoplasma genitalium* genome and found that 8% of the genes have serious disagreement [49]. Devos and Valencia analyzed the different functional descriptions in genes of *M. genitalium*, *Haemophilus influenzae*, and *Methanococcus jannashii* relative to the sequence identity and estimated the error rate of annotations [50]. A recent study by Schnoes et al. [51] analyzed public databases for misannotations. Their results indicate that there are significantly less potential misannotations in Swiss-Prot [41], which is manually curated, as compared with GeneBank [42], TrEMBL [41], and KEGG [1] for the six superfamilies they studied.

The main problem of erroneous annotations is that they will be reused in annotating newer genes and thus will be propagated in the databases [8]. A model of error propagation throughout the database shows that it can significantly degrade overall quality of annotations [52]. Then, how can we avoid the catastrophic deterioration of annotation of databases? First, it is important to examine the validity of annotations by experts of each protein and organism. Researchers of *E. coli* K-12 have held a meeting to examine annotations of this important model organism [53]. A recent attempts to use wiki [54] as a tool for community annotation are along the same direction [55, 56]. Another important direction is to make information and procedure transparent, which are used to make individual annotation. The aforementioned evidence codes available in the GO database provide such important information. Also, as a future direction, the architecture of biological database may need to be improved so that the lineage of annotation, i.e. the software or evidences used to make a particular annotation, homologous sequences from which the annotation are transferred, etc. can be dynamically tracked [57].

Critical Assessment of Function Prediction Methods

For the last section of this chapter, we would like to introduce community efforts for objective assessment of protein function prediction methods. As observed in the structural bioinformatics field, namely, the protein structure prediction and the protein docking prediction, evaluating methods by a quantitative score using blind prediction targets can help assessing the status of the field and also stimulates researchers' motivation for method development. In the protein structure prediction field, the Critical Assessment of Techniques for Protein Structure Prediction (CASP, http://predictioncenter.org/) while the Critical Assessment of Predictions of Interactions (CAPRI, http://www.ebi.ac.uk/msd-srv/capri/capri.html) for the protein docking prediction have served well for these purposes.

For the protein function prediction, there are two such critical assessments. The first one is as a Special Interest Group (SIG) held alongside the Intelligent Systems in Molecular Biology (ISMB) meetings. In 2005, the first meeting for the Automatic Function Prediction Special Interest Group (AFP-SIG) (http://biofunctionprediction.org/) was held at the ISMB conference at Detroit, Michigan; later meetings were followed in 2006, 2007 and 2008. The meetings are focused on exchanging ideas for automatic function predictions, which use protein sequence similarity, motifs, structures, protein-protein interactions, phylogeny, and combined data sources [58]. In 2005, they had set up a blind prediction contest where each participating research group had to provide a web interface where query sequences can be submitted and prediction results were evaluated by the organizers (thus fully automatic function prediction). The predictions were made in terms of GO terms, which were evaluated by using Eq. (7). The subsequent past AFP-SIG meetings consisted of only presentations but it was recently announced that the critical assessment of the methods will be held in the meeting of 2011.

The CASP has also started the function prediction category from CASP6 in 2004 [59]. In CASP6, predictors were allowed to provide GO terms from all three categories, binding site, binding, residue role and posttranslational modifications for each of the targets. As an exploratory category, the prediction groups were not scored and ranked at that time. In the subsequent CASP7 (2006), predictions were accepted for GO molecular function terms, EC numbers, and binding sites [21]. The aforementioned Eq. (3) in the previous section was used to assess the GO term predictions. In the CASP8 (2008) and CASP9 (2010), the function prediction is only restricted to ligand binding residue prediction, mainly because binding residues can be obtained from protein structures solved by experiments and thus can be easily assessed. In future there are many challenges in front of such blind prediction competitions: First of all, there should be availability of new functional knowledge from experimental data to evaluate the results. Also better automatic evaluation techniques may need to be developed to compare predictions with actual annotations. Finally, there should be good consensus on what types of functions will be predicted.

Summary

This chapter started with stating the motivation for development function prediction methods. Then, we overviewed fundamental technical issues for function prediction methods, including the functional ontologies and metrics for assessing accuracy for function prediction. Although steady continuous works are needed, these frameworks, especially functional ontologies, have made it possible to handle protein function computationally and also have opened up ways to for bioinformatics researchers to enter this field.

Acknowledgements MC is supported by grants from Purdue Research Foundation and Showalter Trust. DK also acknowledges a grant from National Institutes of Health (GM075004) and National Science Foundation (DMS800568, EF0850009, IIS0915801).

References

1. Kanehisa, M., Goto, S. KEGG: Kyoto encyclopedia of genes and genomes. Nucleic Acids Res. **28**(1): 27–30 (2000).
2. Flicek, P., Birney, E. Sense from sequence reads: methods for alignment and assembly. Nat. Methods **6**(11 Suppl): S6–S12 (2009).
3. Reeves, G.A., Talavera, D., Thornton, J.M. Genome and proteome annotation: organization, interpretation and integration. J. R. Soc. Interface **6**(31): 129–147 (2009).
4. Bujnicki, J.M. *Prediction of protein structures, functions, and interactions.* Chichester, West Sussex: Wiley. xiv, 287p., [2] p. of plates (2009).
5. Eisenberg, D., et al. Protein function in the post-genomic era. Nature **405**(6788): 823–826 (2000).
6. Friedberg, I. Automated protein function prediction – the genomic challenge. Brief Bioinform. **7**(3): 225–242 (2006).
7. Hawkins, T., Chitale, M., Kihara, D. New paradigm in protein function prediction for large scale omics analysis. Mol. Biosyst. **4**(3): 223–231 (2008).
8. Karp, P.D. What we do not know about sequence analysis and sequence databases. Bioinformatics **14**(9): 753–754 (1998).
9. Altschul, S.F., et al. Basic local alignment search tool. J. Mol. Biol. **215**(3): 403–410 (1990).
10. Pearson, W.R. Rapid and sensitive sequence comparison with FASTP and FASTA. Methods Enzymol. **183**: 63–98 (1990).
11. Pearson, W.R., Lipman, D.J. Improved tools for biological sequence comparison. Proc. Natl. Acad. Sci. USA **85**(8): 2444–2448 (1988).
12. Harris, M.A., et al. The Gene Ontology (GO) database and informatics resource. Nucleic Acids Res. **32**(Database issue): D258–261 (2004).
13. Nomenclature committee of the international union of biochemistry and molecular biology (NC-IUBMB), Enzyme Supplement 5 (1999). Eur. J. Biochem. **264**(2): 610–650 (1999). http://www.ncbi.nlm.nih.gov/pubmed/10491110
14. Ruepp, A., et al. The FunCat, a functional annotation scheme for systematic classification of proteins from whole genomes. Nucleic Acids Res. **32**(18): 5539–5545 (2004).
15. Saier, M.H., Jr. A functional-phylogenetic classification system for transmembrane solute transporters. Microbiol. Mol. Biol. Rev. **64**(2): 354–411 (2000).
16. Mao, X., et al. Automated genome annotation and pathway identification using the KEGG Orthology (KO) as a controlled vocabulary. Bioinformatics **21**(19): 3787–3793 (2005).
17. Ashburner, M., et al. Gene ontology: tool for the unification of biology. The Gene Ontology Consortium. Nat. Genet. **25**(1): 25–29 (2000).
18. Kanehisa, M., et al. KEGG for representation and analysis of molecular networks involving diseases and drugs. Nucleic Acids Res. **38**(Database issue): D355–360 (2010).
19. Smith, B., et al. The OBO Foundry: coordinated evolution of ontologies to support biomedical data integration. Nat. Biotechnol. **25**(11): 1251–1255 (2007).
20. Sheehan, B., et al. A relation based measure of semantic similarity for Gene Ontology annotations. BMC Bioinformatics **9**: 468 (2008).
21. Lopez, G., et al. Assessment of predictions submitted for the CASP7 function prediction category. Proteins **69**(Suppl 8): 165–174 (2007).
22. Vinayagam, A., et al. GOPET: a tool for automated predictions of Gene Ontology terms. BMC Bioinformatics **7**: 161 (2006).
23. Tversky, A. Features of similarity. Psychol. Rev. **84**(4): 327–352 (1977).
24. Hawkins, T., Luban, S., Kihara, D. Enhanced automated function prediction using distantly related sequences and contextual association by PFP. Protein Sci. **15**(6): 1550–1556 (2006).
25. Wass, M.N., Sternberg, M.J. ConFunc – functional annotation in the twilight zone. Bioinformatics **24**(6): 798–806 (2008).

26. Chabalier, J., Mosser, J., Burgun, A. A transversal approach to predict gene product networks from ontology-based similarity. BMC Bioinformatics **8**: 235 (2007).
27. Chagoyen, M., Carazo, J.M., Pascual-Montano, A. Assessment of protein set coherence using functional annotations. BMC Bioinformatics **9**: 444 (2008).
28. Resnik, P. Using information content to evaluate semantic similarity in a taxonomy. Proceedings of International Joint Conference on Artificial Intelligence **1**: 448–453 (1995).
29. Lin, D. An information-theoretic definition of similarity. Proceedings of the 15th International Conference on Machine Learning **1**: 296–304 (1998).
30. Lord, P.W., et al. Investigating semantic similarity measures across the Gene Ontology: the relationship between sequence and annotation. Bioinformatics **19**(10): 1275–1283 (2003).
31. Schlicker, A., et al. A new measure for functional similarity of gene products based on Gene Ontology. BMC Bioinformatics **7**: 302 (2006).
32. Martin, D., et al. GOToolBox: functional analysis of gene datasets based on Gene Ontology. Genome Biol. **5**(12): R101 (2004).
33. Pehkonen, P., Wong, G., Toronen, P. Theme discovery from gene lists for identification and viewing of multiple functional groups. BMC Bioinformatics **6**: 162 (2005).
34. Huang da, W., et al. The DAVID Gene Functional Classification Tool: a novel biological module-centric algorithm to functionally analyze large gene lists. Genome Biol. **8**(9): R183 (2007).
35. Carmona-Saez, P., et al. GENECODIS: a web-based tool for finding significant concurrent annotations in gene lists. Genome Biol. **8**(1): R3 (2007).
36. Pandey, J., Koyuturk, M., Grama, A. Functional characterization and topological modularity of molecular interaction networks. BMC Bioinformatics **11**(Suppl 1): S35 (2010).
37. Zheng, B., Lu, X. Novel metrics for evaluating the functional coherence of protein groups via protein semantic network. Genome Biol. **8**(7): R153 (2007).
38. Curtis, R.K., Oresic, M., Vidal Puig A. Pathways to the analysis of microarray data. Trends Biotechnol. **23**(8): 429–435 (2005).
39. Draghici, S., et al. Global functional profiling of gene expression. Genomics **81**(2): 98–104 (2003).
40. Altschul, S.F., et al. Gapped BLAST and PSI-BLAST: a new generation of protein database search programs. Nucleic Acids Res. **25**(17): 3389–3402 (1997).
41. Boeckmann, B., et al. The SWISS-PROT protein knowledgebase and its supplement TrEMBL in 2003. Nucleic Acids Res. **31**(1): 365–370 (2003).
42. Benson, D.A., et al. GenBank. Nucleic Acids Res. **37**(Database issue): D26–31 (2009).
43. Devos, D., Valencia, A. Practical limits of function prediction. Proteins **41**(1): 98–107 (2000).
44. Valencia, A. Automatic annotation of protein function. Curr. Opin. Struct. Biol. **15**(3): 267–274 (2005).
45. Bork, P., Koonin, E.V. Predicting functions from protein sequences – where are the bottlenecks? Nat. Genet. **18**(4): 313–318 (1998).
46. Tian, W., Skolnick, J. How well is enzyme function conserved as a function of pairwise sequence identity? J. Mol. Biol. **333**(4): 863–882 (2003).
47. Galperin, M.Y., Koonin, E.V. Sources of systematic error in functional annotation of genomes: domain rearrangement, non-orthologous gene displacement and operon disruption. In Silico Biol. **1**(1): 55–67 (1998).
48. Jeffery, C.J. Moonlighting proteins – an update. Mol. Biosyst. **5**(4): 345–350 (2009).
49. Brenner, S.E. Errors in genome annotation. Trends Genet. **15**(4): 132–133 (1999).
50. Devos, D., Valencia, A. Intrinsic errors in genome annotation. Trends Genet. **17**(8): 429–431 (2001).
51. Schnoes, A.M., et al. Annotation error in public databases: misannotation of molecular function in enzyme superfamilies. PLoS Comput. Biol. **5**(12): e1000605 (2009).
52. Gilks, W.R., et al. Modeling the percolation of annotation errors in a database of protein sequences. Bioinformatics **18**(12): 1641–1649 (2002).

53. Riley, M., et al. Escherichia coli K-12: a cooperatively developed annotation snapshot – 2005. Nucleic Acids Res. **34**(1): 1–9 (2006).
54. Hu, J.C., et al. The emerging world of wikis. Science **320**(5881): 1289–1290 (2008).
55. Florez, L.A., et al. A community-curated consensual annotation that is continuously updated: the Bacillus subtilis centred wiki SubtiWiki. Database (Oxford) **2009**: bap012 (2009).
56. Huss, J.W., 3rd, et al. The Gene Wiki: community intelligence applied to human gene annotation. Nucleic Acids Res. **38**(Database issue): D633–639 (2009).
57. Zhang, M., Kihara, D., Prabhakar, S. Tracing lineage in multi-version scientific databases. Proceedings of IEEE 7th International Symposium on Bioinformatics & Bioengineering (BIBE) **1**: 440–447 (2007).
58. Friedberg, I., Jambon, M., Godzik, A. New avenues in protein function prediction. Protein Sci. **15**(6): 1527–1529 (2006).
59. Soro, S., Tramontano, A. The prediction of protein function at CASP6. Proteins **61**(Suppl 7): 201–213 (2005).

Enhanced Sequence-Based Function Prediction Methods and Application to Functional Similarity Networks

Meghana Chitale and Daisuke Kihara

Abstract After reviewing the underlying framework required for computational function prediction in the previous chapter, we discuss two advanced sequence-based function prediction methods developed in our group, namely the Protein Function Prediction (PFP) method and the Extended Similarity Group (ESG) method. PFP extends the traditional homology search by incorporating functional associations between pairs of Gene Ontology terms based on the frequencies of co-occurrences in annotation of the same proteins in the database. PFP also considers very weakly similar sequences to the query, thereby increases its sensitivity and ability to predict low resolution functional terms. On the other hand, ESG recursively searches the sequence similarity space around the query to find consensus annotations in the neighborhood. The last part of the chapter discusses the network structure of gene functional space built by connecting proteins with functional similarity. Function annotation was enriched by predictions by PFP. Similarity to structures of protein-protein interaction networks and metabolic pathway networks is discussed.

Introduction

In the previous chapter we have seen that there is a strong need to develop accurate function prediction techniques to deal with the explosive growth of newly sequenced genomes. The basic approach used for more than a decade is based on homology based annotation transfer. The assumption underneath this approach is that proteins that are evolutionarily related are also functionally related [1]. In this chapter we describe two advanced function prediction techniques, PFP [2, 3] and ESG [4], developed by our group, which extend the conventional homology search methods.

D. Kihara (✉)
Department of Biological Sciences; Department of Computer Science; Markey Center for Structural Biology, College of Science, Purdue University, West Lafayette, IN 47907, USA
e-mail: dkihara@purdue.edu

D. Kihara (ed.), *Protein Function Prediction for Omics Era*,
DOI 10.1007/978-94-007-0881-5_2, © Springer Science+Business Media B.V. 2011

Conventional Sequence-Based Function Prediction Methods

Conventionally, computational protein function prediction is largely based on transferring the functional knowledge from sequences similar to the one being searched. A typical procedure would be to first use sequence homology searches, such as BLAST [5] FASTA [6], or SSEARCH [7] to identify similar sequences from a sequence database. Functional annotations of these homologous sequences were transferred to the query sequences based on the E-value of the searches. SSEARCH [7] is the implementation of the rigorous Smith Waterman algorithm [8], and thus is the most accurate among the three methods [9, 10]. Due to computational complexity of this task, two faster algorithms, BLAST [5] and FASTA [6], that work on approximating the search without missing obvious homologs, are more popular in the research community. PSI-BLAST [11] is another method, which is more sensitive than the aforementioned three methods, which iterates searches by using a sequence profile computed from a multiple sequence alignment obtained from the search from the previous round. Following the homology search, it is common to identify functional domains and motifs in the query sequences by searching against domain databases, like BLOCKS [12], InterPro [13], Pfam [14], PRINTS [15], ProDom [16], PROSITE [17], SMART [18], SUPERFAMILY [19], TIGRFams [20], and PROSITE [17]. For more details, refer to recent review articles [21–23].

However, as discussed in Chapter 1, there are many cases that open reading frames in newly sequenced genomes do not find close homologs in the database, which will result in no annotation to the proteins. This situation has motivated the development of advanced techniques for function prediction. These methods are designed to use sequence search results in a more complex setting for obtaining larger annotation coverage yet maintaining or improving the accuracy. A class of methods extend homology search tools to extract function information in terms of Gene Ontology (GO) terms from retrieved sequences. These include Goblet [24], OntoBlast [25], GOFigure [26], Gotcha [27], GOPET [28], and ConFunc [29]. Using controlled vocabulary is essential for computationally retrieving and summarizing functional terms from a database search.

Protein Function Prediction (PFP) Method

Our group has developed two function prediction methods, the Protein Function Prediction (PFP) method [2, 3] and the Extended Similarity Group (ESG) method [4], both of which predict GO terms from PSI-BLAST search results. There are some technical commonalities between the two methods, however, they are different in their design concepts.

PFP (http://kiharalab.org/pfp.php) is designed to extend the conventional PSI-BLAST search to consider very weakly related sequences. In a conventional use of (PSI-) BLAST searches, only significantly similar sequences to the query, which have a similarity score (e.g. E-value) above a predefined threshold value (typical E-value threshold values are 0.001 or 0.01), are considered for extracting function information. However, there are frequently cases where weakly similar sequences have common function to the query, even if they appear below the threshold value in a search result [2]. Common functions between weakly similar sequences may be of "low resolution", which are less specific terms and are generally at shallower positions in the hierarchical structures of functional vocabularies. Such functions might not be useful for designing biochemical experiments but will be valuable information in large-scale functional analysis, e.g. analyses of microarray data or protein-protein interaction data, when functional information is not available otherwise.

The main advantage of PFP is that it can predict low resolution functions even in the absence of apparent sequence similarity with the query sequence. It extracts functional information (GO terms) from weakly similar proteins with weights derived from the E-value and combines them to form consensus about function of a query protein. PFP also uses an association mining tool called Function Association Matrix (FAM) that captures the relations between pairs of GO terms in term of conditional probabilities of observing one annotation provided that the protein has another annotation.

PFP Algorithm

PFP takes a query sequence as an input and predicts GO terms that are likely to annotate the sequence with a confidence score. It predicts GO terms in all the three categories, Molecular Function (MF), Biological Process (BP), and Cellular Component (CC). PFP first uses PSI-BLAST [11] to obtain similar sequences from a database with the E-value cutoff of 100. For each of the retrieved sequences, GO annotations are obtained from the PFPDB database, which combines GO annotations from Gene Onotology Association (GOA) [30] database, HAMAP [31], InterPro [13], Pfam [14], PRINTS [15], ProDom [16], PROSITE [17], SMART [18], and TIGRFams [20]. GO terms taken from each sequence are weighted and summed as follows:

$$s(f_a) = \sum_{i=1}^{N} \sum_{j=1}^{Nfunc(i)} \left((-\log(E\_value(i)) + b) P(f_a|f_j) \right), \qquad (1)$$

where $s(f_a)$ is the final score assigned to the GO term f_a, N is the number of similar sequences retrieved by PSI-BLAST, $Nfunc(i)$ is the number of GO terms annotating sequence i, $E\_value(i)$ is the E-value given to the sequence i, f_j is a GO term annotating sequence i, and b is the constant value, $2 = (\log_{10} 100)$, which keeps the score positive. $P(f_a|f_j)$ is the association score for f_a given f_j obtained from the function association matrix (FAM).

FAM captures the co-occurrence of pairs of GO terms annotating the same protein in UniProt [32] database as the form of the conditional probability. FAM captures knowledge that is obvious to biologists but not reflected to annotations in the database. For example, a GO term in the MF category, *DNA binding* is frequently related to another GO term in the BP category, *regulation of transcription*. Thus, if we obtain a sequence hit with annotation *DNA binding* then the term *regulation of transcription* will obtain a share of the score from the association. Importantly, the relationship of these two GO terms cannot be captured by considering the GO hierarchy, because the two terms are on different trees.

FAM conditional probability score is obtained as follows:

$$P\left(f_a|f_j\right) = \frac{c(f_a,f_j) + \varepsilon}{c(f_j) + \mu \cdot \varepsilon'} \tag{2}$$

where $c(f_a,f_j)$ is number of times f_a and f_j are assigned simultaneously to each sequence in UniProt, and $c(f_j)$ is the total number of times f_j appeared in UniProt, μ is the size of one dimension of FAM (i.e. the total number of unique GO terms), and ϵ is the pseudo-count.

Thus, the association strategy allows PFP to explore the functional space further from annotations obtained directly from sequence hits using PSI-BLAST, which helps developing consensus about low resolution function in the absence of strong hits.

Additionally, PFP makes use of the hierarchy of the GO terms (directed acyclic graph, DAG) by propagating scores to each parent term based on the number of gene products associated with parent term as compared to the child term, as shown in Eq. (3). Due to this scheme some low resolution functions can get high scores by summing the scores propagated from multiple child nodes, and thus helping PFP predict some annotations where no strong sequence similarity exists.

$$s(f_p) = \sum_{i=1}^{N_c} \left(s(f_{ci}) \left(\frac{c(f_{ci})}{c(f_p)} \right) \right), \tag{3}$$

where $s(f_p)$ is the score of the parent term f_p, N_c is the number of child GO terms which belong to the parent term f_p, $s(f_{ci})$ is the score of a child term ci, and $c(f_{ci})$ and $c(f_p)$ is the number of known genes which are annotated with function term f_{ci} and f_p in the annotation database.

Finally, we compute the p-value significance scores for each prediction using the raw score distribution of each GO term obtained from a benchmarking dataset. Each of the p-values is associated with an expected accuracy score calculated at three different levels (correct predictions within 0, 2 and 4 edge distance on GO DAG) using the benchmarking dataset [2]. Since raw scores from Eq. (1) tend to be large for less specific terms, p-values and expected accuracy should be considered when selecting predictions done by PFP.

PFP Performance and Benchmarking

PFP has been benchmarked in two papers. In the first paper [3], we have studied the prediction performance of PFP at different cutoff levels of E-value of PSI-BLAST search. Namely, sequence hits above each cutoff value were ignored mimicking situations that there are no significant hits up to the E-value. A dataset of 2,000 randomly selected proteins from UniProt was used for the benchmark. The prediction accuracy was measured in terms of the sequence coverage, which is the percentage of sequences in the benchmark set which are annotated with correct predictions. At all E-value cutoffs, PFP showed a significantly higher coverage over a simple transfer of annotations from the top scoring sequence retrieved by PSI-BLAST (top PSI-BLAST). At the E-value cutoff of 10 (i.e. only sequences with an E-value of 10 or larger are used), PFP showed nearly five times more coverage (50%) as compared to the top PSI-BLAST method. It was also shown that the FAM improved the sequence coverage by 5–20%. Interestingly, PFP showed a better coverage over the top PSI-BLAST even when no sequence hits were ignored. This indicates that taking consensus functions among sequence hits yields better prediction in general, since often sequences of significant similarity have different functions. These results indicate that PFP can very well utilize weakly similar proteins which do not share apparent sequence similarity with a query protein.

In another study [2] using a benchmark dataset of 120,260 proteins from 11 genomes, performance of PFP has been compared against two protein function prediction methods, GOtcha [27] and InterProScan [33], as well as the top PSI-BLAST in terms of the three GO category version of the funSim score [34] (Eq. (12) in Chapter 1). It was observed in the head to head comparison among PFP, Gotcha [27], and the top PSI-BLAST [11] that PFP significantly outperformed both methods at all E-value cutoffs used, winning around 60% of cases. We have also tested different parameter values thoroughly in the paper. In addition, the p-value of the raw PFP score and the relationship between the p-value and the accuracy was examined.

As discussed above, one of the main advantages of PFP lies in its ability to increase the annotation coverage as compared to conventional homology searches. Annotations to fifteen genomes showed that more than two third of unknown proteins in each genome were assigned molecular function term at a high confidence with an expected accuracy level of 80%.

The effect of PFP's annotation to less annotated genomes can be quite dramatic. As an illustration, functional enrichment by PFP for the protein-protein interaction (PPI) network of *Plasmodium falciparum* (malaria) is shown in Fig. 1. In the original annotations in the database 664 interactions have both interacting proteins annotated (fully annotated), one of the proteins is annotated in 1,358 interactions, while 824 have neither of interacting nodes annotated. Using PFP predictions with the expected accuracy of over 90%, the number of fully annotated interactions increased to 2,674. And the number of interactions where both interacting partners are unknown was dramatically reduced to 4. These annotations will be useful for biological understanding of protein interactions in the PPI network.

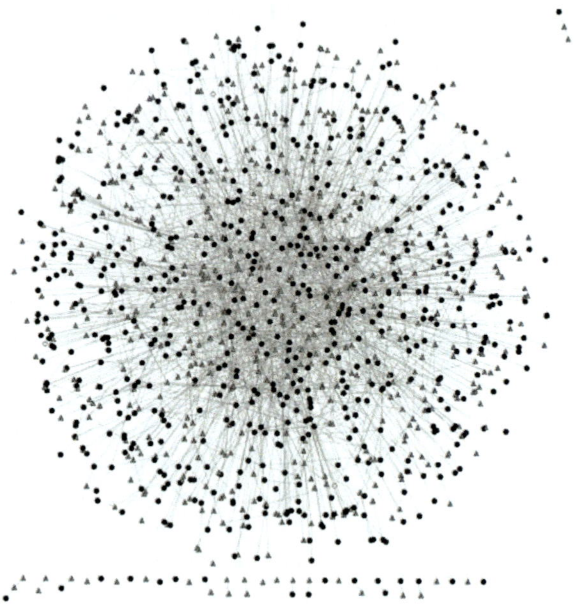

Fig. 1 *P. falciparum* PPI network with following color coding for nodes – *black circles*: proteins annotated by PFP at high confidence (>80% confidence) in at least one GO category, *gray triangles*: proteins previously annotated in the database in at least one GO category, *white diamonds*: un-annotated proteins

Extended Similarity Group (ESG) Method

The Extended Similarity Group (ESG) method [4] (http://kiharalab.org/esg.php) iterates PSI-BLAST searches by using sequences retrieved in a previous round as queries for the next round of search. GO terms taken from a retrieved sequence are weighted in a similar way as PFP, considering the E-value of the sequence. Since there are multiple rounds of searches, each round is weighted by another parameter.

The ESG Algorithm

ESG begins with an initial PSI-BLAST [11] search from the query sequence Q, which will retrieve N sequence hits, $S_1, S_2, \ldots S_N$ each with E-value $E_1, E_2, \ldots E_N$, respectively. The sequences are weighted by W_i, which considers the significance of E-value of sequence S_i relative to the other sequences:

$$W_i = \frac{-\log(E_i) + b}{\sum_{j=1}^{N} \left\{-\log(E_j) + b\right\}}, \tag{4}$$

where score, $-\log(E_i)$, is shifted by a constant value b, which makes the score a non-negative value. Using the Eq. (4) assures that the weights to the N sequences sum up to 1. Using the weights W_i assigned to each sequence, the probability of the

GO term f_a annotating the query sequence Q is defined as the sum of weights of f_a that come from sequences annotated with f_a:

$$P_Q^d(f_a) = \sum_{i=1}^{N} W_i \cdot I_{Si}(f_a) \tag{5}$$

The function I indicates whether the given sequence S_i has annotation f_a:

$$I_{S_i}(f_a) = \begin{cases} 1 & \text{if } S_i \text{ has } f_a \text{ annotation} \\ 0 & \text{otherwise} \end{cases} \tag{6}$$

The index d on the left side of Eq. (5) denotes that function information comes from direct annotations to sequences. Later we formulate integration of the FAM, which captures associated GO terms rather than directly assigned GO terms to each sequence, in the ESG framework.

Now we extend this concept to multiple levels of PSI-BLAST searches by sharing the weights between levels using a weight parameter v. In the second round, each of the sequences $S_1, S_2, \ldots S_N$ retrieved in the first round are in turn used as a query. Suppose sequence S_i obtains N sequences by a PSI-BLAST run, each referred as S_{ij}. The weights for S_{ij}, W_{ij} can be computed in a similar manner to Eq. (4). Combining the two level of searches,

$$P_Q^d(f_a) = \sum_{i=1}^{N} W_i \cdot P_{Si}^d(f_a) \tag{7}$$

$$P_{S_i}^d(f_a) = v \cdot I_{S_i}(f_a) + (1 - v) \cdot \sum_{j=1}^{N_i} W_{ij} \cdot I_{S_{ij}}(f_a) \tag{8}$$

Equation (7) is essentially the same as Eq. (5), representing that the score of a GO term f_a for the query Q is contributed by sequences retrieved at the first level (S_1 to S_N). The weight W_i is defined by Eq. (4). Equation (8) defines the score for f_a for sequence S_i as a combination of $I_{S_i}(f_a)$, which is sequence S_i's annotation, and the second level search. The first and the second terms are weighted by a factor ν. The equations can be recursively extended to multiple levels of searches to explore broader space around the query sequence.

The algorithm for the two level of the search is illustrated in Fig. 2. It shows the probability computations as described by Eqs. (7) and (8).

The FAM, which considers association of GO term pairs (Eq. (2)), can be integrated to the ESG algorithm. Equation (7) is replaced with the following equation, which states that now FAM is used for function annotation:

$$P_Q^{\mathrm{FAM}}(f_a) = \sum_{i=1}^{N} W_i \cdot P_{S_i}^{\mathrm{FAM}}(f_a) \tag{9}$$

For ESG with the second level search, Eq. (8) is modified to

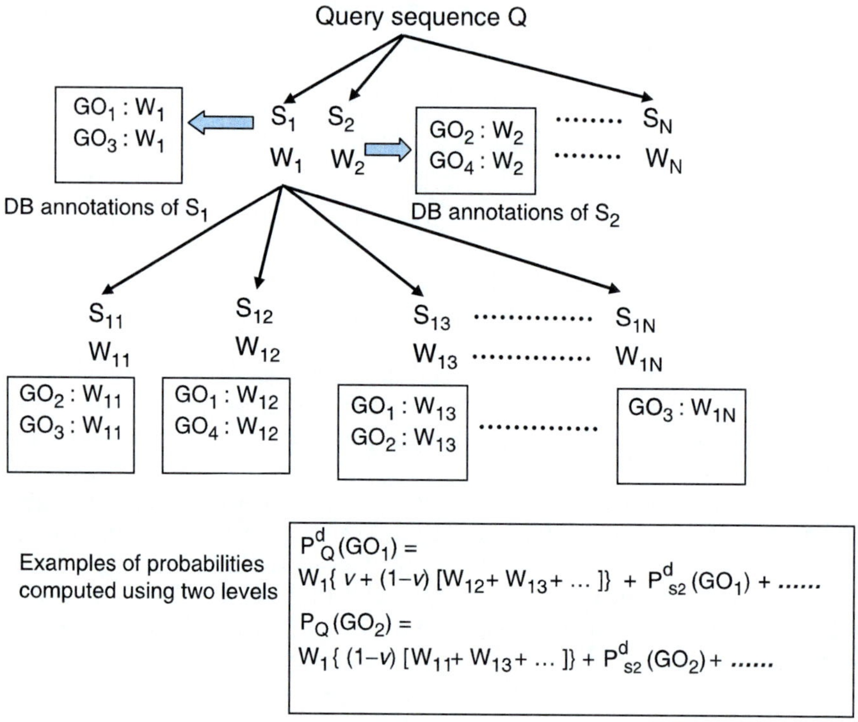

Fig. 2 Probability computation in ESG for two levels

$$P_{S_i}^{\mathrm{FAM}}(f_a) = v\left\{ I_{S_i}(f_a) + (1 - I_{S_i}(f_a)) \cdot \max\left(\sum_{j=1}^{N_{S_i}} P(f_a|f_j), 1 \right) \right\} + (1 - v)\left\{ \sum_{j=1}^{N_{ij}} W_{ij} \cdot P_{S_{ij}}^{\mathrm{FAM}}(f_a) \right\}, \tag{10}$$

where N_{Si} is the number of GO terms annotating sequence S_i. The first and the second level searches are weighted by a factor ν. The first term shows that in case S_i is not directly annotated with f_a, the FAM is used to consider association of each GO term annotating S_i to function f_a. The max operation is used to not to let the FAM-based score exceed 1. $P_{Sij}^{\mathrm{FAM}}(f_a)$ in the second term is expanded in the same way as the first term:

$$P_{S_{ij}}^{\mathrm{FAM}}(f_a) = I_{s_{ij}}(f_a) + (1 - I_{S_{ij}}(f_a)) \cdot \max\left(\sum_{j=1}^{N_{ij}} P(f_a|f_j), 1 \right) \tag{11}$$

The formulation of the score (Eqs. (4), (5), (6), (7), (8), (9), (10), and (11)) provides a value ranging from 0 to 1.

Performance of ESG

Performance of ESG has been benchmarked on a set of 2,400 protein sequences, which consists of 200 randomly selected proteins from twelve different genomes. The results of using two score cutoff values, 0.35, and 0.15, which were shown to provide a good balance of precision and recall, are shown in Fig. 3. Predicted GO terms were evaluated in terms of the funSim score with the three GO categories (Eq. (12) in chapter "Computational Protein Function Prediction: Framework and Challenges"). The FAM was not used and the search was iterated for two levels for these results. It was observed for all but one genome that the funSim scores of ESG are better than PFP. The average score of ESG was around 0.7 while that of PFP was around 0.6. Both of the methods showed superior performance as compared to the top PSI-BLAST method, which showed the average funSim score around 0.2.

Further, it was observed that ESG shows far better performance than PFP in terms of precision. ESG predicts a smaller number of GO terms as compared with PFP (average 7 GO terms are predicted by ESG while 60 terms by PFP), which generally reduces false positives, and results in an increased precision. The average precision for ESG was observed approximately 0.7 while that for PFP and top PSI-BLAST was around 0.10 on this benchmark dataset. Moreover, ESG showed a slightly better recall value than PFP, with 0.6 for ESG and 0.5 for PFP, respectively.

ESG has also been extended to incorporate the FAM, which has been shown to improve prediction recall with slightly reduced precision. For 200 *E. coli* proteins in the dataset, the recall increased from 0.773 to 0.810 by incorporating the FAM but

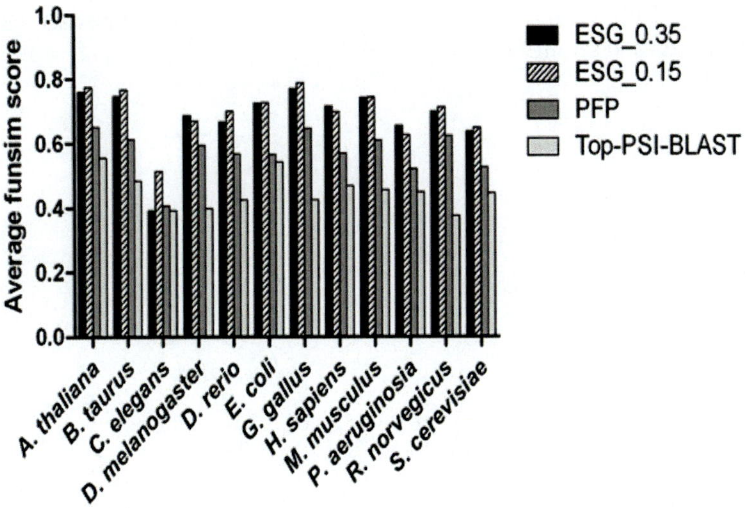

Fig. 3 The average semantic similarity score on the benchmark dataset. Two probability cutoff values are used for ESG, 0.35 and 0.15. For the Top PSI-BLAST, GO terms are extracted from sequence hits with E-value of 0.01 or smaller (better). For the PFP, GO terms with no less than 80% expected accuracy were considered. (This figure is modified from fig. 3 in [4])

the precision decreased from 0.794 to 0.566. This effect is due to the increase of the average number of predicted terms by using FAM. Overall the results indicate that PFP and ESG have considerably improved the prediction accuracy for automated function prediction using sequence similarity search.

Difference between PFP and ESG

Figure 4 illustrates difference of PFP and ESG with a conventional PSI-BLAST search. In the conventional PSI-BLAST search, only significantly similar sequences to the query (shown as the filled circle), e.g. within the E-value of 0.001 or 0.01 (dashed circle), are considered. In contrast, PFP extends the search to the E-value of 100 in the sequence similarity space, which results in more sensitive prediction. On the other hands, ESG iterates searches around the query and takes GO terms that consistently appear among the searches. Thus, ESG is designed to increase the precision of prediction. Both PFP and ESG outperform the conventional PSI-BLAST search in general, because annotations in some of closely similar sequences, which do not apply for the query, can be discarded by considering consensus among a larger number of sequences.

There is also a significant difference in the design of the score of PFP and ESG. In PFP, the raw score for each GO term is simply the sum of the scores computed from each sequences retrieved in the search. Thus, the range of the raw score is practically not pre-determined. Therefore we normalized it to compute the p-value for each GO term individually and further computed the expected accuracy by examining correlation between the p-value and the accuracy. On the other hand, ESG computes probability values varying between 0 and 1, which can be used directly for comparison and setting cutoffs.

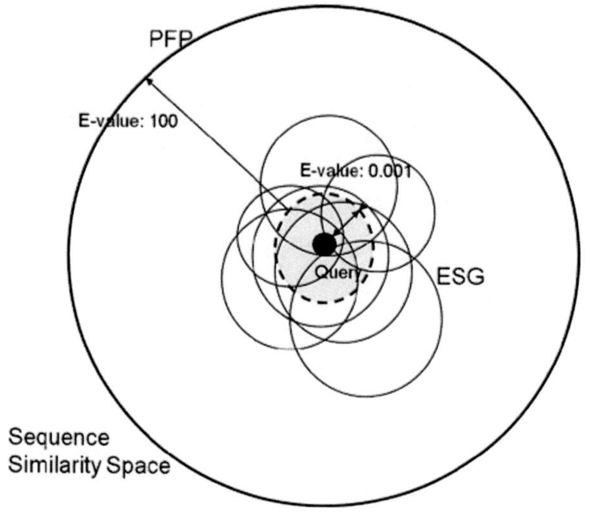

Fig. 4 Conceptual difference of PFP and ESG

PFP and ESG Web Server

Both PFP and ESG are available as web servers at http://www.kiharalab.org/software.php. Figure 5 shows the job submission page of the ESG web server. Users can enter FASTA format sequences in the text box or upload a file containing one or more amino acid sequences. The parameter "number of hits per stage" indicates the number of PSI-BLAST hits considered at each stage of the ESG algorithm (N in Eq. (7) and N_i in Eq. (8)). Another parameter "number of stages" indicates the number of levels considered, e.g. we considered two levels in the previous section. On the right side panel tutorials are provided for PFP and ESG which explain in detail how to format input and how to download and interpret the results. The web servers also provide ability to create a login account for users which can be used for maintaining users' private jobs and also for checking job status or access results from old jobs. Users can choose to be informed about job completion by receiving an email update.

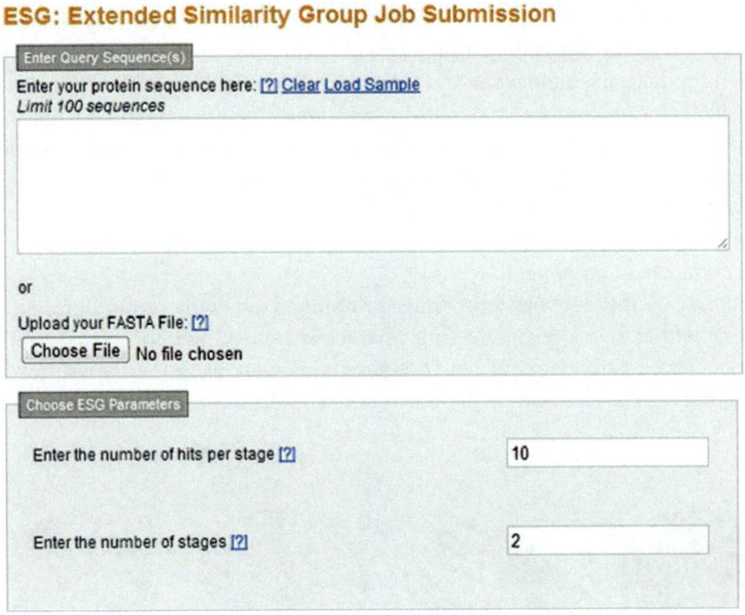

Fig. 5 ESG web server's job submission page at http://kiharalab.org/esg.php

Structure of the Gene Functional Space

In the previous sections we have discussed that PFP can significantly increase the annotation coverage of genomes. The larger annotation coverage can benefit biological research in two ways: obviously, functional clues are provided to a larger number of individual genes. Secondly, we can obtain an overview of the organization of functional space occupied by genomes. And we can further investigate relationship between functions and other important properties of genes, proteins, genomes, and organisms, such as the tertiary structure of proteins, pathways, and gene location in a genome.

To enhance our understanding of the structure of gene functional space, we introduced *functional similarity networks* [35]. We used three genomes, *Escherichia coli* (4,381), *Saccaromyces cerevisiae* (yeast) (6,690), and *Plasmodium falciparum* (malaria) (5,270) for this study. The number of protein genes is shown in the parentheses. *E.coli* and *S. cerevisiae* are well studied model organisms, where over 83.2 and 82.2% of genes, respectively, have been annotated with at least a GO term in the database. *P. falciparum* is an example of less annotated genomes, where only 41.9% of genes have annotation. To the all three genomes, PFP provided a significant number of high confidence predictions, increasing the annotation coverage to 95.2, 96.1, and 90.8%, respectively for *E. coli*, yeast, and the malaria genome.

Using annotated GO terms both in the database and those assigned by PFP, we represented functional similarity of genes in each genome as a network, where genes of similar function are connected with edges. The similarity of sets of GO terms from two genes are quantified using Eq. (11) in chapter "Computational Protein Function Prediction: Framework and Challenges", which compares GO terms in the three categories separately, and also by the three-category version of the *funSim* score (Eq. (12) in chapter "Computational Protein Function Prediction: Framework and Challenges"). Thus, four functional similarity networks, BP-score, MF-score, CC-score, and *funSim*-score networks are computed for each genome (Fig. 6). In all of the functional similarity networks, a majority of the genes are included in the largest connected component.

Analyses of the network properties in comparison with protein-protein interaction networks revealed interesting characteristics of the functional similarity networks. Three parameters of network structures were examined. First, we examined the degree distribution of the networks. The degree distribution concerns the

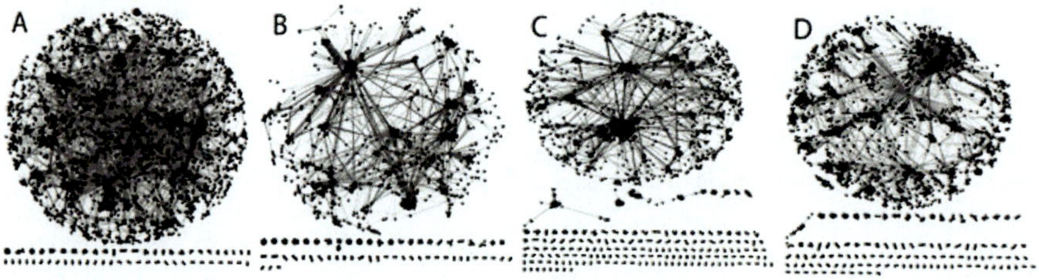

Fig. 6 Functional similarity networks of yeast genome. (**a**) Similarity of biological process terms in the Gene Ontology are used; (**b**) cellular component terms; (**c**) molecular function terms; (**d**) funSim score is used to define functional similarity. (This figure was modified from fig. 3 in Hawkins et al. [35])

probability of nodes with each number of degree k (edges or connections). If the degree distribution follows the power-law, i.e. $P(k) \sim k^{-\gamma}$, where γ is around 1.0, it indicates that the network has few nodes with a large number of connections while the majority of nodes have a small number of connections. In the case of yeast functional similarity networks (Fig. 6), all of them showed a γ value close to 1.0, namely, 1.22, 0.83, 0.96, and 1.31, for the BP-score, CC-score, MF-score, and funSim score networks, respectively. It is known that protein-protein interaction (PPI) networks follow the power-law [36]. Indeed, the yeast PPI network has the γ value of 1.80. Thus, in general both PPI and the functional similarity networks follow the power-law.

Next examined was the clustering coefficient of the networks. The clustering coefficient of a node indicates how well nodes neighboring to the central nodes are connected to each other. It is defined in the following way:

$$C = \frac{n}{\frac{k(k-1)}{2}} \tag{12}$$

k is the number of neighboring nodes connected to the central node and n is the number of pairs of the neighboring nodes that are directly connected. We consider that a network has high *modularity* if it has a large average clustering coefficient [36, 37]. It turned out that the functional similarity networks distinguish themselves from the PPI networks by having higher clustering coefficient, thus they are highly modular compared to the PPI networks. The clustering coefficient value for the yeast PPI network is 0.10, while the BP-, CC-, MF-, and funSim-score networks showed values of 0.63, 0.77, 0.72, and 0.46, respectively.

We also discussed the network hierarchy based on the network model by Ravasz et al. [37]. A network is considered to be hierarchical if the clustering coefficient, C(k) follows the scaling law, $C(k) \sim k^{-1}$. The clustering degree exponent value β, $C(k) \sim k^{-1}$ obtained for the functional similarity networks revealed that only the funSim score network has a β value close to 1: 0.11, –0.05, 0.40, and 1.39, for the BP-score, CC-score, MF-score, and funSim score networks, respectively. Therefore, interestingly, hierarchy is observed in funSim network (Fig. 7) but not in individual GO-score networks. The network hierarchy was first observed in metabolic pathways [37]. It is an interesting observation that hierarchy of the network arises for the funSim score that integrates single GO-scores, which do not show hierarchy individually. This might imply that the funSim score somewhat captures properties of metabolic pathway networks.

In summary, we studied the landscape of the functional space of genes as the functional similarity networks. Analysis of topological properties of these networks revealed different network properties as compared with the PPI networks. This analysis demonstrates that applying annotations by PFP can have a significant impact in investigating biological systems in an omics scale.

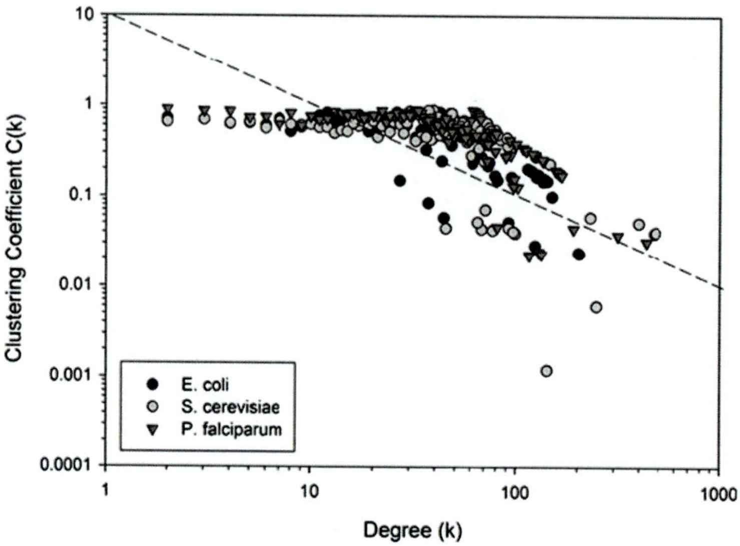

Fig. 7 Hierarchical modularity of funSim score networks of the three organisms. Clustering coefficient is plotted relative to the degree (k) of nodes. The *dotted lines* shows C(k) ~ k^{-1}. (This figure is modified from fig. 5 in Hawkins et al. [35])

Summary

In this chapter, we introduced two sequence-based function prediction methods developed in our group, PFP and ESG. In contrast to conventional sequence-based function prediction methods, the two methods effectively capture function information in weakly similar sequences. Biological implication by the success of PFP and ESG is that there exist functional commonalities among genes with are not traditionally considered as homologous, and such common functions can be captured by making use of very weakly similar sequences. As the number of sequenced genomes is rapidly increasing, there is even stronger need for sensitive and accurate function prediction methods. These two methods show a new direction for function prediction, which is to explore the twilight zone or even lower sequence similarity, rather than sticking with high sequence similarity or conservation.

Acknowledgements MC is supported by grants from Purdue Research Foundation and the Showalter Trust. DK also acknowledges a grant from National Institutes of Health (GM075004) and National Science Foundation (DMS800568, EF0850009, IIS0915801).

References

1. Ofran, Y., et al. Beyond annotation transfer by homology: novel protein-function prediction methods to assist drug discovery. Drug Discov. Today **10**(21): 1475–1482 (2005).
2. Hawkins, T., et al. PFP: automated prediction of gene ontology functional annotations with confidence scores using protein sequence data. Proteins **74**(3): 566–582 (2009).

3. Hawkins, T., Luban, S., Kihara, D. Enhanced automated function prediction using distantly related sequences and contextual association by PFP. Protein Sci. **15**(6): 1550–1556 (2006).
4. Chitale, M., et al. ESG: extended similarity group method for automated protein function prediction. Bioinformatics **25**(14): 1739–1745 (2009).
5. Altschul, S.F., et al. Basic local alignment search tool. J. Mol. Biol. **215**(3): 403–410 (1990).
6. Pearson, W.R. Rapid and sensitive sequence comparison with FASTP and FASTA. Methods Enzymol. **183**: 63–98 (1990).
7. Pearson, W.R., Lipman, D.J. Improved tools for biological sequence comparison. Proc. Natl. Acad. Sci. USA **85**(8): 2444–2448 (1988).
8. Smith, T.F., Waterman, M.S. Identification of common molecular subsequences. J. Mol. Biol. **147**(1): 195–197 (1981).
9. Brenner, S.E., Chothia, C., Hubbard, T.J. Assessing sequence comparison methods with reliable structurally identified distant evolutionary relationships. Proc. Natl. Acad. Sci. USA **95**(11): 6073–6078 (1998).
10. Hulsen, T., et al. Testing statistical significance scores of sequence comparison methods with structure similarity. BMC Bioinformatics **7**: 444 (2006).
11. Altschul, S.F., et al. Gapped BLAST and PSI-BLAST: a new generation of protein database search programs. Nucleic Acids Res. **25**(17): 3389–3402 (1997).
12. Pietrokovski, S., Henikoff, J.G. Henikoff, S. The Blocks database – a system for protein classification. Nucleic Acids Res. **24**(1): 197–200 (1996).
13. Hunter, S., et al. InterPro: the integrative protein signature database. Nucleic Acids Res. **37**(Database issue): D211–215 (2009).
14. Finn, R.D., et al. Pfam: clans, web tools and services. Nucleic Acids Res. **34**(Database issue): D247–251 (2006).
15. Attwood, T.K., et al. PRINTS and its automatic supplement, prePRINTS. Nucleic Acids Res. **31**(1): 400–402 (2003).
16. Bru, C., et al. The ProDom database of protein domain families: more emphasis on 3D. Nucleic Acids Res. **33**(Database issue): D212–215 (2005).
17. Hulo, N., et al. The 20 years of PROSITE. Nucleic Acids Res. **36**(Database issue): D245–249 (2008).
18. Letunic, I., et al. SMART 4.0: towards genomic data integration. Nucleic Acids Res. **32**(Database issue): D142–144 (2004).
19. Wilson, D., et al. The SUPERFAMILY database in 2007: families and functions. Nucleic Acids Res. **35**(Database issue): D308–313 (2007).
20. Haft, D.H., Selengut, J.D., White, O. The TIGRFAMs database of protein families. Nucleic Acids Res. **31**(1): 371–373 (2003).
21. Hawkins, T., Chitale, M., Kihara, D. New paradigm in protein function prediction for large scale omics analysis. Mol. Biosyst. **4**(3): 223–231 (2008).
22. Chitale, M., Hawkins, T., Kihara, D. Automated prediction of protein function from sequence. *Prediction of protein strucutre, functions, and interactions*. Bujnicki, J.M. (ed.). New York, NY: Wiley, pp. 63–86 (2009).
23. Kaminska, K.H., Milanowska, K., Bujnicki, J.M. The basics of protein sequence analysis. *Prediction of protein structures, functions, and interactions*. Bujnicki, J.M. (ed.). New York, NY: Wiley, pp. 1–38 (2009).
24. Hennig, S., Groth, D., Lehrach, H. Automated Gene Ontology annotation for anonymous sequence data. Nucleic Acids Res. **31**(13): 3712–3715 (2003).
25. Zehetner, G. OntoBlast function: from sequence similarities directly to potential functional annotations by ontology terms. Nucleic Acids Res. **31**(13): 3799–3803 (2003).
26. Khan, S., et al. GoFigure: automated Gene Ontology annotation. Bioinformatics **19**(18): 2484–2485 (2003).
27. Martin, D.M., Berriman, M., Barton, G.J. GOtcha: a new method for prediction of protein function assessed by the annotation of seven genomes. BMC Bioinformatics **5**: 178 (2004).
28. Vinayagam, A., et al. GOPET: a tool for automated predictions of Gene Ontology terms. BMC Bioinformatics **7**: 161 (2006).

29. Wass, M.N., Sternberg, M.J. ConFunc – functional annotation in the twilight zone. Bioinformatics **24**(6): 798–806 (2008).
30. Barrell, D., et al. The GOA database in 2009 – an integrated Gene Ontology Annotation resource. Nucleic Acids Res. **37**(Database issue): D396–403 (2009).
31. Gattiker, A., et al. Automated annotation of microbial proteomes in SWISS-PROT. Comput. Biol. Chem. **27**(1): 49–58 (2003).
32. The Universal Protein Resource (UniProt) in 2010. Nucleic Acids Res. **38**(Database issue): D142–148.
33. Zdobnov, E.M., Apweiler, R. InterProScan – an integration platform for the signature-recognition methods in InterPro. Bioinformatics **17**(9): 847–848 (2001).
34. Schlicker, A., et al. A new measure for functional similarity of gene products based on Gene Ontology. BMC Bioinformatics **7**: 302 (2006).
35. Hawkins, T., Chitale, M., Kihara, D. Functional enrichment analyses and construction of functional similarity networks with high confidence function prediction by PFP. BMC Bioinformatics **11**: 265 (2010).
36. Barabasi, A.L., Oltvai, Z.N. Network biology: understanding the cell's functional organization. Nat. Rev. Genet. **5**(2): 101–13 (2004).
37. Ravasz, E., et al. Hierarchical organization of modularity in metabolic networks. Science **297**(5586): 1551–5 (2002).

Gene Cluster Prediction and Its Application to Genome Annotation

Vikas Rao Pejaver, Heewook Lee, and Sun Kim

Abstract Improvements in sequencing technology have made whole-genome sequencing a lot more accessible to researchers in the life sciences. There has been a huge explosion in genomic sequence data over recent years and automated genome-wide function annotation has become a great challenge. The most popular approaches for gene function assignment have been based on sequence similarity. However, homology-based methods are limited in cases where novel sequences show no significant sequence similarity to known genes. This has led to the exploration of innovative methods that make use of additional information such as co-localization, co-evolution and fusion to assign functions computationally. In the case of prokaryotic genomes, functionally related genes tend to be physically clustered together due to evolutionary pressure. Thus, such gene clusters provide effective clues for gene function assignment in prokaryotes. In this chapter, we survey a few of the prominent techniques in this area of research. We also perform simple experiments to detect gene clusters across a given set of genomes. Finally, we provide a few examples from the results of these experiments to show how gene cluster information can be applied to genome annotation and can resolve ambiguities in function assignment.

Introduction

Next generation sequencing technology has made it possible to sequence genomes at a fraction of the cost incurred by using the traditional Sanger sequencing method. Thus, the number of genomes available to research community is growing rapidly and analysis of such a large number of genomes will be a significant challenge. A key issue that needs to be dealt with is that of accurate genome annotation. Experimentally, it is a huge challenge to attempt to identify functions for every

S. Kim (✉)
School of Informatics and Computing, Indiana University, Bloomington, IN, USA
e-mail: sunkim2@indiana.edu

D. Kihara (ed.), *Protein Function Prediction for Omics Era*,
DOI 10.1007/978-94-007-0881-5_3,

gene in a given genome. The process of manual or semi-automated annotation is time-consuming and laborious. Although, automated methods tend to alleviate this situation, accuracy of annotation still remains an issue.

Comparative genomics has played an important role in advancing function prediction and has contributed to great improvements in the accuracy of gene function prediction. The most common methods for automated function assignment rely on sequence homology. The idea is that genes with similar sequences would code for proteins with similar structures and thus, would possess similar functions. This approach has been fairly successful but is ineffective when a given sequence shows no significant sequence similarity to existing genes. This has given rise to methods where sequence similarity information is augmented by genomic context information that allows for function assignment based on a "guilt-by-association" principle. Thus, one important problem is to discover gene sets that are common and/or unique to a subset of genomes as this information can aid in understanding the characteristics of organisms in terms of gene content.

Conserved gene clusters provide effective means to obtain clues about a gene's function based on its neighboring genes. This has been very popular in the case of prokaryotes as physical clustering of functionally related genes is a prominent phenomenon in their genomes. Clusters of genes can be thought of as patterns of genes in terms of their physical proximity, i.e., on the chromosomes. Mining sequential patterns of genes in a number of genomes is a data mining problem as one does not have any prior knowledge about the existence of patterns, the size of patterns, and the genomes in which a gene pattern exists, as shown in Fig. 1. In addition, other challenges in pattern mining from genomic data include: the lack of family definition, the distorted order of genes, missing or inserted genes in pattern occurrences,

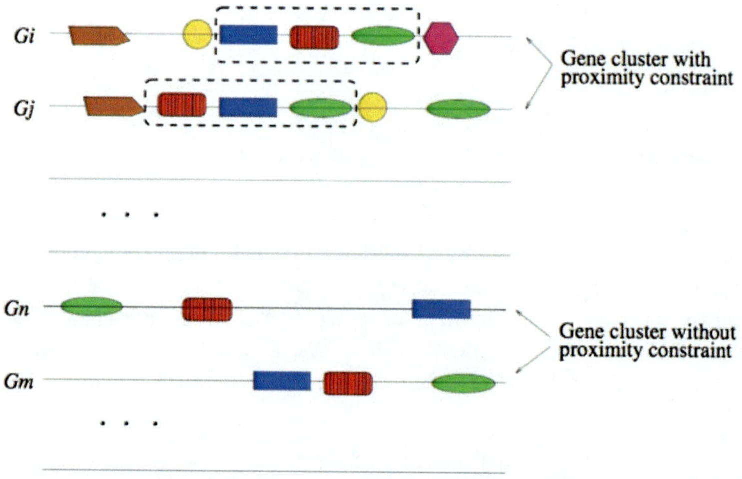

Fig. 1 Illustration of gene cluster mining. We do not know which gene clusters occur in which subset, G_i, G_j, G_n and G_m in this case, of many genomes, say 100 genomes

and different contexts of patterns such as sequences of genes (metabolic pathway or protein-protein interaction networks).

It is worth noting that the gene cluster prediction problem is fundamentally different from the widely studied synteny prediction problem. Synteny prediction aims at finding conserved genomic regions in multiple genomes. Also, it is assumed that all genomes in a given genome set share common syntenic regions. On the contrary, the gene cluster prediction problem seeks to *mine* gene clusters in an *unknown subset* of the genomes, with various constraints as shown in Fig. 1.

Mining gene clusters conserved in multiple genomes can be a useful method for characterizing genomes. For example, Overbeek et al. [1] showed that conserved gene clusters corresponded to biological pathways. Mining conserved gene clusters can help us to characterize subsystems that can be used for genome annotation [2]. In this chapter, we will summarize computational methods for predicting conserved gene clusters and show some examples how gene clusters can be used for genome annotation.

The detection of gene clusters generally follows two broad approaches. In the first approach, sequence similarity information is used as a starting point for cluster detection. This augments the homology-based methods as discussed earlier and can be regarded as a systematic combination of sequence similarity with proximity information. The second approach begins with the assumption that gene families are known. This implies that in order to use this approach one has to have prior knowledge of gene families (such as COG categories [3]). Note that we make a clear distinction between these two approaches and focus only on the first in this chapter. Therefore, we have not included computational methods [4–6] that use pre-classified family information.

Description of Existing Techniques

A number of methods have been proposed to detect conserved gene clusters and can broadly be grouped as those that detect clusters within a single genome, a pair of genomes or within multiple genomes. Single genome methods are generally trivial but become challenging when they take information from other genomes into consideration. The pairwise gene cluster detection problem is well-studied and therefore we discuss some of the more recent approaches on this topic. In the case of multiple genomes, the problem is non-trivial and is much more difficult to address than the detection of gene clusters from within a single genome or conserved between a pair of genomes. It is interesting to note that since most multi-genome methods use pairwise gene cluster information as a starting point, the number of genomes used for the detection of gene clusters may not necessarily be a good way to distinguish between these methods. Therefore, in this section, we cover a few methods across all these categories and provide a brief description of how they work. We focus on the innovative approaches used to solve this problem, ranging from graph-based approaches to query-based strategies. For a detailed study of each of these approaches, we refer the reader to their respective papers.

Graph-Based Approach

One of the first attempts at identifying conserved gene clusters from multiple genomes was made by Fujibuchi et al. [7]. This approach involved the use of two graph algorithms, that had been developed by the group previously [8, 9]. First, locally similar regions were detected across genomes based on a novel graph comparison algorithm. Second, clustering was performed based on a graph linkage feature called P-quasi complete linkage. The resulting information would be analogous to a "multiple genome alignment", but at the level of gene clusters, rather than genes.

The methodology involves an automated analysis pipeline that is summarized in Fig. 2. and consists of three major steps:

1. Application of the graph comparison algorithm to obtain gene clusters conserved in two genomes.
2. Incorporation of related clusters from multiple genomes by means of P-quasi complete linkage analysis.
3. Resolution of issues involving orthology, paralogy and gene fusion by P-quasi can COG clustering methods to generate unambiguous gene cluster tables.

The first step entails the representation of a pair of genomes as a pair of one-dimensionally connected graphs whose vertices correspond to their respective genes. Irrespective of the direction of transcription, two adjacent genes on a chromosome are regarded to be connected to each other by an edge. A sequence similarity matrix is created for both the genomes, based on Smith-Waterman alignment [10] of all pairs of genes in both the genomes. The matrix element is 1 if the optimized score of SSEARCH is 100 [10]; in other cases it is 0. Thus, an $m \times n$ matrix is generated where m is the number of genes in one genome and n is the number of genes in the other genome. This serves as a representation of the correspondences between genes in one genome and genes in the other. A dynamic programming algorithm that is a minor modification of the Floyd-Warshall algorithm is then applied to the similarity matrix to detect all pairwise shortest paths (shorter than a given gap parameter). A detailed description of this graph comparison algorithm can be obtained from [9]. For the detection of conserved gene clusters from a pair of genomes, the authors have allowed gap lengths of up to two for each genome and clusters containing at least two homologous gene pairs with or without rearrangements. This heuristic pairwise gene cluster detection algorithm has been shown to work really well and effectively filters out noise from gene similarity data.

The next step involves the use of correlated gene cluster information from all pairwise analyses to obtain clusters conserved across multiple genomes. Initially, for a pair of genomes, each conserved cluster pair is assigned a similarity score. This similarity score can be defined as the number of best hit gene pairs within a cluster (Fig. 2). In order to avoid overcounting, in cases of paralogy, multiple pairs involving the same node are combined. Grouping of gene clusters is then done based on linkage of similar cluster pairs across multiple genomes by means of a

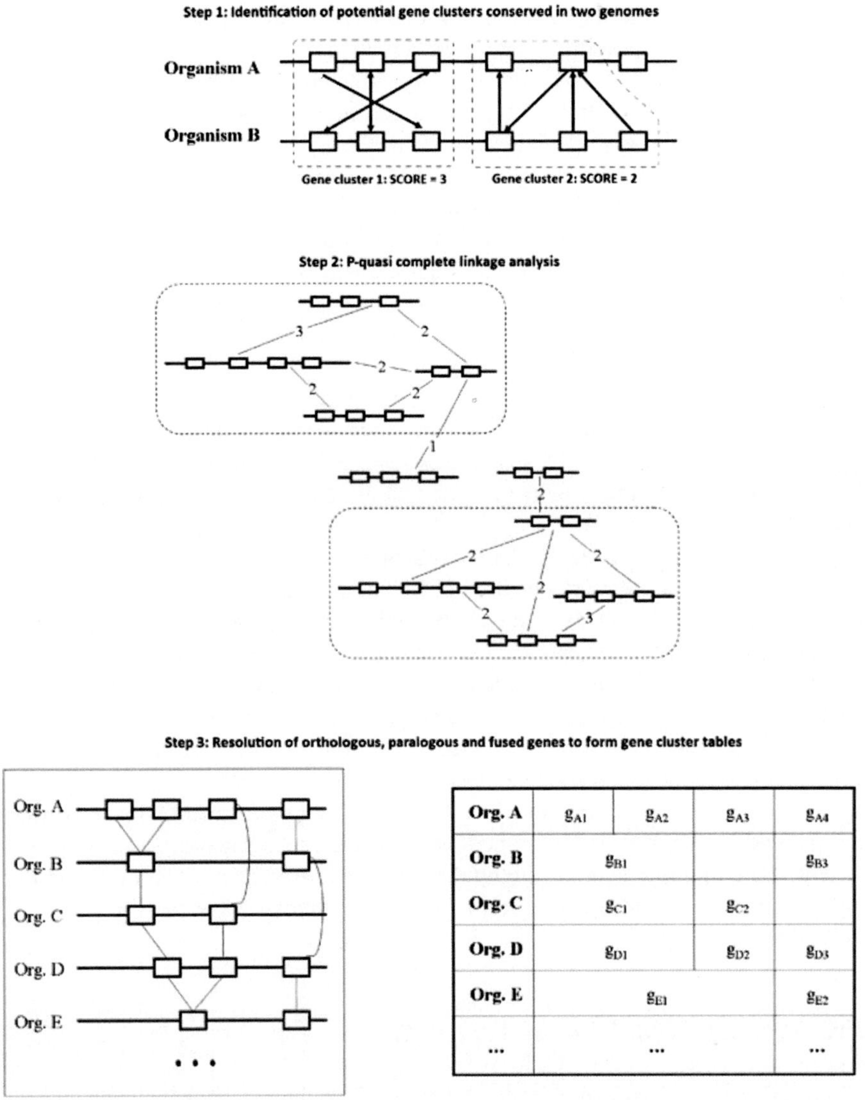

Fig. 2 A schematic description of the three steps involved in the graph-based algorithm proposed by Fujibuchi et al. [7]

clustering algorithm. Typically, clustering is done either by single linkage or complete linkage algorithms. However, Fujibuchi et al. noted that for the detection of clusters conserved across multiple pairs of genomes, these algorithms had certain limitations. The single linkage algorithm would result in a smaller number of genes clustered in larger groups. The complete linkage would cause the creation of groups that are too fine and would be relatively uninteresting in the context of biological inference. This motivated the application of a P-quasi complete linkage algorithm.

This method allows for the creation of groups where any member in one group is connected to $\geq P\%$ of all the other members within the group. The authors have tested different values of the completeness parameter P and have found that generally, as P increases, the number of gene clusters detected increases. They also note that computation of P-quasi linkages is very memory-intensive and time-consuming. We refer the reader to a more comprehensive discussion of the algorithm in [8].

The final step involves the refinement of these merged gene clusters at the gene level. Each gene is defined either as an ortholog, a fused gene or a paralog, based on a set of criteria. For this analysis, the P-quasi complete linkage method is adopted again to assign groups of homologous genes. In cases where one gene in a genome is homologous to more than one gene in another genome, a fused gene is defined based on individual sequence similarity scores and other criteria. If these criteria are not satisfied, the homologs are assumed to be paralogs and are further divided based on the COG triangle method. We refer the reader to the original article for the more specific criteria set for all cases. On the whole, this step results in the formation of the multiple genome alignments discussed earlier and a gene cluster table as illustrated in Fig. 2.

Fujibuchi et al. analyzed 17 completely sequenced microbial genomes, at a completeness parameter 40% and obtained 2313 clusters. Approximately 25% of these contained at least two genes in metabolic and regulatory pathways in the KEGG database [11]. Although a quantitative validation was not performed, qualitative analyses showed that clusters tended to contain functionally related genes. It is interesting to note that the authors found very low cluster conservation, even at reasonable phylogenetic distances. This can be explained by the fact that this method does not incorporate any phylogenetic information. The method has since then been applied to more genomes as and when they have been sequenced. All results from this method can be obtained from the KEGG gene clusters database [12].

Note: The multiple genome alignments are considered as a rough draft and the final results that can be accessed from the KEGG database have been subjected to manual curation.

Evolutionary Model-Based Approach

A few years later, Zheng et al. argued that gene proximity conservation in microbial genomes could be a result of various factors other than functional selection or vertical inheritance [13]. Thus, there would be a tendency to inaccurately estimate the significance of a cluster. Zheng et al. proposed a robust phylogenetic method to detect clusters that contained functionally related genes and to provide a measure of significance that indicated whether they were a likely result of functional selection. Another issue that this method was expected to tackle was that of biases in genomic databases. Microbial genomic data, even today, is not uniform and there are experimental preferences towards model organisms, pathogenic species and easily culturable strains. In several cases, the use of phylogenetic information from genomes that are not properly sampled from databases, could result in errors.

The method involves the use of a stochastic evolutionary model to describe conserved gene clusters. At the uppermost level, this method makes use of two conservation scores to determine whether a pair of genes should be included in a longer conserved cluster or not. For the gth gene on a given chromosome, these scores are defined as:

$$C_u(g) = \sum_{i=1}^{k+1} s(g-i, g)$$

$$C_d(g) = \sum_{i=1}^{k+1} s(g, g+i)$$

where $C_u(g)$ is the upstream conservation score for the gth gene, $C_d(g)$ is the downstream conservation score for the same gene and k is the number of intervening genes allowed in a conserved pair (typically $k = 1$). In general, the term $s(g_1, g_2)$, is the tree-based conservation score for a gene pair consisting of genes g_1 and g_2. This tree-based conservation score represents the overall probability that a gene pair is conserved in a given phylogenetic tree. This score can be explained with the following example:

Consider a simple phylogenetic tree as shown in Fig. 3. Let the leaf nodes represent extant genomes G_1, G_2, G_3, G_4 and G_5. The internal nodes A_0, A_1, A_2, and A_3 represent the inferred ancestor genomes. The idea is to model the evolution of a gene cluster as a stochastic process on this tree. A gene cluster can be regarded as a group of gene pairs and thus, for Fig. 3, the basic unit that needs to be considered is a gene pair. The phylogenetic tree in Fig. 3 can be treated as a Bayesian network in a tree form. If a gene pair is present in a particular genome as a pair of neighboring genes, we assign it a value 1 and if it is absent, a value of 0 is assigned. These values are determined by the definition of conserved neighboring gene pairs – a gene pair is said to be neighboring if each of these genes is separated by no more than k open reading frames as mentioned previously. A conserved neighboring gene pair is a gene pair where the orthologs of its members form a neighboring gene pair in

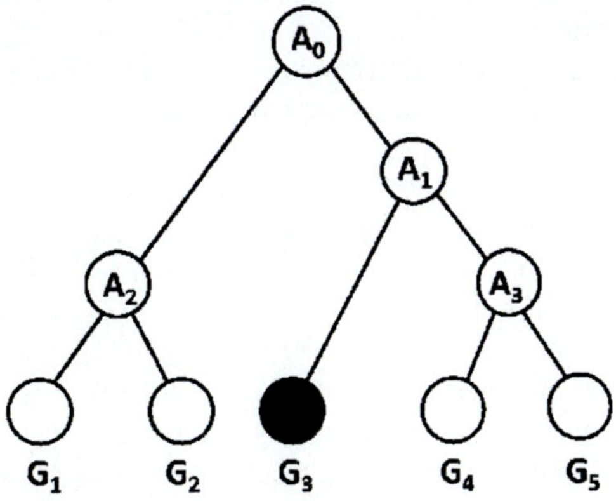

Fig. 3 A simplified example of a phylogenetic tree where the filled node represents a genome in which a gene pair is not conserved

another genome. Note that this definition results in gene pairs (and thus, clusters) that do not take strand direction into account and are solely based on orthology or similarity data (BLAST E-value $< 1E-5$) [14].

In order to model the probability that a gene pair in one of the leaf genomes is conserved across the phylogenetic tree in Fig. 3, the binary random variables discussed previously, are used. Consider a situation where a gene pair in query genome G_1 is conserved in genomes G_2, G_4, and G_5, but not in G_3. It becomes obvious that the vertical inheritance along any path on the tree is a generative probabilistic process and the inheritance of a gene pair depends on the immediate ancestor. However, since the probability of a gene pair existing in a genome, but not in its immediate ancestor is negligible, A_0 can be assigned the value 1. Thus, the following expression can be deduced for a phylogenetic tree T:

$$\begin{aligned} P(\text{conservation of a gene pair}) &= P(G_1 = 1, G_2 = 1, G_3 = 0, \\ &\qquad G_4 = 1, G_5 = 1 | A_0 = 1) \\ &= \prod_{X,Y \epsilon T} P(X = 1 | Y = 1) \end{aligned}$$

where Y is an immediate evolutionary ancestor of X in T. For a detailed derivation of this expression, we refer the reader to the original article. The product on the right hand side of this expression can be converted into a summation context by taking the negative logarithm on both sides. A key assumption of this methodology is that $\log(P(X = 1|Y = 1))$ is proportional to the phylogenetic distance between X and Y. Therefore,

$$\log(P(\text{conservation of a gene pair})) \sim \sum_{X,Y \epsilon T} d(X, Y)$$

where $d = -\ln(s)$ and s is a measure that is used to derive pairwise distances for the construction of the phylogenetic tree. s is defined as the ratio of the number of shared orthologs to the average of the total numbers of genes in a pair of genomes. Thus, for each gene pair, a genome phylogenetic tree can be constructed, irrespective of the query genome and conservation scores can be calculated based on the branch lengths.

Once the upstream and downstream conservation scores are calculated, longer clusters are formed on the basis of the following criteria:

- In order to define the "boundary genes" of a long cluster, either C_u or C_d (not both), for the genes should exceed a predetermined threshold.
- All genes in between these two genes would then be considered as being a part of the cluster.

The threshold cutoff value for the conservation scores is determined based on their P-values so that only statistically significant clusters are preserved. These P-values are calculated from bootstrap simulations where each genome is randomly shuffled and conservation scores are calculated on these genomes to obtain the null distribution.

The authors tested this method on a set of 345 operons from *E. coli* from the RegulonDB database [15] and it was found that a sensitivity of 65% and a specificity of 85% could be achieved when starting with orthology data. When gene similarity data was used as a starting point, the sensitivity improved but the specificity worsened. This could be explained due to the conservative nature of the reciprocal BLAST hit method to determine orthology. This method was extended to other genomes and it was generally found that 10-40% of the total number of genes in a genome lie in gene clusters. Conserved gene clusters have now been computed for over 200 microbial genomes and have been stored in a database called GeneChords [16].

Note: Although, in practice, the above method predicts clusters on a given query genome, it successfully captures information from multiple genomes and projects that information on to the query genome. Therefore, we have included its discussion in this section.

EGGS: Gene Pattern Prediction Based on Genome Context

Kim et al. developed a gene pattern prediction algorithm [17], called EGGS (Extraction of Gene clusters by iteratively using a Genome context-based Sequence matching technique). Given all pairwise gene similarity data, EGGS predicts a set of gene patterns in two genomes. The most widely used approach to gene pattern prediction is to model it as an optimization problem where all gene matches are treated "equally" although the optimal score considers interdistance between genes. However, it is obvious that some gene matches are more accurate than others. The simplest way to discriminate significance of gene matches is to use match scores such as E-value, Zscore, or bitscore. Adding genome context to gene similarity improves the correctness (specificity) of gene matches. The motivation for EGGS is to utilize *genome context* to distinguish more reliable gene matches from less reliable ones, which was inspired by the seminal work from Overbeek et al. [1].

First, EGGS defines a four-level gene matching technique and uses it in an *iterative constraint relaxation* fashion from least to most significant. At *Level 1*, gene matches are defined by simply using standard pairwise sequence match tools. By default, EGGS uses FASTA [18] with a Z-score cutoff score of 200. At *Level 2*, two genes are matches if they are bi-directional best hits (BBHs). This uses genome context since the constraint of being the best hit requires the best match on the whole genome. At *Level 3*, two pairs of matching genes (a quartet of genes) are considered and they are called *pair of close homologs (PCHs)*. The definition of being "close" means that two genes in one genome should be in the same run of genes. A run of genes is defined by Overbeek et al. [1] and it can be seen as a cluster of genes where distance between any adjacent genes on the same strand is within a certain threshold, say 300 bp (Fig. 4). At *Level 4*, two pairs of BBHs are considered and they are called *pair of close BBH (PCBBH)* (Fig. 4).

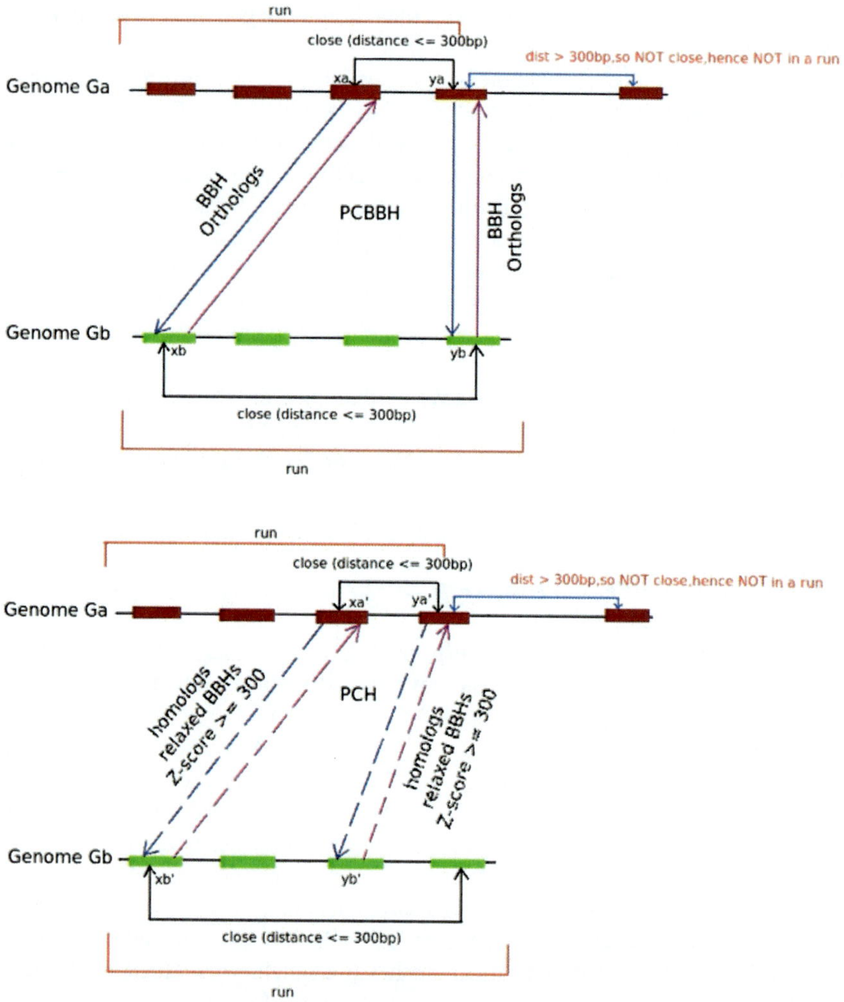

Fig. 4 Graphical representation of PCBBH and PCH

First, gene clusters are defined at the level 4 with the PCBBH criterion and then at level 3 with the PCH criterion. The basic idea to determine clusters first by merging PCBBH and PCH quartets when two quartets share a common side and then by adding more genes using BBH and pairwise matches. This can be viewed as relaxing iteratively two constraints *distance*(proximity) and *similarity* to achieve highly accurate gene cluster predictions. Below are the steps of the algorithm.

1. Compute PCBBHs and PCHs.
2. Construct a *graph* where nodes are PCBBH and PCH quartets and edges are defined when two quartets share a common side, i.e., a pair of genes.

3. Clusters are determined by computing connected components of the graph.
4. Relax the distance constraint (say < 1000 bp) and merge adjacent clusters if two clusters are within the new distance criterion.
5. Create a *hyper graph* where nodes are gene clusters and edges are created between adjacent clusters.
6. For each adjacent pair of nodes (edge of the hyper graph), introduce new BBH and pairwise matches above a certain threshold (say Z-score > = 200) if they are consistent with the clusters in terms of distance. This will create a *local graph* by defining edges between "close"(< 3000 bp) nodes.
7. Gene clusters are predicted by computing connected components of the local graph.
8. Post-process the clusters and remove those smaller than a preset threshold (say, > 3 gene pairs)

Kim et al. claimed that the gene cluster prediction based on genome context works well for distantly related genomes. For example, it was shown that EGGS was able to produce 99 gene clusters of 2268 genes between distantly related genome pairs, *Sulfolobus tokodaii* (NC_003106) and *Sulfolobus solfataricus* (NC_002754) while an optimization method FISH [19] was not able to find gene clusters.

Gene Cluster Prediction Based on a Mutable Pattern Model

Hu et al. have developed a gene cluster model called mutable patterns [20, 21]. The idea behind this method is to detect gene clusters by extending pattern mining techniques that are widely used in the data mining community. The pattern mining problem is to look for a set of items that appear frequently in many records where a record is a set of items. By "frequently," it means that the set of items occurs together at least in a certain number of records; the number of records is called the *support* of the pattern. Searching for frequent patterns is done by exhaustively looking for all maximal patterns. The maximality condition is that the pattern cannot be a sub-pattern of any pattern without changing the support value. There are several challenges involved while applying pattern mining methods to the gene cluster prediction problem.

1. There are no item labels in the gene cluster prediction problem; the protein family labels can be seen as item labels but family classification is not used since only a fraction of genes in many genomes are classified as families. Thus matching genes across multiple genomes are based on sequence similairty, which can lead to errors in matching genes.
2. The order of matchings genes in genomes can be different.
3. Gene clusters typically appear only in a subset of genomes, within a given genome set. Thus, we have to enumerate all possible combinations of genomes to find gene clusters exhaustively according to the maximality constraint, which can be computationally very expensive.

The mutable pattern method utilizes a two-scan approach to deal with these challenges. First it uses a concept of *interchangeable gene set* that is a set of genes that share sequence similiarity above a preset threshold. The first scan of genomes constructs interchangeable gene sets where a gene can belong to multiple interchangeable gene sets. Then in the second scan, it uses a concept of *reachability*. Two genes g_a and g_c are reachable if and only if there are genes $g_{c_1}, \ldots, g_{c_k}$ between g_a and g_c such that no adjacent genes in a gene sequence, $(g_a, g_{c_1}, \ldots, g_{c_k}, g_c)$, are distant in bp no more than a preset threshold, say 200 bp. This simple concept of reachability can be used to handle *distorted pattern* based on the mutable sets. In the traditional sequential pattern model, a sequence supports a pattern only when the total order defined by the pattern is contained by the total order defined by the sequence. However, in the gene cluster prediction problem, we are interested in finding groups of genes appearing in proximity of each other on a certain number of genome sequences. The exact order is not important since the order of genes in genomes could be distorted. To deal with the challenge, Hu et al. used a new pattern mining concept called *mutable order-distorted pattern*. This method has been shown to perform significantly better in terms of the COG database [3] when compared with the widely used method based on the bi-directional best hit concept.

Query-Based Approach

More recently, Yang et al. have suggested that a query-based strategy could be used to identify gene clusters conserved across a given genome set that contains hundreds of genomes [22]. Genome sequence data has greatly increased over the recent years and scalability is an important issue for the conserved gene cluster detection problem. Since the time complexity of the GCQuery algorithm, proposed by Yang et al. is $O(n^2)$, it is expected to perform really well for larger genome sets. Moreover, by querying experimentally confirmed gene clusters, it is also possible to achieve a relatively higher accuracy as well. Another key aspect of this approach is that it does not depend on gene densities or orientations within a cluster and utilizes only proximity information. This contributes greatly to its flexibility and robustness.

GCQuery assumes that genes from a cluster are provided and first, finds all the locations of all their related genes on each chromosome (genome) for a given set. The basic ideas behind this algorithm are that a window-based approach can be adopted to model the distribution of related genes on each chromosome and that they can be modeled by a hypergeometric probability distribution. In order to identify gene clusters, the expectation value (E-value) is calculated based on this distribution. An appropriate E-value cutoff across a list of windows would result in accurate gene cluster detection.

Consider a query cluster Q and chromosome c from a given set, such that each chromosome c is represented by an ordered sequence of genes $(g_1, g_2, g_3, ..., g_n)$. The set of all genes related to Q on c are first identified by considering BLASTP matches between genes from Q and c with E-values less than 10^{-7}. Thus, a subsequence $c' = (g'_1, g'_2, g'_3, ..., g'_{n'})$ is defined, where each g'_i is related to at least one

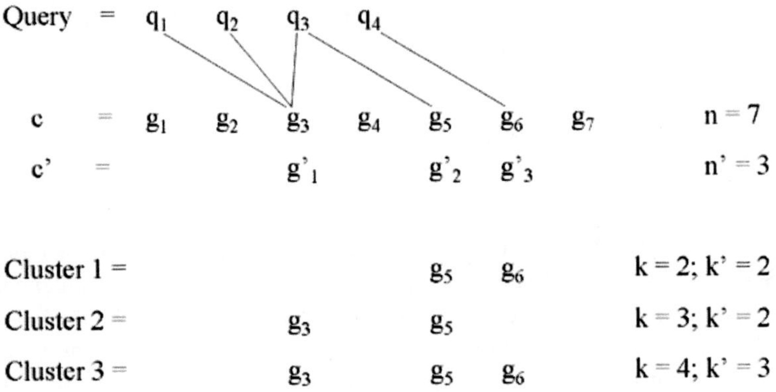

Fig. 5 An illustration depicting the basics of the GCQuery algorithm

gene in the query cluster. GCQuery then estimates E-values for each substring on c' between the jth gene and the $(j + k' - 1)$th gene, i.e., $(g'_j, g'_{j+1}, g'_{j+2}, \ldots, g'_{j+k'-1})$. These substrings are considered as potential gene clusters spanning the window on c between the ith gene and the $(i + k - 1)$th gene, i.e., $(g_i, g_{i+1}, g_{i+2}, \ldots, g_{i+k-1})$, where $g_i = g'_j$ and $g_{i+k-1} = g'_{j+k'-1}$. An example of such a window has been illustrated in Fig. 5. The E-value is estimated from the following equation for a circular chromosome c:

$$e(n, n', k, k') = n.p(n, n', k, k')$$

The left hand side of this equation represents the expected number of clusters spanning a window k. The probability on the right hand side of the equation is modeled by a hypergeometric distribution and represents the probability of finding a cluster of size at least k', that spans a window of k. This is given by:

$$p(n, n', k, k') = \sum_{i=k'}^{k} \binom{n'}{i}\binom{n - n'}{k - i} / \binom{n}{k}$$

For more details on the actual GCQuery algorithm and its time complexity analysis, we refer the reader to the original article.

Yang et al. have applied this algorithm to study gene clustering in 400 bacterial genomes spread across 18 different phylogenetic groups. For validation purposes, GCQuery was also applied to the *B. subtilis* and the *E. coli* genomes and the results were compared to experimentally verified operons. In both cases, the query clusters were 123 known operons from *E. coli*. Evaluation was done considering two criteria: $s_{\min}$ and $s_{\max}$. These terms represent the ratios of the overlap of genes between actual operons and predicted gene clusters to the minimum and maximum, respectively, of the sizes of the operon and the predicted cluster. In the case of *B. subtilis*, for a given GCQuery cutoff of 10^{-5}, the average $s_{\min}$ score was found to be 0.59 and $s_{\max}$ score was 0.31. Apart from this quantitative comparison, the authors have also discussed detailed qualitative analyses of certain operons across different bacterial groups. Sections have also been dedicated to the discussion of rearrangements within and across clusters.

Experiments

Researchers are often interested in looking at or finding gene clusters that are conserved over multiple genomes. However, this problem poses several challenges that have to be overcome. One of the biggest challenges is that the problem size grows rapidly as the number of genomes increases. In order to tackle this problem, we need an effective gene cluster prediction algorithm. Availability of gene clusters from given genomes allows scientists to further explore functionally coupled genes conserved over multiple genomes [1]. Also, predicting gene clusters that are conserved over well studied genomes and draft genomes provides a starting point for annotation. In this section, we propose an algorithmic framework which predicts gene clusters and demonstrate its ability and application by conducting a small experiment.

Formulation and Algorithm

Due to the rapidly growing problem size, an algorithm to solve this problem must be able to effectively cut down the problem size and still be able to compute a set of gene clusters. One way of addressing this issue is to utilize phylogeny information. Assuming that phylogenetic distance reflects the rate of divergence between two organisms, we can incorporate the phylogenetic tree with distance, of n genomes as a guide to hierarchically compute gene clusters. Then, we can give the following problem formulation. Given all pairwise gene cluster sets and a phylogenetic tree of n genomes, for each internal node, compute a set of gene clusters over all of the child genomes of the node. Let's call this algorithm PhyloEGGS (Fig. 6) and it is as follows:

Input: A set of all pairwise gene clusters sets and a phylogenetic tree of n genomes.
Output: A set of gene clusters conserved over n genomes.
Tree Traversing Direction: left to right and bottom-up

For each internal node,

1. Compute the intersection between the gene clusters of both subtrees of the current node based on the pairwise gene clusters of two closest genomes defined by phylogenetic distance.
2. Expand the result by rescuing based on sequence similarity results.

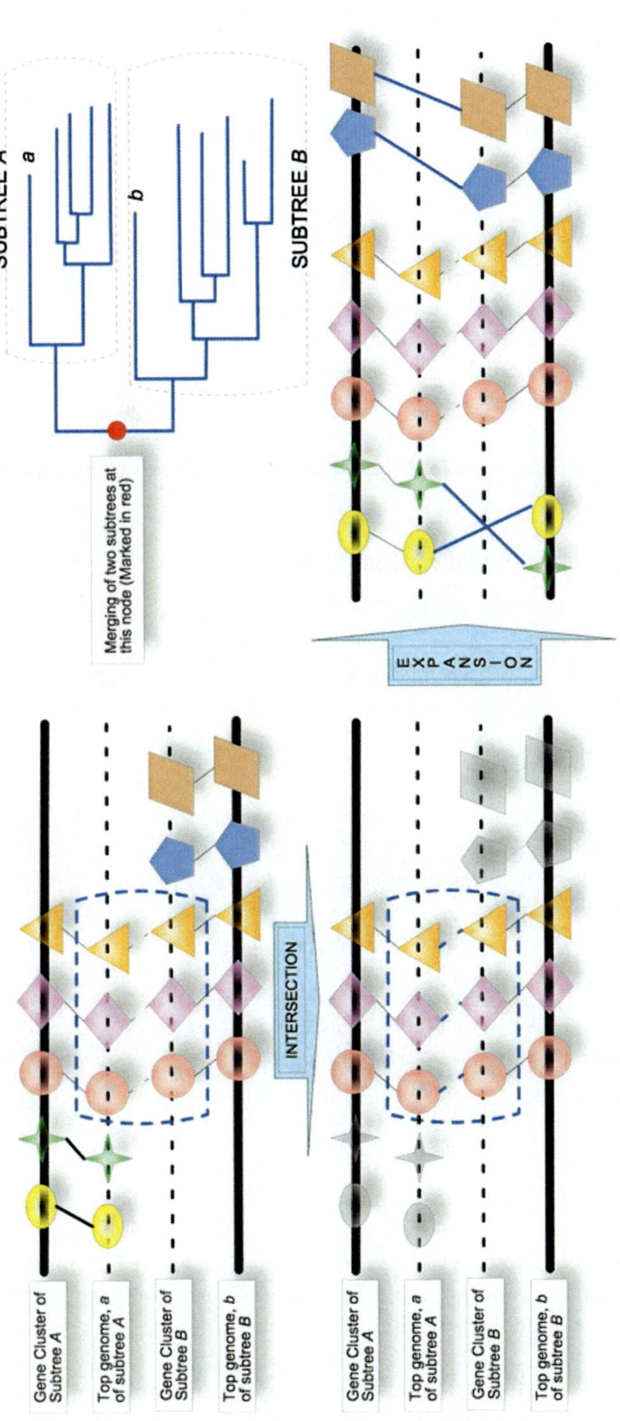

Fig. 6 Pictorial representation of PhyloEGGS algorithm: the figure demonstrates how PhyloEGGS computes clusters given a phylogenetic tree along with gene clusters from both subtrees. The cluster is computed at the node labelled *red*

Demonstrative Experiments

Gene Cluster Prediction

This experiment demonstrates how gene clusters of multiple genomes are predicted by PhyloEGGS.

Methods

1. Selected the following 6 genomes: *Escherichia coli* str. K-12 substr. MG1655, *Shigella flexneri* 2a str. 301, *Salmonella typhimurium* LT2, *Enterobacter* sp. 638, *Yersinia pestis* CO92, and *Haemophilus influenzae* Rd KW20.
2. Using EGGS, computed all pairwise gene cluster sets.
3. Constructed a phylogenetic tree using 16s rRNA sequences from the six genomes.
4. Applied PhyloEGGS to compute the set of gene clusters over the genomes.

Annotation Assignment based on Gene Cluster

In this experiment, we used Annotation Confidence Score (ACS) [23] to assign annotation to a given gene cluster. ACS is a annotation scoring system for existing genome annotation. ACS computes a confidence score to each annotation of a target genome by comparing annotations of a set of selected reference genomes.

Methods

1. Ran ACS with each genome as a target genome and the rest of the genomes as reference genomes.
2. Assumed that there is no annotation for the *i*th genome from the gene cluster prediction experiment
3. For each gene of the *i*th genome of each gene cluster,
   a. Compared ACS scores of the corresponding genes of the rest of the genomes.
   b. Selected the annotation with the highest ACS score.
   c. Assigned the selected annotation to the gene.

Results

Given a draft or newly sequenced genome, one of the most common ways to infer functions of genes is based on sequence homology. BLAST often serves as a popular tool to make this homology based inference. However, often in this type of inference, there are occasions where it is difficult to assign annotation with confidence due to weak homology or multiple matching genes. The context of gene clusters provides more confidence in such cases. Consider a case where certain genes are observed as a tightly conserved cluster over multiple genomes and one of the genes has a weak match in another genome. In this case, we can infer, with more confidence, that the weakly matched gene in another genome has similar function. Alternatively, when there is a tightly conserved gene cluster over multiple genomes, one or more genes in the cluster may have multiple matching genes in another genome. In this case, since the proximity information is utilized when computing gene clusters, the algorithm is able to pick the best gene; Hence resulting in more confident assignment of annotation. Here we provide a few examples of such clusters reported in the previously described experiment.

Case 1 : Gene clusters with weak pairwise matches

Here is a gene cluster of the sigma E heat shock sigma factor cluster [24] conserved over the six genomes, *rpoE-rseABC*. These four genes are tightly conserved in *E. coli*, *S. flexneri*, *S. typhimurium*, *Enterobacter*, and *Y. pestis* and BLAST E-values are very small. However, in *H. influenzae*, BLAST E-value of *rseC* to its orthologous genes in the rest of genome is much higher for this specific gene. Also, unlike in the other five genomes, *rseC* is remotely located from the rest of the genes on this genome. Despite the high e-value and remoteness, the context of gene cluster suggests a clear indication that they should be clustered as functionally coupled genes. To verify, a profile HMM for *rseC* genes from the five genomes was built using HMMER [25]. Then, the program, hmmsearch was invoked to search rseC from *H. influenzae* against the model and the e-value of the hit provided by HMMER was significantly low, hence, supporting the case.

Case 2: Gene clusters with multiple matches

dmsABC is a gene cluster responsible for encoding anaerobic dimethylsulphoxide reductase [26]. All of the three genes are very tightly conserved among the six

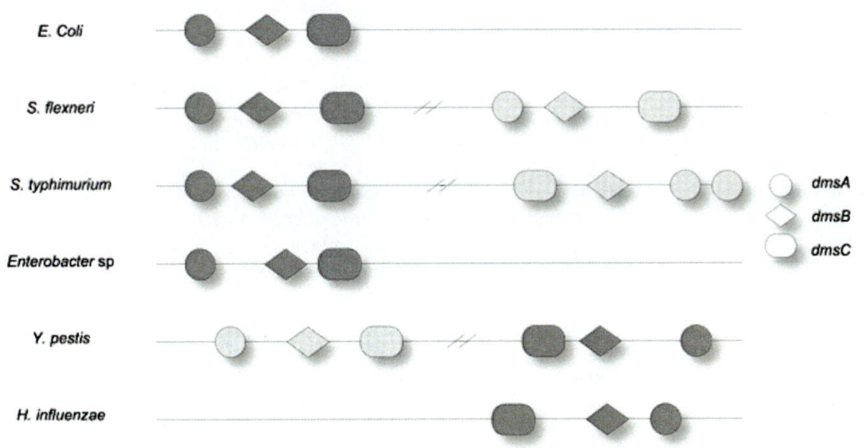

Fig. 7 Pictorial representation of *dmsABC* cluster over six genomes. *Darker shaded shapes* represent the orthologous copies and *lighter shaded shapes* represent the paralogous copies. The figure is just a pictorial representation and it is not drawn to scale

genomes in terms of sequence similarities and relative positions. Interestingly, these genomes happen to have one or more paralogs to at least one of *dmsA*, *dmsB*, and *dmsC*. *S. flexneri*, *S. typhimurium*, and *Y. pestis* actually have a paralogous cluster that is tightly conserved and most of the paralogs have high sequence similarities (Fig. 7). In case of the *dmsB* genes of *E. coli* and *S. typhimurium*, they have strikingly high sequence similarity both in terms of identity (over 95%) and E-value. When there are multiple matches that have such high sequence similarity, it is often very difficult to select a correct pair with a simple homology based approach. Consideration of gene cluster context can often help in this situation. The proposed algorithm was able to predict *dmsABC* cluster with only orthlogous copies over the six genomes.

Summary

In this chapter, we have briefly discussed methods that enable the prediction of gene clusters. Each of these methods have unique approaches that provide unique advantages. They have been summarized in Table 1 The approach for the KEGG database as outlined by Fujibuchi et al. is deterministic in nature and is rigorous in its approach. However, it suffers from a lack of scalability. This problem is circumvented by the GCQuery algorithm. However, a query-based strategy assumes the knowledge of known clusters to begin with and may thus, be a limiting factor when it comes to the discovery of novel gene clusters. The probabilistic approach prescribed by Zheng et al. seems to strike a balance between the above approaches. However, probabilistic approaches are based upon assumptions that may result in

Table 1 Summary of existing approaches to detecting gene clusters

Method	Section	Overview
KEGG graph-based approach [7]	Graph-based approach	Starts with gene clusters conserved across a pair of genomes (obtained by graph comparison) and extends them to multiple genomes by P-quasi linkage analysis
GeneChords [13]	Evolutionary model-based approach	Uses a stochastic evolutionary model for the conservation of gene clusters to identify potentially related genes
EGGS [17]	EGGS: gene pattern prediction based on genome context	Detects gene clusters conserved across a pair of genomes by iteratively merging PCBBHs, PCHs, BBHs and pairwise matches
Mutable pattern model approach [20, 21]	Gene cluster prediction based on a mutable pattern model	Uses a pattern mining technique to predict gene clusters across multiple genomes
GCQuery [22]	Query-based approach	Uses a hypergeometric distribution to model the occurrence of conserved gene clusters within a given window, across multiple genomes

conservative results. Therefore, the use of a method to detect conserved gene clusters across multiple genomes greatly depends on the requirements of the researcher. The decision to use an appropriate method for cluster detection solely depends on whether a researcher prefers accurate coverage or a lesser runtime for a larger experiment or the application of evolutionary constraints. However, with all their merits and demerits, the above methods have certainly paved the way for breakthroughs in structural, functional and comparative genomics and have proved to be useful tools in the study of the co-localization of genes in bacterial genomes.

The detection of conserved gene clusters is still an open area of research and we still have not reached the full potential of this domain yet. Apart from shedding light on genomic structure and evolutionary patterns, gene clusters provide useful insights into gene function and serve as good additions to known annotation methods. We have shown that gene clusters can be really effective when resolving ambiguities obtained by the simple homology method. When genes show really weak sequence similarity or show matches to multiple genes, gene clusters provide greater confidence in assigning function. Additionally, novel genes found within known clusters can also be annotated based on their neighbors. Thus, improved solutions to the gene cluster detection problem can be regarded as major comparative genomics approaches for genome annotation and genome-wide function prediction.

We have discussed some of the benefits of using genome context information for annotation by mainly focusing on the problem of gene cluster detection. It becomes obvious that methods utilizing genome context can serve as a complementary alternative to homology-based methods. A good example of using such genome context information for characterizing gene functions is MSOAR [27], which makes ortholog assignments based on genome rearrangement information. We expect such methods to be more prominent and widely used for gene function assignment in the years to come.

References

1. Overbeek, R., Fonstein, M., D'Souza, M., Pusch, G.D., Maltsev, N. The use of gene clusters to infer functional coupling. Proc. Nat. Acad. Sci. **96**(6): 2896–2901 (1999).
2. Overbeek, R., et al. The subsystems approach to genome annotation and its use in the project to annotate 1000 genomes. Nucleic Acids Res. **33**(17): 5691–5702 (2005).
3. Tatusov, R.L., Koonin, E.V., Lipman, D.J. A genomic perspective on protein families. Science **278**(5338): 631–637 (1997).
4. He, X., Goldwasser, M. Identifying conserved gene clusters in the presence of orthologous groups. Proceedings of RECOMB, San Diego, CA, USA, pp. 272–280 (2004).
5. Kim, S., Choi, J., Saple, A., Yang, J. A hybrid gene team model and its application to genome analysis. J. Bioinform. Comput. Biol. **4**(2): 171–196 (2006).
6. Kim, S., Choi, J., Yang, J. Gene teams with relaxed proximity constraint. IEEE Comput. Syst. Bioinform. CA, USA, 44–55.
7. Fujibuchi, W., Ogata, H., Matsuda, H., Kanehisa, M. Automatic detection of conserved gene clusters in multiple genomes by graph comparison and P-quasi grouping. Nucleic Acids Res. **28**(20): 4029–4036 (2000).
8. Matsuda, H., Ishihara, T., Hashimoto, A. Classifying molecular sequences using a linkage graph with their pairwise similarities. Theor. Comput. Sci. **210**(2): 305–325 (1999).

9. Ogata, H., Fujibuchi, W., Goto, S., Kanehisa, M. A heuristic graph comparison algorithm and its application to detect functionally related enzyme clusters. Nucleic Acids Res. **28**(20): 4021–4028 (2000).
10. Smith, T.F., Waterman, M.S. Identification of common molecular subsequences. J. Mol. Biol. **147**(1): 195–197 (1981).
11. Kanehisa, M., Goto, S. KEGG: Kyoto encyclopedia of genes and genomes. Nucleic Acids Res. **28**(1): 27–30 (2000).
12. http://www.genome.jp/kegg/ssdb/
13. Zheng, Y., Anton, B.P., Roberts, R.J., Kasif, S. Phylogenetic detection of conserved gene clusters in microbial genomes. BMC Bioinform. **6**(243) (2005).
14. Altschul, S.F., Gish, W., Miller, W., Myers, E.W., Lipman, D.J. Basic Local Alignment Search Tool. J. Mol. Biol. **215**(3): 403–410 (1990).
15. Gama-Castro, S., et al. RegulonDB (version 6.0): gene regulation model of *Escherichia coli* K-12 beyond transcription, active (experimental) annotated promoters and Textpresso navigation. Nucleic Acids Res. **36**: D120–D124 (2008).
16. http://genomics10.bu.edu/cgi-bin/GeneChords/GeneChords.cgi
17. Kim, S., Bhan, A., Maryada, B.K., Choi, K., Brun, Y.V. EGGS: extraction of gene clusters by iteratively using genome context based sequence matching techniques. IEEE International Conference on Bioinformatics and Biomedicine, Silicon Valley, CA, USA, pp. 23–28 (2007).
18. Pearson, W.R., Lipman, D.J. Improved tools for biological sequence comparison. Proc. Nat. Acad. Sci. **85**(8): 2444–2448 (1988).
19. Calabrese, P., Chakravarty, S., Vision, T.J. Fast identification and statistical evaluation of segmental homologies in comparative maps. Bioinformatics **19**: 74–80 (2003).
20. Hu, M., Choi, K., Su, W., Kim, S., Yang, J. A Gene Pattern Mining Algorithm using mutable sets for prokaryotes. BMC Bioinform. **9**: 124 (2008).
21. Hu, M., Yang, J., Su, W. Permu-pattern: discovery of mutable permutation patterns with proximity constraint. Proceedings of the ACM SIGKDD International Conference on Knowledge Discovery and Data Mining, Las Vegas, NV: USA, pp. 318–326.
22. Yang, Q., Sze, S. Large-scale analysis of gene clustering in bacteria. Genome Res. **18**: 949–956 (2008).
23. Yang, Y., Gilbert, D., Kim, S. Annotation confidence score for genome annotation: a genome comparison approach. Bioinformatics **26**(1): 22–29 (2010).
24. Raina, S., Missiakas, D., Georgopoulos, C. The *rpoE* gene encoding the sigma E (sigma 24) heat shock sigma factor of Escherichia coli. The EMBO Journal **14**(5): 1043–1055 (1995).
25. http://hmmer.org/
26. Bilous, P.T., Cole, S.T., Anderson, W.F., Weiner, J.H. Necleotide sequence of the dmsABC operon encoding the anaerobic dimethylsulphoxide reductase of Escherichia coli. Mol. Microbiol. **2**(6): 785–795 (1998).
27. Fu, Z., Chen, X., Vacic, V., Nan, P., Yang, Z., Jiang, T. MSOAR: a high-throughput ortholog assignment system based on genome rearrangement. J. Comput. Biol. **14**(9): 1160–1175 (2007).

Functional Inference in Microbial Genomics Based on Large-Scale Comparative Analysis

Ikuo Uchiyama

Abstract By virtue of the recently accumulated microbial genomic sequence data, we can utilize this large amount of information for predicting and understanding microbial gene functions and microbial life through comparative analysis of variously related genomes. Here, I introduce basic concepts and issues in microbial comparative genomics including ortholog analysis and core genome analysis, and introduce our approach to large-scale comparative genomics focusing on our database, Microbial Genome Database for Comparative Analysis (MBGD) and related methods and tools.

Introduction

Since the completion in 1995 of the first and second whole-genome sequences of cellular organisms, i.e., those of *Haemophilus influenzae* and *Mycoplasma genitalium* [1, 2], comparative genome analysis has played a central role in understanding microbial life from a genomic perspective. Indeed, the first comparison of the completed *H. influenzae* genome and the then incomplete *Escherichia coli* genome was conducted soon after the first whole-genome sequence became available [3]. Subsequently, the minimal-gene-set concept was proposed on the basis of comparisons between the genome sequences of *H. influenzae* and *M. genitalium* [4]. During these initial studies, the problem of distinguishing orthologs from paralogs was recognized as being important for identifying equivalent genes between genomes through exhaustive homology analysis.

The terms "orthology" and "paralogy" were originally introduced into the molecular evolution field by Fitch in 1970, in order to denote different types of homologous relationships [5]. Orthology is a type of homology that is derived from speciation, while paralogy is a type of homology that is derived from duplication.

I. Uchiyama (✉)
Laboratory of Genome Informatics, National Institutes of Natural Sciences, National Institute for Basic Biology, Nishigonaka 38, Myodaiji, Okazaki, Aichi 444-8585, Japan
e-mail: uchiyama@nibb.ac.jp

D. Kihara (ed.), *Protein Function Prediction for Omics Era*,
DOI 10.1007/978-94-007-0881-5_4, © Springer Science+Business Media B.V. 2011

This distinction is essential in molecular phylogenetics, since the phylogeny of a given set of orthologs, by definition, always coincides with the species phylogeny. More importantly in the genomics context, gene functions are typically conserved between orthologs, whereas paralogs generally acquire different functions; this is a consequence of the famous concept of "evolution by gene duplication" proposed by Ohno in 1970 [6]. Therefore, ortholog identification is crucial for identifying a pair of genes that have the same function in different genomes, which is a central task in genome annotation.

As of the end of 2009, more than 1,000 genomic sequences have been determined, and the number of completed sequences continues to grow. Ortholog identification has been proved to be effective and is now routinely used in the genome annotation process. However, ortholog identification is not just similarity-based prediction of individual gene functions; it can also serve as a basis for any kind of comparative genomics study, whose goals include extracting useful information for understanding the diversity and evolution of living systems.

Here, I focus on microbial (mainly prokaryotic) genome comparison. Thanks to their small genome sizes with a relatively simple organization, prokaryotic genome sequences have accumulated rapidly, and although more complex eukaryotic genomic sequences have also accumulated recently, prokaryotic genomics is still at the forefront of large-scale comparative genomics studies. This is partly because of the large diversity of microbes whose habitats span a broad range of environments, and partly because of their dynamic genome evolution due to horizontal gene transfers between distantly related organisms. Moreover, the emerging field of metagenomics, which analyzes sequences of collective microbial genomes obtained from environmental samples, is now addressing the issue of elucidating the structure, function and evolution of microbial communities in their natural habitats.

In this chapter, I introduce some of the recent problems of microbial genomics as well as the basic concepts and methodologies of comparative genomics, including ortholog analysis, and then I introduce our approach to this problem, focusing on our database, the Microbial Genome Database for Comparative Analysis (MBGD) [7, 8], and some related methods and tools.

General Schemes in Comparative Genomics

As described above, the number of completed microbial genome sequences has been increasing during the past decade; more than 1,000 genomes are now completed, and the number continues to grow (Fig. 1a). Reflecting the variety of relationships between microbes and human life, the motivations behind the sequencing projects have also been diverse, including medical, biotechnological, agricultural and environmental concerns (Fig. 1b). In addition, there are thousands of ongoing sequencing projects, and those recently launched include large-scale microbial genome projects, such as the Human Microbiome Project (HMP) [9] and the Genomic Encyclopedia of Bacteria and Archaea (GEBA) [10], both of which are

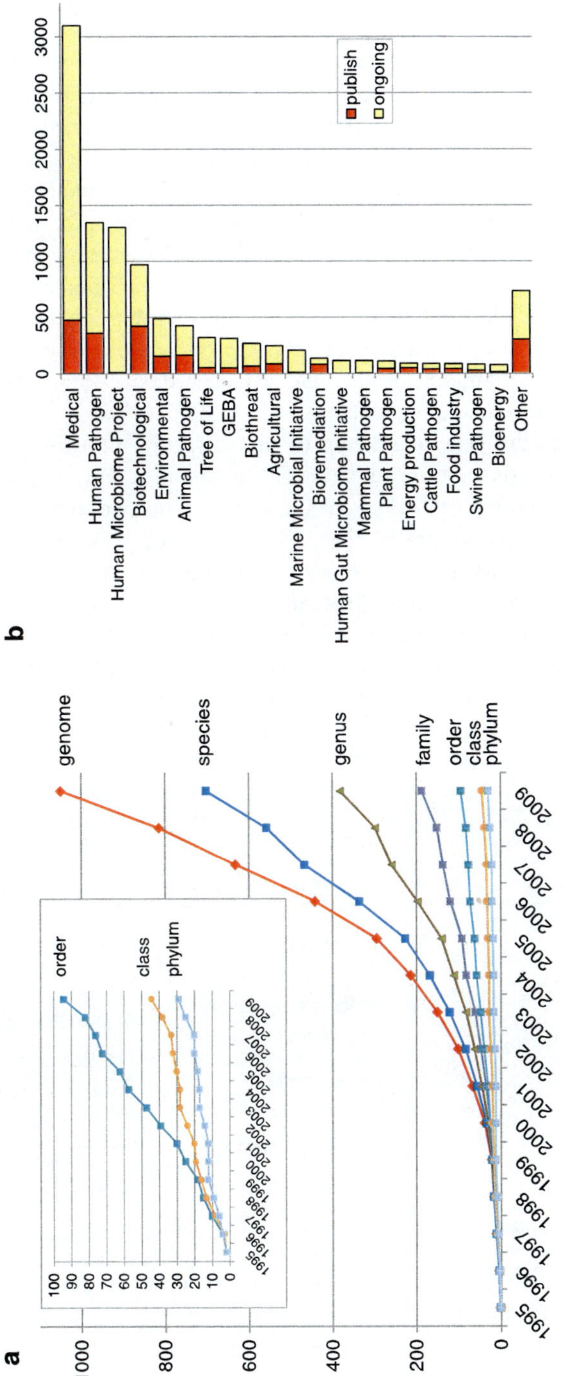

Fig. 1 The current status of microbial genome sequencing projects. The data were taken from the GOLD database [54]. (**a**) Growth of the (cumulative) number of published microbial genomes and the numbers of unique taxonomic classes thereof. The *inset* shows a magnified view of the highest three ranks. (**b**) Relevance of the published and ongoing genome projects. Note that multiple relevance values may be assigned to each genome, so the total number of this graph does not coincide with the total number of published/ongoing genomes

generating sequence data on tens or hundreds of genomes covering a broad range of organisms (Fig. 1b).

However, when we look at the taxonomical breakdown of the completed genomes, the number of unique species is about 700, and the number of unique genera is less than 400 (Fig. 1a). Therefore, the majority of the newly sequenced genomes are closely related to some genomes which have already been determined. On the other hand, the higher taxonomic ranks, i.e., phyla, classes and orders, appear to be almost saturated. However, when we look at the figure on a smaller scale (Fig. 1a, inset), we can see that the number of phyla is still increasing gradually. In fact, it is considered that more than 99% of microbes from environmental samples are unculturable, and our knowledge of the extent of microbial diversity even on the phylum level is still very limited. Therefore, the challenge is to extract useful information from the vast amount of genomic sequence data, taking into account the taxonomic relatedness among organisms.

Figure 2 illustrates the general scheme of the extraction of various types of information from comparisons of differently related genomes. For comparisons of closely related genomes, we can construct nucleotide sequence alignments from which we can enumerate polymorphic sites in order to understand the elementary processes of genome evolution or extract well-conserved regions in order to infer functionally important sites. For more distantly related genomes, it may become difficult to align nucleotide sequences correctly, but we can still identify the conservation of gene order on a chromosome (hereafter, we use the term "synteny" to refer to the concept of gene order), from which we can infer mid-term evolutionary process. For genomes which are further distantly related, it becomes difficult to identify syntenic conservation, but we can still identify orthologous relationships and

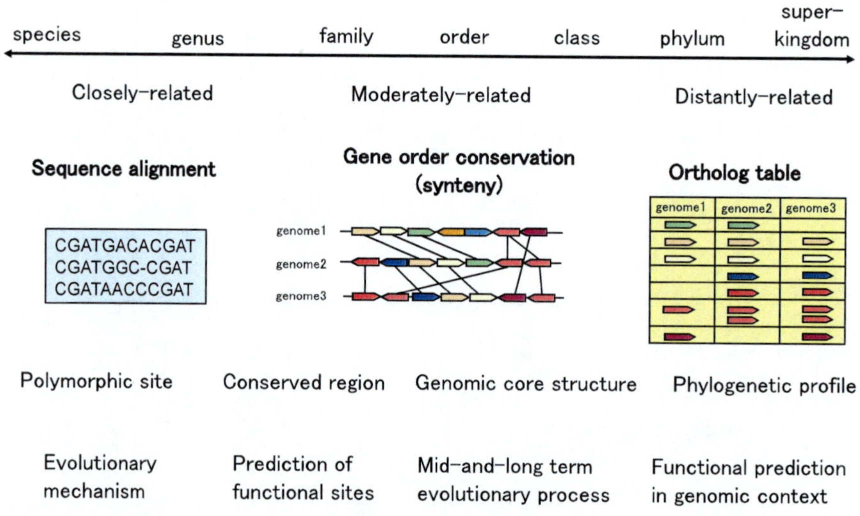

Fig. 2 Various approaches to comparative genomics for variously related organisms

use this information to infer gene functions. Moreover, gene neighborhood relationships among several genes may be conserved even between such distantly related organisms, which strongly suggests functional coupling among these genes.

There are several methods and tools for calculating [11–14] and visualizing [15–17] a genomic sequence alignment (including ours [18]), but this is beyond the scope of this chapter. Here, I focus on the issue of functional inference based on genome comparisons among distantly or moderately related organisms.

Functional Inference Based on Comparative Genomics

Comparative sequence analyses, including similarity searches (e.g., by BLAST [19]) and motif searches (e.g., by InterProScan [20]), are the most commonly used approach for gene function prediction. In fact, although protein or gene function was traditionally described only in natural languages that have quite complex semantics, the sequence-similarity-based approach has successfully been applied to the problem of functional inference as an almost universal tool. The basic logic behind this approach is quite simple: transferring the annotation from similar sequences identified in the database to the query sequence. However, determining what information can be allowed to be transferred and what cannot is generally not easy, and the actual annotation process usually requires a more complex manual task called "curation." Recent developments in functional genomics resources, such as GeneOntology (GO) [21] and the Kyoto Encyclopedia of Genes and Genomes (KEGG) [22], have greatly improved the situation and facilitated the computerization of the annotation process. In GO, the terms (or concepts) required to describe various gene functions are arranged in a directed acyclic graph, in which each node represents a term and each edge represents a relationship between terms, where types of relationships include `is_a`, `part_of` and `regulates`. To address the ambiguity of the term "gene function," there are three top-level categories in the GO hierarchy: molecular function, biological process and cellular component. This semantic structure helps overcome the difficulty in dealing with complex gene functions. On the other hand, KEGG represents molecular functions using pathway map diagrams, which are more intuitive for human cognition. KEGG also provides a hierarchical gene function classification scheme that is much simpler than that of GO, which is, in many cases, practically more useful for microbial genome annotation and comparison.

Advanced similarity-based methods for functional inference have also been developed. Traditionally, many research endeavors have focused on the issue of improving the sensitivity of similarity searches; i.e., the possibility of detecting more subtle similarities. Methods using a scoring system based on profiles [23], the position-specific scoring matrix (PSSM) [19, 24] and the profile hidden Markov Model (HMM) [25] are the most important advancements in this direction and have been successfully applied to the identification of distantly related proteins of similar folds. However, although the use of such a sensitive method can allow the prediction of functional similarities even between weakly similar sequences, the more distantly related two proteins are, the more difficult it generally becomes to infer the exact function, because homologous proteins can acquire versatile functions during the course of evolution. Nonetheless, identifying locally conserved motifs that are likely to be related to functionally important sites, such as ligand-binding sites and catalytic active sites, often allows researchers to infer the functions of distantly related proteins in detail. In any case, however, this approach is applicable only to the inference of biochemical molecular functions, and is difficult to apply to the prediction of biological functions in the context of cellular processes.

In this regard, comparative genomics based on orthology analysis is another important approach to this issue. The basic assumption behind this approach is that orthologs have an equivalent function in different organisms. This assumption allows the user to transfer annotations from one gene to its orthologs, not only in terms of biochemical functions, but also in terms of biological functions. Moreover, this scheme is simple enough to be easily extended to the whole genome, and whole-genome comparative analysis provides additional "genomic context" information, which allows more powerful functional inference. For example, if a genome contains the entire set of genes encoding a certain biochemical pathway, this strongly indicates that this genome contains the entire pathway, whereas if the genome contains only one of the genes in the pathway, it is unlikely that the genome contains that pathway, and therefore it is likely that the gene has lost its function under this pathway and probably plays a different role, even if the similarity to the gene of known function is sufficiently high. Alternatively, if the genome contains all but one of the genes in the pathway, it is likely that the genome has this pathway, and some gene that is not orthologous to the missing gene may play the role of this missing gene [26]. Such a phenomenon, called "non-orthologous gene displacement" [27], apparently represents one of the limits of ortholog analysis, but this limit can often be overcome by an inference using the genomic context described above, together with a more detailed analysis based on motif or profile analysis (see ref. [28] for various examples of this type of analysis).

The basic idea behind the above consideration is that gene function can be defined in the context of molecular interactions, and that genes which function together should, in principle, coexist in organisms that have this function. This idea has led to the establishment of a function prediction method called the "phylogenetic pattern (or profile) method" (Fig. 3a) [29]. A phylogenetic pattern is defined as a binary vector for each orthologous group that represents the presence (1) or absence (0) of genes in each genome, and, in this method, two orthologous groups are predicted to have a related function if they have similar phylogenetic patterns. Several other methods that predict functional linkages between genes have also been investigated, including the domain fusion method [30, 31] and the gene neighborhood method [32]. In the domain fusion method (also known as the "Rosetta Stone method") (Fig. 3b), when two proteins (domains) are fused into a single gene in a genome, the fusion protein is considered to be evidence (Rosetta Stone sequence) of a functional linkage between the two fused proteins. In the gene neighborhood

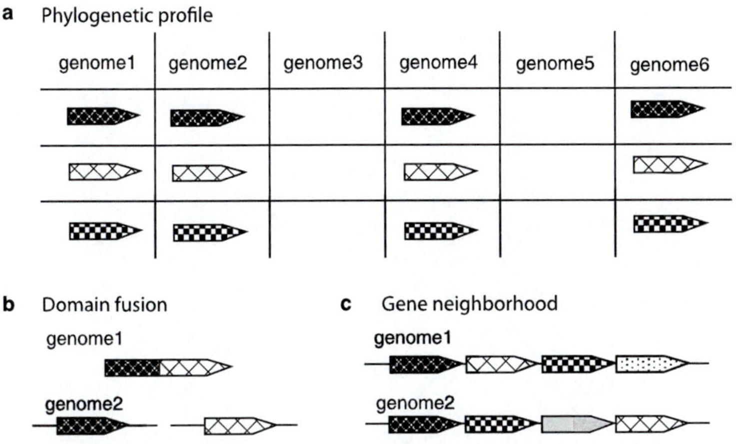

Fig. 3 Prediction of functional links between genes by comparative genomics approaches. (**a**) Phylogenetic profile method, (**b**) domain fusion method, (**c**) gene neighborhood method. Genes with the same pattern represent orthologs among different organisms

method (Fig. 3c), two genes that are identified as neighbors in several different genomes are considered to have related functions. This method is especially effective in comparisons of prokaryotic genomes, where functionally related genes often form operons. The prediction accuracy of this type of analysis can be further improved by combining it with multiple methods. For example, STRING [33] uses the above three methods along with additional methods, including co-expression (genes that have similar expression patterns), text-mining (co-occurrence of gene names in scientific texts) and simple homology (homologs of interacting genes), and evaluates the reliability of the interaction by combining multiple sources of evidence.

The inference methods using the genomic context are sometimes called "non-homology methods" [34], since they do not directly use homology to infer functions, as in the traditional homology method. However, they do use homology in an indirect manner. In fact, these methods are based on the assumption that functional linkages between proteins in one organism are conserved in different organisms, and ortholog identification is a crucial part of identifying corresponding genes between organisms.

Ortholog Classification Problem

As described in the Introduction, "ortholog" and "paralog" are terms that are defined on the basis of evolutionary events that generate homologous genes, i.e., speciation and duplication, respectively [5]. Therefore, orthology and paralogy are binary relationships between two genes, and whether two given homologous genes are

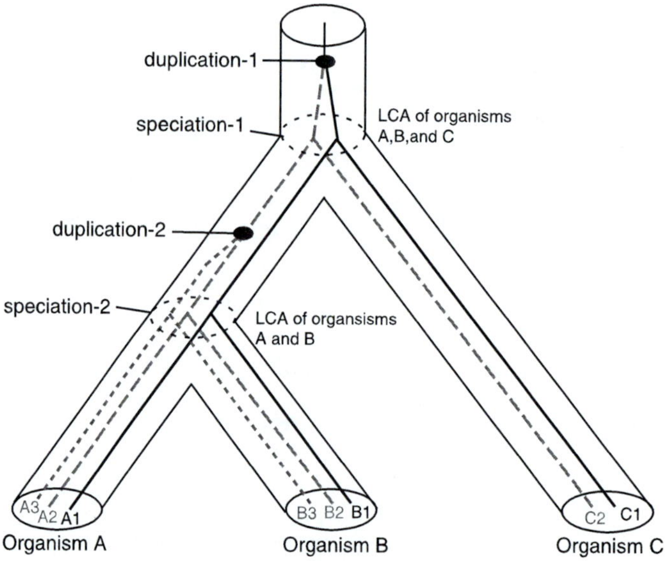

Fig. 4 The ortholog-paralog distinction problem. Pairs of genes indicated by the same line type, such as (A1, B1), (B2, C2) and (A3, B3), are orthologs, since they originated as a result of a speciation event. On the other hand, (B2, C1) and (B2, A3) are paralogous, since they originated from a duplication-1 or -2 event, respectively, while (C2, A3) is orthologous, since it originated from the speciation-1 event. To define orthologous relationships among multiple organisms, one must consider the speciation of the last common ancestor (LCA) of the target organisms as the reference speciation event and separate the paralogous lineages generated by duplications before this speciation. In this case of the three organisms A, B, and C, speciation-1 is the reference speciation event, and there are two orthologous groups, (A1, B1, C1) and ([A2, A3], [B2, B3], C2). The paralogous genes (B2, C1) are called *outparalogs*, since they were separated by a duplication *before* the given speciation event, while the (B2, A3) genes are called *inparalogs*, since they were separated by a duplication *after* the given speciation event [36]. In general, outparalogs are classified into different orthologous groups, whereas inparalogs are classified into the same orthologous group and form many-to-many orthologous relationships. Note that these definitions are dependent upon the set of target organisms

orthologs or paralogs can be determined unambiguously by examining whether the type of event that generated them is speciation or duplication, provided that the phylogenetic gene tree and species tree are exactly known (Fig. 4). For example, in Fig. 4, the gene pairs (A1, C1) and (B2, C2) are orthologs because both of them originated from a speciation event (speciation-1), whereas (A1, A2) and (B2, C1) are paralogs because both of them originated from a duplication event (duplication-1).

Since orthology is a binary relationship, it is natural that orthology is defined in pairwise genome comparison, and the simple-criterion bidirectional best-hit (BBH) is often used in such a situation: two genes, a and b in genomes A and B, respectively, are BBH if a is the best-hit of b in genome A and b is the best-hit of a in genome B. This strategy for identifying orthologs is very simple and does not require any knowledge of species phylogeny, and yet it typically gives results in good agreement with those of more complicated methods. Therefore, the BBH strategy has been widely used for the genome-wide identification of orthologs.

However, there are some cases where the BBH criterion misidentifies the correct orthologs. First, the BBH strategy generally assumes one-to-one orthologous relationships, whereas many-to-many orthologous relationships can be generated by duplication in each lineage after speciation. The BBH strategy can misidentify some orthologous relationships, although in such cases, all pairs of homologous genes between the two species should be orthologous since they all originated from a single speciation event. For example, both (A2, C2) and (A3, C2) in Fig. 4 are orthologs because both originated from speciation, but one of the relationships can be missed by the BBH strategy when one of the scores is significantly smaller than the other. According to the terminology proposed recently [35, 36], paralogs that are generated as a result of duplication before and after a given speciation event are called "outparalogs" and "inparalogs", respectively (Fig. 4); thus, the lineages of outparalogs, but not inparalogs, should be separated in the ortholog identification process. A solution to this problem requires a clustering procedure that incorporates inparalogs into an orthologous group and correctly distinguishes them from outparalogs. Inparanoid [35] was the first to address this problem, and gives one of the best-known solutions.

Second, paralogous genes may become a BBH pair when both of their respective orthologs have been lost, although a single gene-loss event generally does not make them a BBH pair. For example, in Fig. 4, B1 and C2 can become a BBH pair if both B2 and C1 are lost during the course of evolution. This problem is difficult to overcome when using only two genomes, but can be overcome by using a third genome [37].

Third, although orthology is originally defined on a pairwise basis, for simultaneous comparison of multiple genomes, it is necessary to derive sets of orthologs, or orthologous groups, among multiple genomes. The single-linkage clustering algorithm or a related method is often used for constructing homologous gene groups, or families, from pairwise homology relationships. The rationale behind this method is that the homology relationship is transitive, i.e., the homolog of a homolog of a certain gene is a homolog of that gene. However, the orthology relationship is not necessarily transitive [38], and, therefore, it is not clear how to extend binary BBH relationships into an orthologous relationship for multiple genomes. Some sophisticated graph-based clustering algorithms have been developed to improve this situation. For example, the Markov Clustering (MCL) algorithm [39] evaluates transitive similarities with a probability measure, instead of just assuming transitivity, using the equilibrium probabilities of a Markov chain defined on the similarity graph. The MCL algorithm is applied in a protein classification program, TribeMCL [40], which is also successfully applied in an ortholog classification program, OrthoMCL [41].

Alternatively, a phylogenetic tree-based algorithm is generally more accurate than a graph-based algorithm with BBH relationships for identifying ortholog/paralog relationships in multiple genomes. Here, an orthologous group is defined as a set of homologous genes that are derived from the speciation event in the last common ancestor of the target set of organisms. Therefore, one can identify orthologous groups by first assigning a speciation or duplication event to each node of the gene phylogeny and then identifying the nodes corresponding to the target speciation event. The plausible assignment of speciation/duplication events can be done by a procedure for reconciling the gene tree with the species tree [42–44], provided that the exact topologies of both trees are known. One of the drawbacks of this approach is that constructing an accurate phylogenetic tree is generally not easy and requires a long computation time for genome-scale analysis. It is also difficult to assume the knowledge of the exact topology of the species tree. Incorrect assumption of the tree topology for either the gene tree or the species tree leads to incorrect assignment with many more duplication events than in reality. Therefore, although the theoretical advantage of tree-based methods over BBH-based methods is obvious, a straightforward implementation of this approach is not suitable for large-scale automatic classification. Instead of using a precise tree reconciliation approach, some programs [45] including our own, DomClust [46] (see below), uses a simpler "species overlap" criterion where only intraspecific paralogous genes are considered to be excluded from an ortholog cluster. One benchmark test concluded that a species reconciliation method cannot outperform a species overlap method even when a trusted species tree can be assumed [47].

Horizontal gene transfer (HGT) is quite a common event in prokaryotic evolution, but is ignored in the traditional ortholog/paralog distinction scheme. HGT is another event that generates homologous genes within a genome, and the existence of HGT nullifies the above ortholog/paralog assignment rule. In fact, Fitch introduced another term, "xenolog": two homologous genes are defined as xenologs when their gene history since the time of their common ancestor involves an HGT event [38, 48]. However, although several methods have been developed to address the issue [49, 50], it is generally not easy to identify HGT events exactly, even when the true topologies of both the gene tree and the species tree are known.

However, as discussed above, an orthology relationship is typically used for functional inference through "annotation transfer" between orthologous proteins, and knowing the exact evolutionary history may not be so important for this purpose. In fact, two genes sharing highly similar sequences can be considered to share the same function, whether or not these genes experienced horizontal transfers during the course of evolution; such a situation may arise when the original orthologous lineage is replaced by a functionally equivalent xenologous gene [51]. In such cases, a simple BBH strategy or a species overlap strategy may be sufficient, or even more effective, than more complex tree-based classification schemes.

Domain fusion/fission events are frequently seen in both prokaryotic and eukaryotic evolution, and are yet another factor that complicates the ortholog classification problem. Since a domain fusion/fission event violates the one-to-one correspondence between orthologs, a simple BBH strategy cannot identify the correct relationship. Moreover, fusion proteins often cause serious trouble in the clustering procedure by connecting independent, non-homologous domains. To avoid this problem, fusion proteins should be split into domains. Another solution is to consider only global matches (i.e., the alignment coverage must be above a certain level) as orthologous relationships. Interestingly, domain fusion events can also be used for inferring functional relationships between fused proteins [30, 52], and automatic detection of fusion events is also useful for identifying such interesting cases. On the other hand, the strict criterion that takes into consideration only global matches always classifies genes with different domain organizations into different groups. This strategy might be more useful for reliable annotation.

Many methods and databases for orthology identification have been developed so far. In the next section, we introduce some of these databases, mainly focusing on those useful for microbial comparative genomics. A more comprehensive list which includes those useful for eukaryotic genome comparisons can be found in a recent review [53].

Ortholog Databases for Microbial Comparative Genomics

Table 1 summarizes the resources useful for microbial comparative genomics. Besides collections of genome project information such as the Genomes Online Database (GOLD) [54] and primary genomic sequence databases such as NCBI Genomes (Table 1A), many types of databases geared toward comparative genomics have been developed, and these databases require ortholog assignment among organisms as a basis for comparison (Table 1B).

The clusters of orthologous groups (COGs) database [55], established in 1997, is probably the best-known database of orthologous groups in microbial genomes. The BBH strategy is used for constructing the COG database at the step of identifying orthologous gene pairs, which are subsequently extended to orthologous groups among multiple genomes through a unique clustering algorithm based on triangles that consist of three consistent BBH relationships [55]: two triangles are merged if they share a common side. To overcome the problems associated with BBH described in the previous section, however, the overall construction process must be implemented in conjunction with several additional procedures, which include the addition of pre-identified species-specific inparalogs, the splitting of proteins into multiple domains if required, and the splitting of large groups into appropriate small groups by manual inspection of multiple alignments and phylogenetic trees [56]. The latest version of the COG database, released in 2003, was constructed through an incremental process, and comprises 4,373 COGs generated from 66 genomes [57]. Although the COG classification is still widely used for various genome analyses, the database itself has not been updated since 2003.

Several other ortholog databases have been constructed through the curation processes, e.g., TIGRFAMs [58], KEGG Orthology (KO) [59, 60], HAMAP [61] and FIGfams [62]. These databases aim at collecting "functionally equivalent homologs" or "equivalogs" [58] useful for genome annotation, rather than identifying true evolutionary orthologs. Although these databases do not ensure comprehensive classification, they do provide reliable classification.

On the other hand, other databases have been constructed by automated procedures, such as Inparanoid [63], OrthoMCL-DB [64] and our database, MBGD [8].

Table 1 Resources for microbial comparative genomics

Name	URL	Description
A. Genome project database		
GOLD:Genomes OnLine Database	http://www.genomesonline.org/	Comprehensive collection of genome and metagenome projects
Entrez Genome	http://www.ncbi.nlm.nih.gov/sites/entrez?db=genomeprj	NCBI collection of genome projects with links to sequence entries
B. Ortholog database		
COG	http://www.ncbi.nlm.nih.gov/COG/	Well-known resource for microbial ortholog classification
TIGRFAMs	http://www.jcvi.org/cms/research/projects/tigrfams/	Protein classification based on a hidden Markov model
Protein Clusters	http://www.ncbi.nlm.nih.gov/sites/entrez?db=proteinclusters	A resource for protein functional annotation in NCBI RefSeq
HAMAP	http://au.expasy.org/sprot/hamap/	A resource for protein functional annotation in SWISS-PROT
KEGG Orthology	http://www.genome.jp/kegg/ko.html	A resource for protein functional annotation in the KEGG project
FIGfams	http://seed-viewer.theseed.org/seedviewer.cgi?page=FigFamViewer	A resource for protein functional annotation in the SEED project
EggNOG	http://eggnog.embl.de/	Automatic extension of the COG database
OMA	http://omabrowser.org/	Automated large-scale ortholog grouping
C. Integrated comparative genomics system		
MBGD	http://mbgd.genome.ad.jp/	A platform for microbial genome comparison based on ortholog grouping
IMG	http://img.jgi.doe.gov/	An integrated platform for microbial genome analysis at JGI
CMR	http://cmr.jcvi.org/	A comprehensive resource for microbial genome analysis at JCVI
MicrobesOnline	http://www.microbesonline.org/	An integrated resource for microbial comparative and functional genome analysis
D. Pathway-based annotation		
KEGG	http://www.genome.jp/kegg/	A system for integrating genomic data and metabolic pathways
BioCyc	http://www.biocyc.org/	A system for integrating genomic data and metabolic pathways
The SEED	http://www.theseed.org/	A system for annotating microbial genomes based on a subsystem approach
E. Functional-link prediction tools based on comparative genomics		
STRING	http://string.embl.de/	An integrated prediction system for protein–protein interaction based on the genomic context
Phydbac	http://www.igs.cnrs-mrs.fr/phydbac/	A database of predicted functional associations between prokaryotic proteins
Predictome	http://visant.bu.edu/	A database of predicted functional associations in the VisANT project

Inparanoid is a database of orthologous relationships between two genomes identified by the Inparanoid program [35], which is an improvement on the BBH strategy for correctly identifying orthologous groups that contain inparalogs. Although the Inparanoid program can only be used for pairwise genome comparisons, the creators of this program later developed MultiParanoid, a program that can construct orthologous groups among multiple genomes by merging multiple pairwise orthologous groups generated by the Inparanoid program [65]. OrthoMCL-DB [64] is a database constructed by means of the OrthoMCL program [41], which uses an improved BBH strategy in combination with an improved clustering algorithm, TribeMCL [40], and is applicable to multiple genomes. However, these databases classify mainly eukaryotic genomes and contain none (Inparanoid) or a limited set (OrthoMCL-DB) of prokaryotic genomes. Recently, large-scale ortholog databases that contain hundreds of prokaryotic genomes have been developed, including EggNOG [66] and OMA [67]. EggNOG was constructed based on the COG database; the triangular linkage clustering method used in the COG construction procedure was used to extend each COG group and to construct new groups that were not assigned to any COG.

In addition to the general ortholog databases, there are some integrated systems providing various functions for the comparative analysis of microbial genomes (Table 1C). Such databases include CMR [68], IMG [69] and MicrobesOnline [70]. Our database, MBGD [8], can also be classified into this category. Another type of database is an integrated database of genomes and metabolic pathways or interaction networks, such as KEGG [22] and BioCyc [71] (Table 1D). Such databases integrate various resources and are extremely useful for genomic data annotation. The SEED project, which is an annotation system for thousands of genomes based on the subsystem approach [72], is another effort in a similar direction. These systems are also reliant on ortholog identification for integrating genomic data and pathway/subsystem information. Functional prediction methods using genomic context information, such as STRING [33], Phydbac [73] and Predictome [74], also depend on ortholog identification as a fundamental basis (Table 1E). Actually, many of these databases appear not to be actively updated, possibly because of the discontinuance of the COG database updating; the exception is STRING, which is based on the EggNOG database.

Before introducing MBGD in detail, I would like to briefly describe in the following section the ortholog classification method that is used in the MBGD database, named DomClust, which was developed to overcome many of the drawbacks of the BBH-based methods described in the previous section.

DomClust: Hierarchical Clustering Algorithm for Ortholog Group Construction at the Domain Level

As an ortholog grouping method, DomClust adopts an intermediate approach between the graph-based and the tree-based approach. Instead of using BBH, DomClust uses as input all-against-all similarity search results, and applies a traditional hierarchical clustering algorithm, UPGMA. The most significant feature of DomClust is that it can detect domain fusion or fission events during the course of clustering, and splits clusters into domains if required. The subsequent procedure splits the resulting trees, such that intra-species paralogous genes are divided into different groups so as to create plausible orthologous groups. As a result, the procedure can split genes into the domains minimally required for ortholog grouping.

In the DomClust algorithm, similarity search results are represented as a similarity graph, $G=(V,E)$, where V is the set of protein sequences (vertices) and E is the set of homologous relationships identified by BLAST or other methods (edges). The clustering procedure is basically a successive contraction of this graph by UPGMA [75]. At each iteration, the procedure takes the best-similarity edge and replaces the vertices connected by the best edge with a new vertex (a merged cluster). In addition, for each vertex connected to the merged vertices (e.g., S_3 in Fig. 5a), it also merges the edges that join these vertices, assigning the averaged score. The procedure is repeated until the best score becomes worse than the given cutoff, c. While the usual UPGMA requires a complete similarity matrix, many edges are missing in our similarity graph, G, due to the elimination of insignificant similarities. When one of the two edges is missing, we assign a fixed score (parameter), m ($<c$), to the missing edge for calculating the average. This modification reduces the computational cost of UPGMA from $O(|V|^2)$ to $O(|E|)$.

To address domain fusion/fission events, we added a process for domain splitting to the basic procedure outlined above (Fig. 5a, b). At each iteration with the best edge sim(s_1, s_2), a merged vertex was split into at most 5 vertices according to the aligned segment of the best-scoring edge: the aligned segment itself and the left and right overhangs on either of the sequences (or clusters). For each vertex s_3 connected by the best edge, the edges sim(s_1,s_3) and sim(s_2,s_3) were reconnected to one or more of the new segments, and the information for these edges was updated appropriately, by averaging the alignment lengths as well as the scores over all of the relationships included in the merged edges. The resulting structure is a directed acyclic graph representing overlapping trees, as shown in Fig. 5c. On this graph, one can split genes into the domains minimally required for ortholog grouping by specifying a set of internal nodes as roots. To do this, one can carry out tree reconciliation between the gene tree and the species tree [42, 76], but this requires the assumption that there are no horizontal transfers and that both the gene tree and the species tree are exactly known. Instead, here we adopted the simpler "species overlap" criterion: each root node is recursively cut until the two clusters that are merged at that node share no or few intra-species paralogous genes (Fig. 5d). More precisely, a root node with two child nodes, A and B, is cut when $|Ph(A \cap Ph(B)|/\min(|Ph(A)|, |Ph(B)|) > p$ with a given cutoff parameter, p, where $Ph(A)$ denotes the set of species contained in cluster A (phylogenetic pattern). Strictly speaking, the parameter p must be 0 according to the definition of orthologs, but actually we found that a relaxed condition, around $p = 0.5$, often generated a more plausible classification and better coincided with the COG classification.

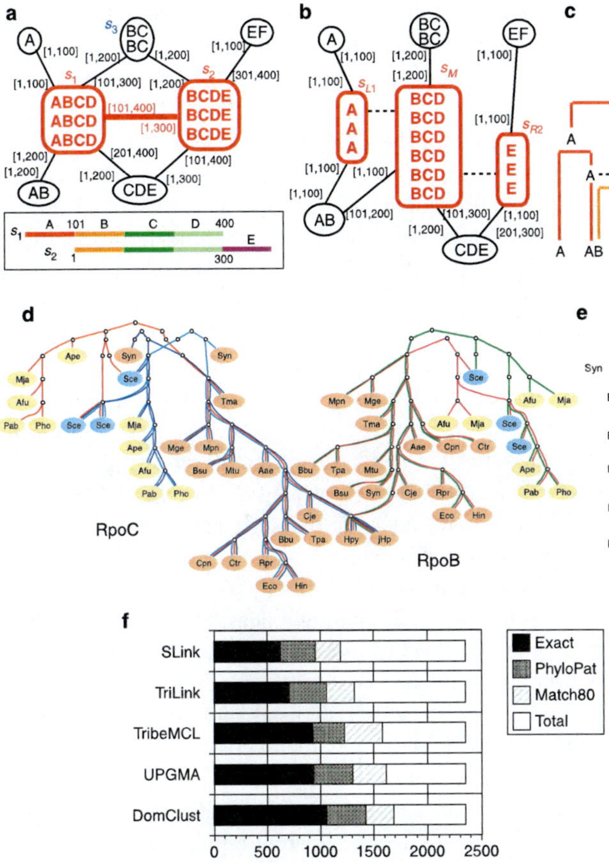

Fig. 5 Overview of the domain-splitting procedure in the DomClust algorithm. (**a**) A similarity graph that has 7 clusters (vertices) containing 12 sequences, which are constituted from six domains, (**A–F**), each of which is 100 residues long. The *numbers in brackets* on each edge indicate the coordinates of the aligned segments. The best similarity edge, sim(s_1,s_2), that is selected for merging is indicated by a *thick line*. At the bottom is a schematic illustration of the alignment between s_1 and s_2. (**b**) A similarity graph, which shows the situation after the two clusters have been merged and split. (**c**) The resulting clustering tree. The process of merging ABCD and BCDE, at the center of the figure, corresponds to the process shown in (**a**) and (**b**). (**d**) Orthologous groups of RNA polymerase beta (RpoB) and beta' (RpoC) subunits, as an example of hierarchical clustering trees created by the DomClust program. Each tree corresponds to an individual domain. (**e**) Schematic illustration of the gene structures of RpoB and RpoC in the selected genomes. (**f**) Comparison of the various clustering methods with the COG recovery test

Figure 5d shows an example of DomClust classification containing domain fusion or fission events: the orthologous groups of RNA polymerase beta (RpoB) and beta' (RpoC) subunits. These subunits are fused into one gene in the genomes of two strains of *Helicobacter pylori,* 26695 (Hpy) and J99 (jHp), while, in most archaea, each subunit is further divided into two genes. The algorithm first joined the fused genes of Hpy and jHp, and then divided the cluster into two domains when joining the cluster with the RpoC ortholog of the *Campylobacter jejuni* (Cje) genome; the remainder domain was subsequently joined with the RpoB ortholog of Cje. By repeating the procedure, DomClust finally identified five orthologous domains (Fig. 5e).

The algorithm was evaluated based on the COG recovery tests [46], where the orthologous groups were regenerated using the same set of sequences that were used for COG construction, and the resulting classification was evaluated according to the agreement with that of the COG database. Three other clustering methods were tested for comparison, including single-linkage clustering (Slink), triangular linkage clustering (TriLink) (our implementation of the basic procedure for constructing the original COG database), and the TribeMCL algorithm (which has been used for the construction of the OrthoMCL database). In addition, we also tested the DomClust algorithm without a domain-splitting procedure (gUPGMA). In each test, we first optimized the parameters to those giving the best agreement with the COG recovery test, and compared the results of different clustering methods. The outcome is shown in Fig. 5f. DomClust recovered 1,060 out of 2,360 (44.9%) well-defined COGs exactly (Exact). If we consider two OGs having the same phylogenetic patterns as equivalent (taking into account only the presence/absence of orthologs and ignoring the number of inparalogs), 1,429 (60.6%) OGs are correctly recovered (PhyloPat); if we consider two OGs sharing at least 80% of each of their member genes as equivalent, 1,685 (71.3%) OGs are correctly recovered (Match80). For any of the equivalency criteria, DomClust showed the best agreement with the COG database, followed by gUPGMA, TribeMCL, TriLink and Slink, although the difference between gUPGMA and TribeMCL was relatively small (Fig. 5f). The performance superiority of DomClust over gUPGMA indicates that the domain-splitting procedure is indeed effective for reconstructing the COG classification. On the other hand, COG groups often include many apparent outparalogs that should not be included in the same ortholog group by definition. This tendency seems to have emerged in the above result as a difference between Exact and PhyloPat, which should reflect the difference in the number of inparalogs between the results of different classification systems. In this sense, we think PhyloPat gives a better indication of the classification performance than Exact.

Actually, benchmarking ortholog classifications is quite a difficult task, in that the way classification results should be evaluated is still a debatable issue. Recently, several assessments of different ortholog classification schemes have been done in terms of functional equivalence and/or phylogenetic relationships [77–79]. In these tests, COG (or its eukaryotic version, KOG) did not necessarily give a better result than other methods. In fact, in some tests, COG gave worse results than the simpler BBH methods [77, 79]. This is partly because of the aforementioned drawback of COG, i.e., the inclusion of many apparent outparalogs that cause many false positive assignments. On the other hand, most of these assessments examined pairwise orthologous relationships rather than the all the members of the orthologous groups, as in our COG recovery test. As a result, these assessments could not evaluate the clustering quality well. In fact, the way of evaluating orthology assignment should depend on the purpose of the analysis: accurate identification of orthologous gene pairs is generally sufficient to ensure an accurate transfer of functional annotation, but is not necessarily sufficient for phylogenetic pattern analysis; the latter requires all the information on the orthologous groups.

MBGD: Microbial Genome Database for Comparative Analysis

MBGD is a microbial genome database for comparative analysis established in 1997. It is not simply an ortholog database where precalculated orthology relationships are stored, but a platform for large-scale comparative genome analysis based on comprehensive ortholog classification [7]. MBGD uses the DomClust program to automatically construct orthologous groups. Hence, MBGD is comprehensive and routinely updated. In addition, unlike other automatically constructed ortholog databases, MBGD allows the user to classify genes dynamically using a specified set of genomes. This feature is especially useful when the user's interest is focused on specific taxonomically related organisms.

In MBGD, a default ortholog table has been precomputed using a default set of organisms that contains one strain from every species. Currently, this set contains less than 40% of all the completed genomes (see Fig. 1), and offers an unbiased sample of the entire set of genomes currently sequenced, which is useful for comparing distantly related genomes. In addition, to cover all the genomes in the database, MBGD also has an "extended" ortholog table, where each gene of the unselected genomes in the default set are assigned to the orthologous group that gives the best average similarity score. In this way, MBGD provides orthology information for every gene stored in the database. On the other hand, MBGD allows the user to choose a set of genomes for ortholog analysis; for this purpose, the user can refer to taxonomic information taken from the NCBI Taxonomy database, or phenotypic information taken from the GOLD database. In this way, MBGD identifies orthologous relationships among the genomes that the user is especially interested in.

The orthologous groups among the selected genomes are not a simple subset of the default ortholog groups, even if the selected species is a subset of the default species set. In fact, as described above, an ortholog group can be defined as a set of homologous genes that are derived from a speciation event in the last common ancestor of the target set of organisms. Therefore, a different partitioning of the same set of genes may result when different sets of organisms are considered (Fig. 6). In general, when one compares the genomes of closely related organisms, the resulting orthologous groups are expected to contain more one-to-one relationships than those created from all of the organisms sequenced to date.

In addition, MBGD provides the MyMBGD functionality [80], which allows the user to add his or her own genome sequences to MBGD to identify orthologs among new and existing genomes. In this mode, the user can submit his or her data either in GenBank format or as amino acid sequence data in FASTA format plus a tab-delimited annotation table. Alternatively, the user can submit a raw genomic sequence without any annotation, in which case a protocol based on GeneMarkS

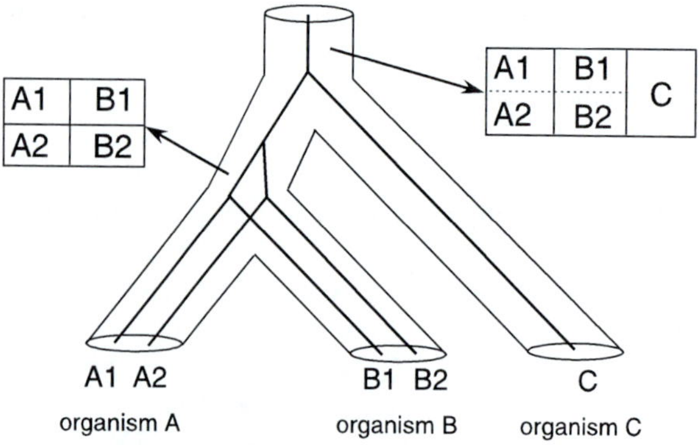

Fig. 6 Ortholog grouping as a mapping from tree structures to a classification table. In this figure, a species tree that contains the organisms A, B and C is drawn with *pipes*, and a gene tree of genes A1, A2, B1, B2, and C is drawn with *lines*. The *left-hand* table represents an ortholog grouping that is created from organisms A and B and contains two ortholog clusters, whereas the *right-hand* table created from organisms A, B and C consists of only one ortholog cluster. The *broken line* in this table indicates the line separating ingroup-specific subgroups, which makes the table a nested table; this type of ortholog table is available in the RECOG system described in the last section

[81] and Glimmer [82] is invoked to identify genes in the query genome. The submitted or identified amino acid sequence data are then subjected to similarity search analysis against the published genomes selected by the user under the same protocol as that for constructing the MBGD database, after which ortholog analysis of those genomes is carried out. In MyMBGD mode, the private data and the public data are logically merged using the "merge table" feature implemented in the MySQL database management system. By this mechanism, the user can use almost every function of MBGD without noticing the differences between the MyMBGD mode and the usual mode.

MBGD provides several methods of retrieving specific orthologous groups from the default or created ortholog table. For example, the user can specify keywords on the top page of MBGD. The system searches for the keywords first in individual gene records, after which it finds ortholog clusters containing the retrieved genes. The user can also specify query sequences for a similarity search. In this function, the system calculates similarities between the query and the database sequences in the same way as all-against-all similarities are calculated in MBGD, and then finds the clusters that contain these genes. Alternatively, MBGD also provides a usual genome map interface if the user wishes to specify a particular locus in a chromosome to retrieve a gene. All information about the retrieved gene is summarized on a gene information page, which includes a link to an ortholog table page showing the orthologs of the gene currently displayed. In any case, the user can see the part of the ortholog table that contains the orthologous groups of their interest.

An ortholog cluster entry page is shown in Fig. 7. Several types of information are added to each cluster entry that are generated from the annotation of

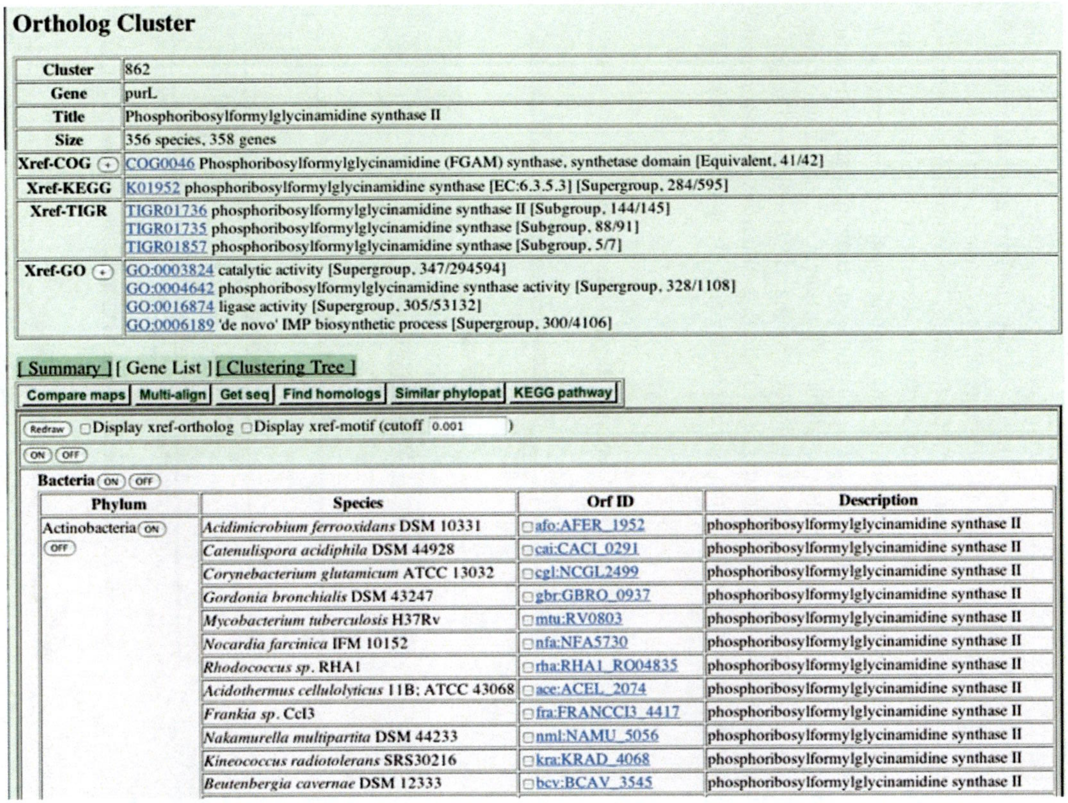

Fig. 7 An example of the ortholog cluster entry page displaying the orthologous group of the *purL* gene in the default cluster set. The page contains a cluster annotation table showing the gene name, title, and cross-references to other databases, and a table showing the list of member genes

each member gene. The title of each ortholog cluster is automatically generated by extracting the best-scoring title line among those of the member genes, where the scores of the title lines are calculated based on the frequency of words appearing in the title lines of the member genes. Each cluster entry also contains cross-references to the corresponding entries of COGs [57], KEGG Orthology [22], TIGRFAMs [83] and Gene Ontology [84] terms. Cross-reference data are also used to assign functional categories: the functional category of each gene is taken from that of the cross-reference entry of COG, KEGG and TIGR, and category assignment to each cluster is based on a majority vote of categories assigned to individual genes.

The cluster entry page also contains several functions for comparing genes within that orthologous group, including multiple sequence alignment for comparing proteins or nucleotide sequences and multiple map comparison for comparing gene orders around orthologous genes (Fig. 8). In addition, functions for comparing different orthologous groups are also available. The function "Find homologous clusters" lists the clusters that are homologous to the target cluster. The function "Similar phylogenetic pattern search" calculates dissimilarities in phylogenetic pattern between each cluster and the target cluster, and orders the cluster table

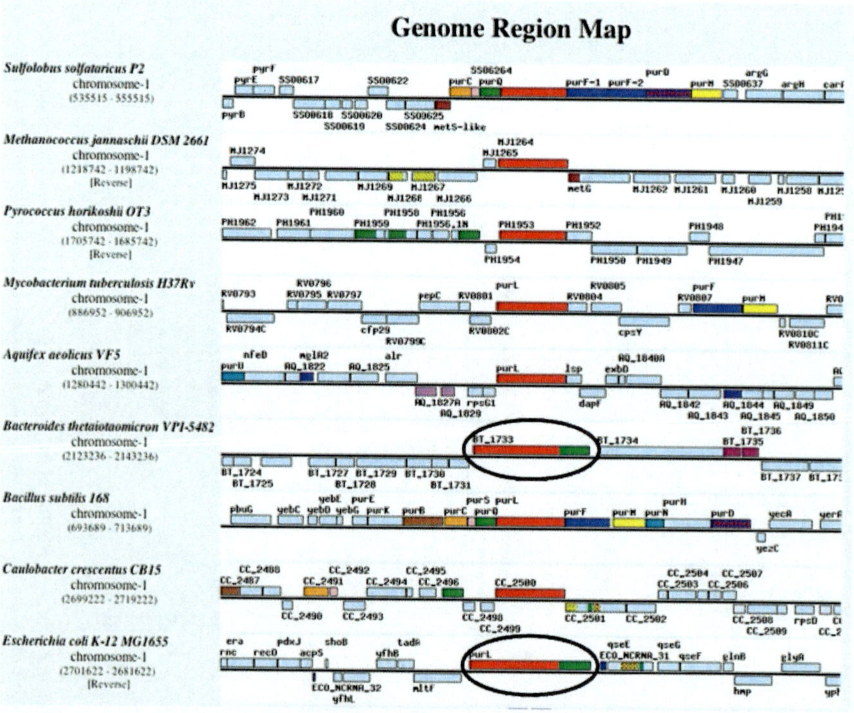

Fig. 8 Comparative genome map of the *purL* orthologs and their vicinity. Genes that belong to the same orthologous group are assigned the same color and pattern. Note that in some genomes, including *E. coli*, the *purL* gene is fused with the *purQ* ortholog (enclosed by *ellipse*)

according to the dissimilarity value, where the dissimilarities are calculated based on a correlation coefficient, the hamming distance or mutual information [85]. This function is useful for predicting functional linkages [29], and similar functions are implemented in certain more specialized databases [33, 86]. In MBGD, the user can combine this type of analysis with a more flexible ortholog analysis.

A Sample MBGD Session: Phylogenetic Pattern Analysis

MBGD also provides a function to search for orthologous groups that have a similar phylogenetic pattern to the one specified by the user. This type of analysis is especially useful when looking for genes that are potentially related to a particular phenotypic trait. In the following, we show an example of this type of analysis.

By clicking the "View Ortholog Table" item on the top page of MBGD, the user can see a summary of the current ortholog table. By default, this is a histogram showing the distribution of cluster size; optionally, it can display the distribution of phylogenetic patterns with colors assigned according to functional categories. The histogram can be redrawn with a restricted set of clusters by specifying phylogenetic patterns or keywords in the input form below. Alternatively, a detailed ortholog table can be shown by pressing the [Show cluster table] button below.

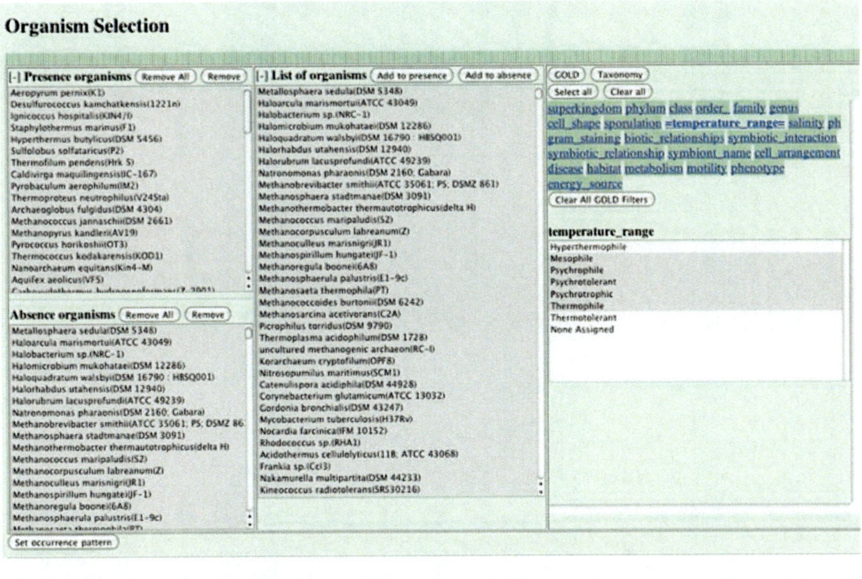

Fig. 9 Organism selection window in MBGD for specifying a phylogenetic pattern using the phenotypic properties defined in the GOLD database. Here a phylogenetic pattern corresponding to "hyperthermophile-specific" is specified

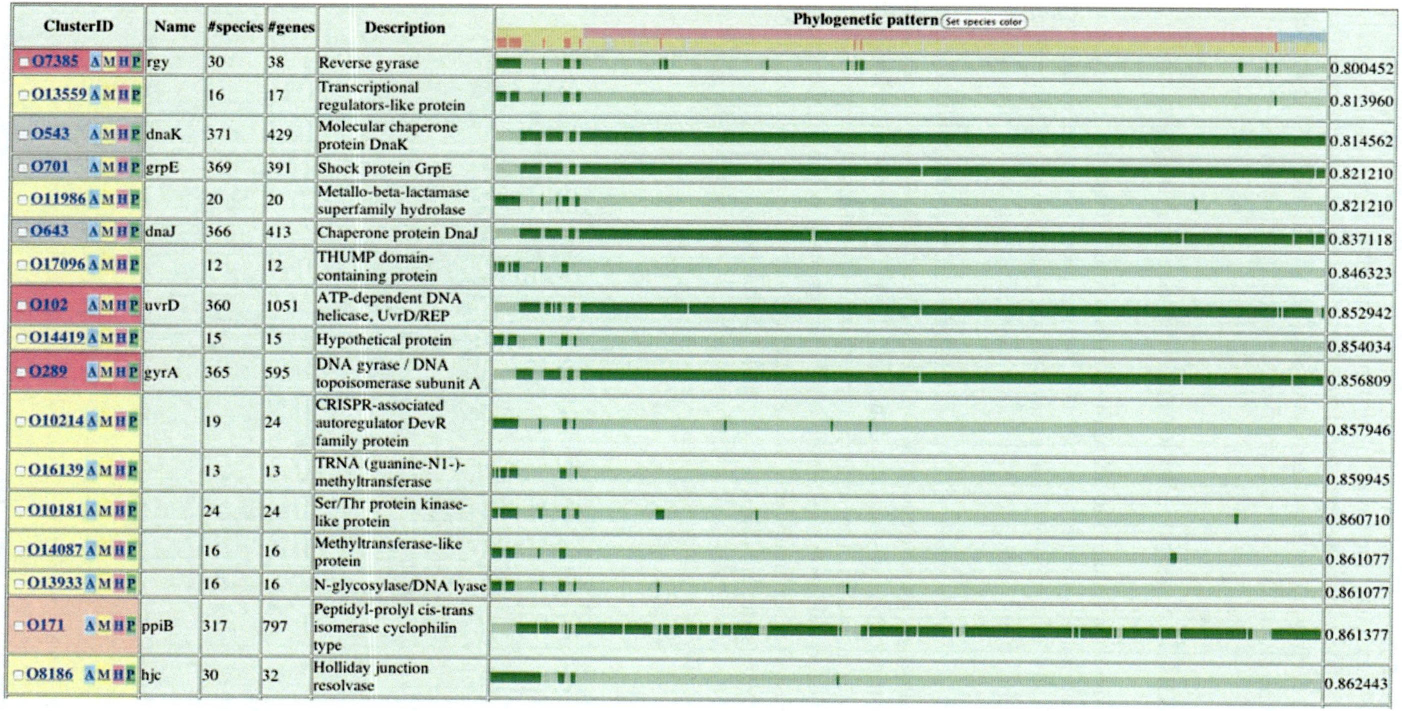

ClusterID	Name	#species	#genes	Description	Phylogenetic pattern	
O7385 A M H P	rgy	30	38	Reverse gyrase		0.800452
O13559 A M H P		16	17	Transcriptional regulators-like protein		0.813960
O543 A M H P	dnaK	371	429	Molecular chaperone protein DnaK		0.814562
O701 A M H P	grpE	369	391	Shock protein GrpE		0.821210
O11986 A M H P		20	20	Metallo-beta-lactamase superfamily hydrolase		0.821210
O643 A M H P	dnaJ	366	413	Chaperone protein DnaJ		0.837118
O17096 A M H P		12	12	THUMP domain-containing protein		0.846323
O102 A M H P	uvrD	360	1051	ATP-dependent DNA helicase, UvrD/REP		0.852942
O14419 A M H P		15	15	Hypothetical protein		0.854034
O289 A M H P	gyrA	365	595	DNA gyrase / DNA topoisomerase subunit A		0.856809
O10214 A M H P		19	24	CRISPR-associated autoregulator DevR family protein		0.857946
O16139 A M H P		13	13	TRNA (guanine-N1-)-methyltransferase		0.859945
O10181 A M H P		24	24	Ser/Thr protein kinase-like protein		0.860710
O14087 A M H P		16	16	Methyltransferase-like protein		0.861077
O13933 A M H P		16	16	N-glycosylase/DNA lyase		0.861077
O171 A M H P	ppiB	317	797	Peptidyl-prolyl cis-trans isomerase cyclophilin type		0.861377
O8186 A M H P	hjc	30	32	Holliday junction resolvase		0.862443

Fig. 10 Ortholog cluster table containing ortholog clusters with a phylogenetic pattern similar to the hyperthermophily-specific pattern as specified in Fig. 9. The table is ordered by the correlation coefficient against the specified phylogenetic pattern. Phylogenetic patterns are graphically shown with *green bars* indicating "present". The value shown in the rightmost column is a dissimilarity value, $d=(1-r)/2$, calculated with the correlation coefficient, r. The specified phylogenetic pattern is shown in the header row with *red bars* indicating "present" and *yellow bars* indicating "absent"

By pressing the [Change pattern] button in the Occurrence pattern box, the organism selection window is displayed for specifying a phylogenetic pattern (Fig. 9). Here, the condition can be set by choosing the sets of present organisms and absent organisms. For this organism selection, the phenotypic properties of each organism provided by the GOLD database [54] can be used. By using this functionality, a user can specify a query such as "What are the genes specifically present in hyperthermophiles?" To find the answer to this question, choose "temperature range" as the property type and "Hyperthermophile" as the property value, and press the [Add to presence] button; next, choose the remaining values other than "None assigned" in the selection box, and press the [Add to absence] button; then press the [Set occurrence pattern] button below. The phylogenetic pattern bar is then updated with the present organisms colored green and the absent organisms colored yellow. After choosing "Mutual information" as the similarity (dissimilarity) measure and pressing the [Show cluster table] button below, the user can see the ortholog table, which is ordered according to the dissimilarity of the phylogenetic pattern to the specified pattern (Fig. 10).

In the resulting table, the ortholog group that is most similar to the specified "hyperthermophilic" phylogenetic pattern is reverse gyrase, which was previously reported as the only hyperthermophile-specific protein [87]. The second best ortholog group is putative transcriptional regulator (COG1318), which was also reported in the same paper as a candidate hyperthermophilic-specific gene, but was eliminated because of its absence in the *Sulfolobus* genomes [87]. Later, the other authors adopted this family as a candidate hyperthermophile-specific transcriptional regulator [88]. Since then, the increasing number of genomes appears to strengthen the hypothesis that this gene is related to hyperthermophily. On the other hand, negatively correlated patterns can also be identified when using the similarity measure "mutual information." In this case, three chaperone proteins, DnaJ, DnaK and GrpE, show patterns that are negatively correlated with the hyperthermophilic pattern, suggesting that this heat-shock system does not work in extremely high temperatures.

Core Genome Analysis: What Is the Essential Part of a Set of Related Genomes?

Another important application of ortholog analysis is to elucidate which genes are shared among related genomes and which are not. The result of this type of analysis is often represented in a Venn diagram (Fig. 11a), in which the commonality and difference among gene sets among genomes are displayed. In this diagram, the intersectional area (area 7 in Fig. 11a), which contains universally conserved genes in all the compared organisms, is particularly interesting because these genes are commonly required for those organisms to survive in their natural habitats, and thus correspond to the most essential part of the respective genomes. This type of analysis can be applied to any set of phylogenetically related organisms, and is related to

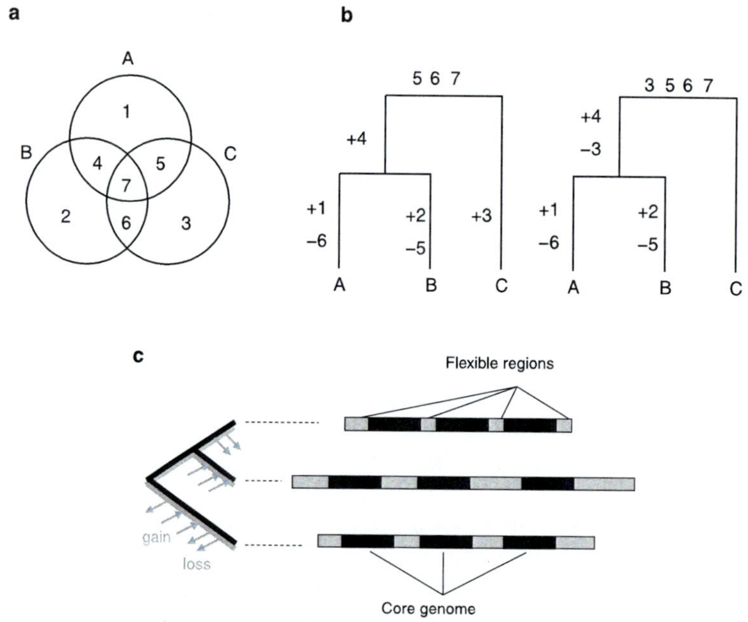

Fig. 11 Definition of a core gene set. (**a**) Venn diagram showing the commonality of genes among three related organisms, A, B and C. (**b**) Phylogenetic trees showing two parsimonious evolutionary scenarios of the three organisms shown in (**a**). Each number represents a gene subset that is included in each region of the Venn diagram; a *plus sign* indicates acquisition and a *minus sign* indicates deletion. (**c**) The concept of a core genome as well-conserved genomic segments mainly transmitted vertically. Here, *black lines* represent core regions and *gray lines* represent flexible regions

the concept of a "core genome" in a broad sense, although the meaning of this term is vague since it can be used in various contexts.

One of the related topics that has attracted broad interest is the problem of identifying the minimal gene set [4], which is defined as the smallest possible set of genes that is sufficient for a cellular organism to survive under the most favorable conditions [89, 90]. Since the universally conserved genes among all the cellular organisms correspond to the genes commonly required under any environmental condition, they are candidates for the minimal gene set. The first version of the minimal gene set was generated by comparison of only two genomes, *H. influenzae* and *M. genitalium*, and comprised 256 genes [4]. Interestingly, this number coincides well with the essential genes that were later identified in *Bacillus subtilis* (271 genes) [91] and *Escherichia coli* (297 genes) [92] in systematic gene knockout experiments. However, when the same authors examined the conservation status of each gene in the minimal gene set using the COG database with 21 genomes, they found that only one third (81 genes) of the genes in the original minimal gene set were still conserved in all organisms [89]. The number of genes appears to be too small for organisms to survive, which was partly explained by the existence of non-orthologous gene displacement [89, 90]. In fact, currently only around 10 genes are commonly conserved among all organisms in the MBGD default table, which are clearly insufficient for maintaining cellular life. There are some trivial reasons for this reduction in number, including the existence of unusually small genomes such as those of endosymbionts, as well as possible annotation errors. In any case, the requirement of 100% conservation will become too strict for this type of analysis as the number of target genomes increases, and a more relaxed criterion is needed to avoid the "extinction" of the core genes [93].

The core genome analysis can also be applied to comparisons of more closely related organisms. In fact, this type of analysis is commonly done for intra-species genome comparisons, and is often referred to in terms of the bacterial species genome concept [94], since, in prokaryotic species, the gene contents can be quite different among strains. In this context, "core genome" corresponds to the intersection of the gene sets, i.e., the set of genes commonly conserved in all the species, and "pan-genome" corresponds to the union of the gene sets, i.e., the set of genes contained in either of the genomes [95, 96]. A similar analysis can also be applied to genus-level comparisons [97], and comparisons of even more distantly related genomes [90, 98]. In any case, the set of genes conserved throughout a given taxonomic group should contain the genes commonly required for these organisms to survive in their natural habitats. However, the number of universally conserved genes may again decrease excessively as the number of target genomes increases.

From the evolutionary viewpoint, the universally conserved genes among related organisms are likely to have been inherited from their common ancestor. However, when applying the parsimony criterion to infer the evolutionary history of multiple genomes, many genes other than the universally conserved genes are also assigned to the last common ancestor (Fig. 11b); such non-universal genes have been lost during the course of evolution. This consideration allows the formulation of a relaxed version of the definition of a core genome: the core genome of phylogenetically related organisms consists of those genes that existed in the genome of their common ancestor and have been inherited by the majority of the current genomes. In this regard, however, the existence of horizontal gene transfers (HGTs) may become a serious problem in prokaryotic genome comparisons, and this has introduced another aspect to the concept of a core genome.

A growing body of evidence is supporting the idea that HGTs have played a significant role in prokaryotic genome evolution [51, 99–103], and these observations have stimulated researchers to develop a new paradigm of HGT-driven reticulate evolution that challenges the traditional tree-based phylogeny concept [104–106]. However, it can be argued that prokaryotic phylogeny can still be inferred using a certain subset of genes that have mainly transferred vertically throughout evolution [107–109]. In fact, the genes constituting a prokaryotic genome appear to be divided into two classes: a "core gene pool" that comprises intrinsic genes encoding the proteins of basic cellular functions, and a "flexible gene pool" that comprises HGT-acquired genes encoding proteins which function under particular conditions, such as genomic islands [110] (Fig. 11c). Combining the conservation criterion with the vertical/horizontal transference criterion, here we consider vertically transferred conserved genes as core genes. Then, the problem is how to identify the core genes among the genes of the given genomes according to this definition.

Several criteria have been developed to identify HGT events, including (1) unusual best-hit relationships or contradictory topologies of phylogenetic trees, and (2) a biased nucleotide composition, which includes a codon usage bias and G+C content at the third-codon positions. However, although these criteria can be successfully applied to identify HGTs as extraordinary events, their precision is not sufficient to exactly discriminate non-HGT genes from HGT genes. As a complementary approach to the identification of the core genome in the above sense, here we take into consideration the information on gene order (or synteny) conservation. In the next section, we introduce our own approach to this issue.

CoreAligner: Multiple Genome Alignment Procedure for Identifying the Core Genome Structure

Here, we consider the "structural core gene set," or simply the "core structure," of closely or moderately related genomes. It is defined as the set of sufficiently long consecutive genomic segments in which gene orders are conserved among multiple genomes so that they are likely to have been inherited from a common ancestor mainly through vertical transfer [111]. The rationale behind this approach is that horizontally transferred genes are unlikely to insert themselves at the same chromosomal position. For this purpose, we developed an algorithm for aligning the conserved regions of multiple genomes, which finds the order of pre-identified gene families that retains to the greatest possible extent the conserved gene orders.

The program, named CoreAligner [111], requires a set of well-conserved orthologous groups (OGs) as input. Here, we compiled them using the DomClust algorithm on the MBGD server, and considered an OG as "conserved" when it was present in at least half of the genomes (the parameter *CONS_RATIO*=0.5). Next, a neighborhood graph was constructed using the set of conserved neighborhood pairs, which are defined as two conserved OGs that are located within 20 genes (the parameter *MAX_GAP*=20) in at least half of the genomes (Fig. 12a). A neighborhood graph, $G=(V,E)$, was then constructed with the set of conserved OGs, V, as nodes and the set of conserved neighborhood relationships, E, as edges. Our algorithm for constructing alignments of the core genome structures is based on finding the longest path of this conserved neighborhood graph. A similar algorithm has been previously developed mainly for identifying much shorter but more widely conserved gene clusters such as operons [112], but, unlike that method, our method considers not only genes in the same direction but also those in the opposite direction as neighboring genes, thereby generally generating longer alignments. In addition, our method uses the dynamic programming (DP) algorithm to calculate the longest path. To apply the DP algorithm, we devised a heuristic scheme comprising a series of preprocessing procedures to convert the initial conserved neighborhood graph into a directed acyclic graph. Then, the extracted longest path is added to the

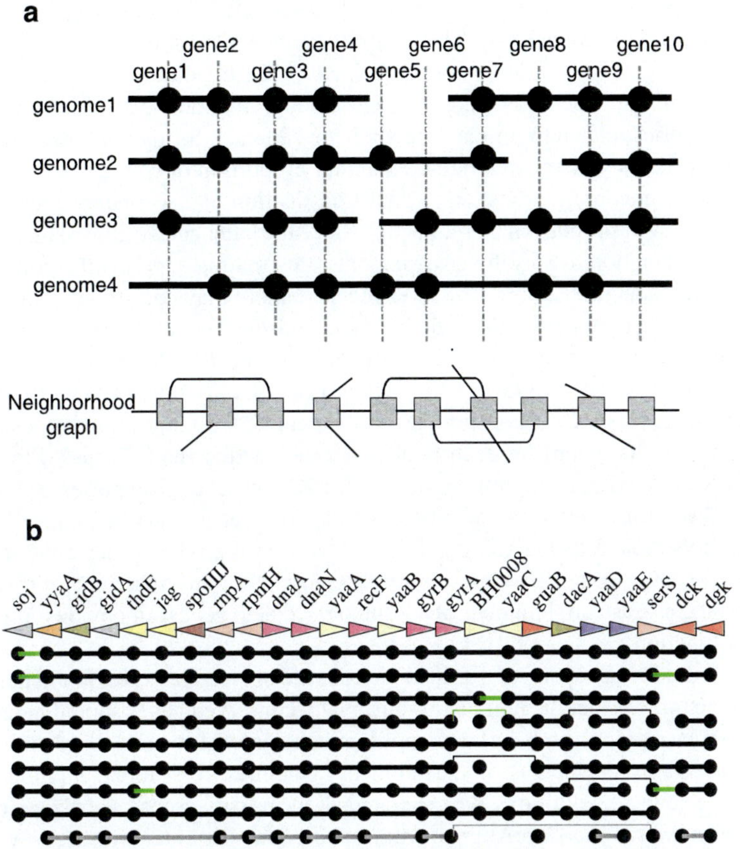

Fig. 12 Schematic illustration of the core genome construction procedure. (**a**) A "core genome alignment" is defined as the order of pre-identified conserved orthologous groups (OGs) (*vertical lines*) that retains to the greatest possible extent the conserved neighborhood relationships on the chromosomes (*horizontal lines*). To determine this order, a neighborhood graph representing the neighborhood relationships is constructed (*bottom*). For simplicity, only OGs that are directly adjacent to each other are connected. (**b**) A part of the core structure constructed from the *Bacillaceae* data set

core structure when the path consists of more than 10 OGs and at least half of the genes (OGs) in that path are present in every genome. The procedure is repeated to find the next longest path in the remaining graph, and the iteration is continued until all such paths are found.

The method was applied to the genome comparison of two well-characterized families, *Bacillaceae* and *Enterobacteriaceae*, and their core structures were found to comprise 1,438 and 2,125 OGs, respectively [111], which corresponds to a third of the number of genes in the *B. subtilis* genes (4,105) and half of the *E. coli* genes

(4,237), respectively (see Fig. 12b, which shows a part of the resulting core structure of the *Bacillaceae* dataset). The core sets contained most of the essential genes (>90%) identified in *B. subtilis* [91] and *E. coli* [92]. In addition, the ratio of core genes to non-core genes was quite different among functional categories: the functional categories related to primary metabolism, genetic information processing and cellular processes generally contained a higher proportion of core genes, while the categories of membrane transport, signal transduction and secondary metabolism contained a lower proportion thereof [111]. Since the latter categories are likely to be related to adaptation to specific environments, this finding supports the notion that the genes included in the core structures indeed tend to play core-functional roles.

Actually, the core gene sets of *Bacillaceae* and *Enterobacteriaceae* share some common orthologs. The two core gene sets share around 700 OGs (class CC: core to core), while the other core OGs are specific to each family (class CO: core to non-core ortholog; class CN: core to none). As expected, the majority of the essential genes (around 200 genes for each family) are included in the CC class (Fig. 13a). We also examined the proportions of the KEGG functional categories and found that the "core-functional" characteristics of the core genes described above appear to be mainly linked to the CC class (Fig. 13b). On the other hand, although the majority of OGs in the CN class are not categorized in any functional class, they also include a substantial number of sporulation-related genes, an obvious *Bacillus*-specific function.

As discussed above, one of the drawbacks of the universality criterion in defining the core genome is that the number of universally conserved genes should monotonically decrease as the number of genomes increases. In other words, this definition is not robust against the increase in the number of genomes. To examine the robustness of our core genome definitions, we generated all the possible subsets of six genomes, with which we ran the CoreAligner program to define the core structures. We also examined the numbers of universal genes (*CONS_RATIO*=1) and genes conserved in at least half of the genomes (*CONS_RATIO*=0.5) for the same genome subsets (Fig. 14). As expected, the number of universal genes decreases monotonically as the number of genomes increases. In contrast, the number of conserved genes fluctuates widely with the change in the number of genomes, probably because of the fluctuation of the actual *CONS_RATIO* values due to the rounding-up effect. On the other hand, the number of core genes shows a relatively stable pattern in both families (Fig. 14). In fact, the magnitude of fluctuation is much smaller than that for the conserved genes described above. These observations suggest that the use of synteny information with a relaxed conservation criterion helps the CoreAligner program to identify robust and reliable core gene sets, although the setting of the *CONS_RATIO* parameter still remains somewhat arbitrary.

To further characterize the structural core genes, we also investigated them in terms of G+C content homogeneity and phylogenetic congruence, which are important indicators of the indigenousness of genes (i.e., non-HGT genes). As a result, we found that the structural core genes primarily had higher homogeneous G+C contents at the third-codon positions (GC3) than non-core genes [111]; more precisely, core genes showed the smallest standard deviation of the GC3 values in almost all

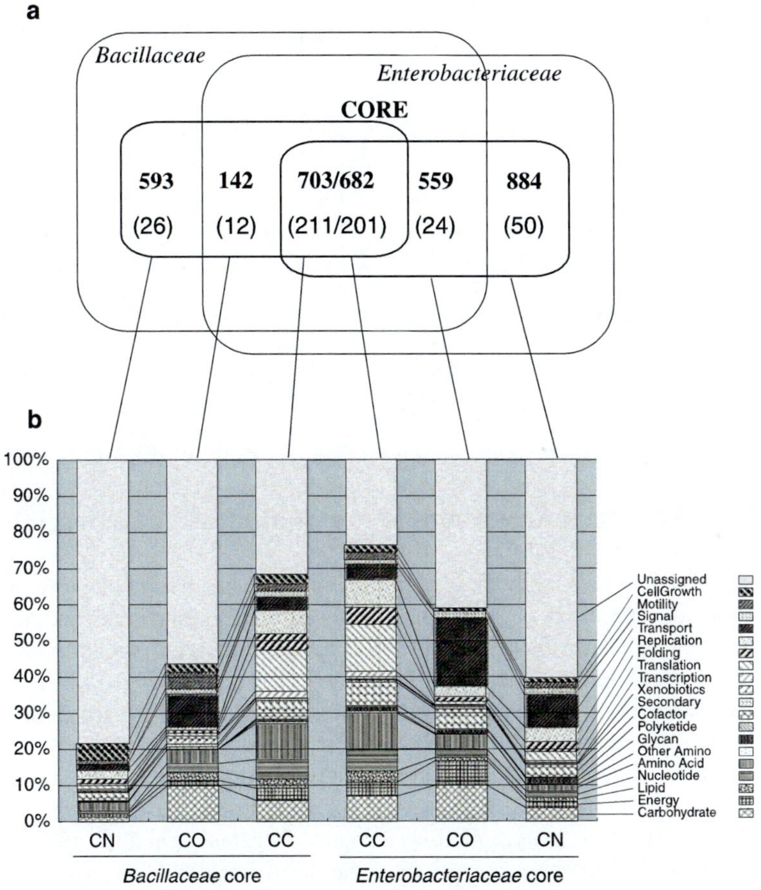

Fig. 13 Overlap between the core OG sets of *Bacillaceae* and *Enterobacteriaceae*. (**a**) A Venn diagram showing the number of core OGs defined in *Bacillaceae* and *Enterobacteriaceae* that overlap each other. For each family, the *outer circle* indicates the entire gene set, and the *inner circle* indicates the core gene set. The *number in parentheses* indicates the number of essential genes in *B. subtilis* and *E. coli*. (**b**) Functional breakdown of each subtype defined in (**a**): CC core to core; CO, core to non-core ortholog; CN, core to none. The functional classification is based on the KEGG functional categories

the organisms examined, followed by non-core-conserved genes and non-conserved genes. To test the phylogenetic congruency, we also compared the topologies of individual gene trees with the reference tree topology created from the concatenated core gene sequences and identified significant incongruent trees using the Shimodaira-Hasegawa (SH) test [113]. The result indicates that the structural core genes (orthologous groups) tended to have less incongruent topologies (rejected by the SH test) than non-core genes in every conservation ratio [111]. These results indicate that structural core genes indeed show the expected characteristic, i.e., being indigenous and sharing the same history in comparison to non-core genes.

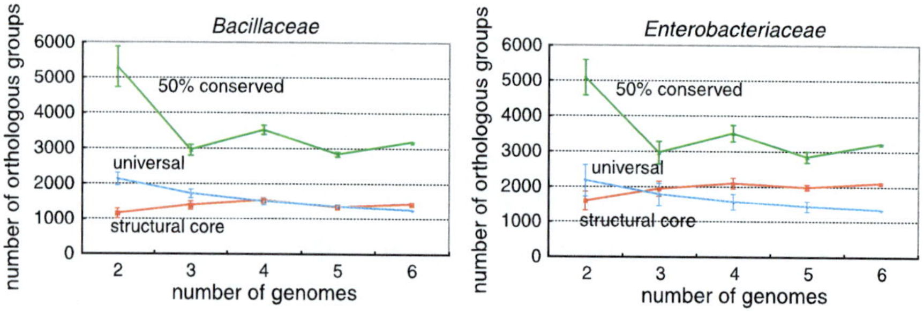

Fig. 14 Test for the robustness of the core genome sizes. The average numbers of the structural core, universally conserved and ≥50% conserved orthologous groups are plotted for each number of input genomes in the test datasets that were generated as subsets of the original dataset of *Bacillaceae* (*left*) and *Enterobacteriaceae* (*right*). The *error bars* show the standard deviations

RECOG: Research Environment for Comparative Genomics

To integrate the various methods of comparative genomics described so far and to incorporate the knowledge of individual genomes and gene functions into one analysis, we are now developing a general workbench for comparative genomics named RECOG (Research Environment for COmparative Genomics) (Fig. 15). The entire RECOG system employs a client-server architecture. The RECOG server program has been developed based on the MBGD server, sharing the same database construction protocol. The RECOG client program is a Java application running locally that receives data from any RECOG server. By default, the user can connect to the public MBGD server to analyze publicly available data, but, optionally, the user can install the RECOG server program on his or her local machine to analyze his or her own genomic data.

As with MBGD, RECOG allows the user to choose a set of organisms to compare, and constructs orthologous groups among those organisms using the DomClust program. However, RECOG also allows the user to specify ingroup species and outgroup species in order to combine a comparison of closely related genomes with a comparison of distantly related genomes, where the ingroup is usually specified as a set of taxonomically related organisms that the user is interested in. The result is displayed as a nested table, where the genes in the outgroup species form an outgroup cluster that corresponds to multiple sub-clusters consisting of genes in the ingroup species (see Figs. 6 and 15b).

The central function of RECOG is to display and manipulate a large-scale ortholog table, but this function is much more enhanced and flexible than that of MBGD. The ortholog table viewer in the central portion of the main window (Fig. 15a) can display the entire ortholog table. Using the zoom in/out function, it can display the entire table or a section of the main table with more detailed information (Fig. 15b). In RECOG, several basic operations on the ortholog table are defined, which are classified into categories such as filtering, sorting and coloring,

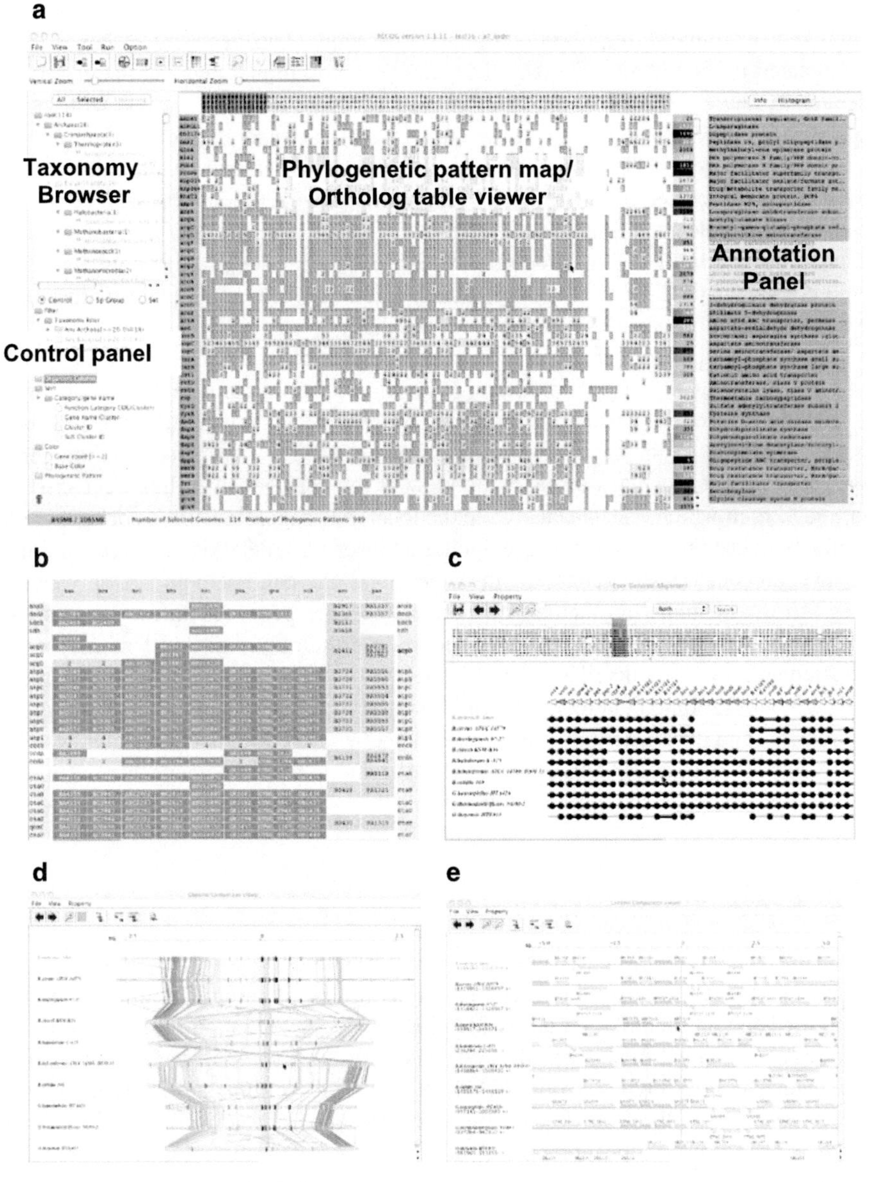

Fig. 15 Screenshots of the RECOG client software. (**a**) The main window of RECOG. (**b**) A nested ortholog table generated from the organisms in *Bacillaceae* as ingroup species and *Escherichia coli* and *Pseudomonas aerugninosa* as outgroup species. (**c**) Core genome alignment viewer showing the core structure of the *Bacillaceae* dataset. (**d**) Genome comparison viewer showing the entire chromosomal maps of the *Bacillaceae* core genes. (**e**) Genome comparison viewer showing a zoomed-in picture of (**d**) around a particular orthologous group

and various comparative analyses can be done by combining these basic operations. For example, "Neighborhood gene clustering" identifies a set of genes that are located in the vicinity of each other in both the ortholog table and the genomic sequence, and assigns the same color to each group. By combining this with various sorting functions, the user can use this function to efficiently identify conserved neighborhood genes. "Phylogenetic pattern clustering" performs hierarchical cluster analysis based on the dissimilarity between phylogenetic patterns, and reorders the ortholog table according to the clustering result. In addition, RECOG allows the user to input arbitrary gene properties such as sequence length, nucleotide/amino acid contents and functional classes, which can be used in conjunction with table operation functions such as filtering and coloring.

For the comparison of closely or moderately related genomes, RECOG also implements the CoreAligner program. With this function, the user can invoke the CoreAligner program using the orthologous groups generated by DomClust as input. The result is displayed in two graphical views. The core genome alignment viewer (Fig. 15c) displays a schematic view of the core genome alignment shown in Fig. 12b, in which one can easily see the local rearrangement of the core genome structure in each individual genome, including insertions, deletions and breakpoints of inversions or translocations. On the other hand, the genome comparison viewer (Fig. 15d, e) displays the actual location of the core genes on linear chromosomal maps with lines connecting corresponding orthologous genes. This map can be zoomed in to see a detailed map of a specific orthologous group and its vicinity (Fig. 15e) and zoomed out to see the entire map (Fig. 15d). The user can manipulate these maps and the cluster table in the main window in a coordinated manner, which facilitates the comprehension of the complex genome rearrangement that occurred in the core genome structure during the course of evolution.

Conclusion and Future Prospects

In this chapter, I reviewed several basic issues in microbial comparative genomics focusing on our own solutions primarily based on our microbial comparative genome database, MBGD. Ortholog identification is the common basis for various comparative genome analyses, although there are still some dificulties in defining plausible orthologous groups. Our DomClust algorithm addresses many of these issues, including the domain-fusion/fission problem as well as the inparalog/outparalog distinction problem. Actually, however, the ortholog identification problem is not well-defined in prokaryotic genome comparison due to the existence of horiontal gene transfers. The use of synteny information, as in our core genome extraction strategy, is a promising approach to this issue for the comparison of closely or moderately related genomes.

With thousands of microbial genome sequences in hand, we can now ask questions like "How prokaryotic genomes are organized?" and "Are there any general rules or principles that underlie this highly diverse prokaryotic world?" From a very simplified viewpoint, a prokaryotic genome is composed of the "core genome" that characterizes the taxonomic groups that that organism belongs to and various functional modules that have been mainly acquired horizontallly depending on the habitat. However, a detailed answer to the above question is yet to be given. Combining various strategies for the genome comparison of both closely related and distantly related organisms may lead us to a deeper understanding of prokaryotic genome evolution, and a comparative genome workbench supporting such analyses, like RECOG, can facilitate the study along this direction. The knowedge gained from such study should further facilitate the next challenges to the understanding of the diversity of microbial life, including microbial community analysis based on metagenomics.

References

1. Fleischmann, R.D., Adams, M.D., White, O., Clayton, R.A., Kirkness, E.F., Kerlavage, A.R., Bult, C.J., Tomb, J.F., Dougherty, B.A., Merrick, J.M., et al. Whole-genome random sequencing and assembly of *Haemophilus influenzae* Rd. Science **269**: 496–512 (1995).
2. Fraser, C.M., Gocayne, J.D., White, O., Adams, M.D., Clayton, R.A., Fleischmann, R.D., Bult, C.J., Kerlavage, A.R., Sutton, G., Kelley, J.M., et al. The minimal gene complement of *Mycoplasma genitalium*. Science **270**: 397–403 (1995).
3. Tatusov, R.L., Mushegian, A.R., Bork, P., Brown, N.P., Hayes, W.S., Borodovsky, M., Rudd, K.E., Koonin, E.V. Metabolism and evolution of *Haemophilus influenzae* deduced from a whole-genome comparison with *Escherichia coli*. Curr. Biol. **6**: 279–291 (1996).
4. Mushegian, A.R., Koonin, E.V. A minimal gene set for cellular life derived by comparison of complete bacterial genomes. Proc. Natl. Acad. Sci. USA **93**: 10268–10273 (1996).
5. Fitch, W.M. Distinguishing homologous from analogous proteins. Syst. Zool. **19**: 99–113 (1970).
6. Ohno, S. *Evolution by gene duplication*. New York, NY: Springer (1970).
7. Uchiyama, I. MBGD: microbial genome database for comparative analysis. Nucleic Acids Res. **31**: 58–62 (2003).
8. Uchiyama, I., Higuchi, T., Kawai, M. MBGD update 2010: toward a comprehensive resource for exploring microbial genome diversity. Nucleic Acids Res. **38**: D361–365 (2010).
9. Turnbaugh, P.J., Ley, R.E., Hamady, M., Fraser-Liggett, C.M., Knight, R., Gordon, J.I. The human microbiome project. Nature **449**: 804–810 (2007).
10. Wu, D., Hugenholtz, P., Mavromatis, K., Pukall, R., Dalin, E., Ivanova, N.N., Kunin, V., Goodwin, L., Wu, M., Tindall, B.J., et al. A phylogeny-driven genomic encyclopaedia of Bacteria and Archaea. Nature **462**: 1056–1060 (2009).
11. Delcher, A.L., Kasif, S., Fleischmann, R.D., Peterson, J., White, O., Salzberg, S.L. Alignment of whole genomes. Nucleic Acids Res. **27**: 2369–2376 (1999).
12. Ma, B., Tromp, J., Li, M. PatternHunter: faster and more sensitive homology search. Bioinformatics **18**: 440–445 (2002).
13. Schwartz, S., Kent, W.J., Smit, A., Zhang, Z., Baertsch, R., Hardison, R.C., Haussler, D., Miller, W. Human-mouse alignments with BLASTZ. Genome Res. **13**: 103–107 (2003).
14. Darling, A.C., Mau, B., Blattner, F.R., Perna, N.T. Mauve: multiple alignment of conserved genomic sequence with rearrangements. Genome Res. **14**: 1394–1403 (2004).
15. Schwartz, S., Zhang, Z., Frazer, K.A., Smit, A., Riemer, C., Bouck, J., Gibbs, R., Hardison, R., Miller, W. PipMaker – a web server for aligning two genomic DNA sequences. Genome Res. **10**: 577–586 (2000).
16. Frazer, K.A., Pachter, L., Poliakov, A., Rubin, E.M., Dubchak, I. VISTA: computational tools for comparative genomics. Nucleic Acids Res. **32**: W273–279 (2004).

17. Carver, T.J., Rutherford, K.M., Berriman, M., Rajandream, M.A., Barrell, B.G., Parkhill, J. ACT: the Artemis comparison tool. Bioinformatics **21**: 3422–3423 (2005).
18. Uchiyama, I., Higuchi, T., Kobayashi, I. CGAT: a comparative genome analysis tool for visualizing alignments in the analysis of complex evolutionary changes between closely related genomes. BMC Bioinformatics **7**: 472 (2006).
19. Altschul, S.F., Madden, T.L., Schaffer, A.A., Zhang, J., Zhang, Z., Miller, W., Lipman, D.J. Gapped BLAST and PSI-BLAST: a new generation of protein database search programs. Nucleic Acids Res. **25**: 3389–3402 (1997).
20. Quevillon, E., Silventoinen, V., Pillai, S., Harte, N., Mulder, N., Apweiler, R., Lopez, R. InterProScan: protein domains identifier. Nucleic Acids Res. **33**: W116–120 (2005).
21. Ashburner, M., Ball, C.A., Blake, J.A., Botstein, D., Butler, H., Cherry, J.M., Davis, A.P., Dolinski, K., Dwight, S.S., Eppig, J.T., et al. Gene ontology: tool for the unification of biology. The Gene Ontology Consortium. Nat. Genet. **25**: 25–29 (2000).
22. Kanehisa, M., Goto, S., Furumichi, M., Tanabe, M., Hirakawa, M. KEGG for representation and analysis of molecular networks involving diseases and drugs. Nucleic Acids Res. **38**: D355–360 (2010).
23. Gribskov, M., McLachlan, A.D., Eisenberg, D. Profile analysis: detection of distantly related proteins. Proc. Natl. Acad. Sci. USA **84**: 4355–4358 (1987).
24. Tatusov, R.L., Altschul, S.F., Koonin, E.V. Detection of conserved segments in proteins: iterative scanning of sequence databases with alignment blocks. Proc. Natl. Acad. Sci. USA **91**: 12091–12095 (1994).
25. Eddy, S.R. Profile hidden Markov models. Bioinformatics **14**: 755–763 (1998).
26. Osterman, A., Overbeek, R. Missing genes in metabolic pathways: a comparative genomics approach. Curr. Opin. Chem. Biol. **7**: 238–251 (2003).
27. Koonin, E.V., Mushegian, A.R., Bork, P. Non-orthologous gene displacement. Trends Genet. **12**: 334–336 (1996).
28. Koonin, E.V., Galperin, M.Y. *Sequence – evolution – function: computational approaches in comparative genomics*. Boston, MA: Kluwer (2002).
29. Pellegrini, M., Marcotte, E.M., Thompson, M.J., Eisenberg, D., Yeates, T.O. Assigning protein functions by comparative genome analysis: protein phylogenetic profiles. Proc. Natl. Acad. Sci. USA **96**: 4285–4288 (1999).
30. Marcotte, E.M., Pellegrini, M., Ng, H.L., Rice, D.W., Yeates, T.O., Eisenberg, D. Detecting protein function and protein-protein interactions from genome sequences. Science **285**: 751–753 (1999).
31. Enright, A.J., Iliopoulos, I., Kyrpides, N.C., Ouzounis, C.A. Protein interaction maps for complete genomes based on gene fusion events. Nature **402**: 86–90 (1999).
32. Overbeek, R., Fonstein, M., D'Souza, M., Pusch, G.D., Maltsev, N. The use of gene clusters to infer functional coupling. Proc. Natl. Acad. Sci. USA **96**: 2896–2901 (1999).
33. Jensen, L.J., Kuhn, M., Stark, M., Chaffron, S., Creevey, C., Muller, J., Doerks, T., Julien, P., Roth, A., Simonovic, M., et al. STRING 8 – a global view on proteins and their functional interactions in 630 organisms. Nucleic Acids Res. **37**: D412–416 (2009).
34. Marcotte, E.M. Computational genetics: finding protein function by nonhomology methods. Curr. Opin. Struct. Biol. **10**: 359–365 (2000).
35. Remm, M., Storm, C.E., Sonnhammer, E.L. Automatic clustering of orthologs and in-paralogs from pairwise species comparisons. J. Mol. Biol. **314**: 1041–1052 (2001).
36. Sonnhammer, E.L., Koonin, E.V. Orthology, paralogy and proposed classification for paralog subtypes. Trends Genet. **18**: 619–620 (2002).
37. Dessimoz, C., Boeckmann, B., Roth, A.C., Gonnet, G.H. Detecting non-orthology in the COGs database and other approaches grouping orthologs using genome-specific best hits. Nucleic Acids Res. **34**: 3309–3316 (2006).
38. Fitch, W.M. Homology a personal view on some of the problems. Trends Genet. **16**: 227–231 (2000).

39. van Dongen, S. Performance criteria for graph clustering and Markov cluster experiments. INS-R0012, Center for Mathematics and Computer Sciences (2000).
40. Enright, A.J., Van Dongen, S., Ouzounis, C.A. An efficient algorithm for large-scale detection of protein families. Nucleic Acids Res. **30**: 1575–1584 (2002).
41. Li, L., Stoeckert, C.J., Jr., Roos, D.S. OrthoMCL: identification of ortholog groups for eukaryotic genomes. Genome Res. **13**: 2178–2189 (2003).
42. Goodman, M., Czelusniak, J., Moore, W.M., Romero-Herrera, A.E., Matsuda, G. Fitting the gene lineage into its species lineage, a parsimony strategy illustrated by cladograms constructed from globin sequences. Syst. Zool. **28**: 132–163 (1979).
43. Page, R.D., Charleston, M.A. From gene to organismal phylogeny: reconciled trees and the gene tree/species tree problem. Mol. Phylogenet. Evol. **7**: 231–240 (1997).
44. Zmasek, C.M., Eddy, S.R. A simple algorithm to infer gene duplication and speciation events on a gene tree. Bioinformatics **17**: 821–828 (2001).
45. Jothi, R., Zotenko, E., Tasneem, A., Przytycka, T.M. COCO-CL: hierarchical clustering of homology relations based on evolutionary correlations. Bioinformatics **22**: 779–788 (2006).
46. Uchiyama, I. Hierarchical clustering algorithm for comprehensive orthologous-domain classification in multiple genomes. Nucleic Acids Res. **34**: 647–658 (2006).
47. van der Heijden, R.T., Snel, B., van Noort, V., Huynen, M.A. Orthology prediction at scalable resolution by phylogenetic tree analysis. BMC Bioinformatics **8**: 83 (2007).
48. Gray, G.S., Fitch, W.M. Evolution of antibiotic resistance genes: the DNA sequence of a kanamycin resistance gene from *Staphylococcus aureus*. Mol. Biol. Evol. **1**: 57–66 (1983).
49. MacLeod, D., Charlebois, R.L., Doolittle, F., Bapteste, E. Deduction of probable events of lateral gene transfer through comparison of phylogenetic trees by recursive consolidation and rearrangement. BMC Evol. Biol. **5**: 27 (2005).
50. Beiko, R.G., Hamilton, N. Phylogenetic identification of lateral genetic transfer events. BMC Evol. Biol. **6**: 15 (2006).
51. Koonin, E.V., Makarova, K.S., Aravind, L. Horizontal gene transfer in prokaryotes: quantification and classification. Annu. Rev. Microbiol. **55**: 709–742 (2001).
52. Yanai, I., Derti, A., DeLisi, C. Genes linked by fusion events are generally of the same functional category: a systematic analysis of 30 microbial genomes. Proc. Natl. Acad. Sci. USA **98**: 7940–7945 (2001).
53. Kuzniar, A., van Ham, R.C., Pongor, S., Leunissen, J.A. The quest for orthologs: finding the corresponding gene across genomes. Trends Genet. **24**: 539–551 (2008).
54. Liolios, K., Chen, I.M., Mavromatis, K., Tavernarakis, N., Hugenholtz, P., Markowitz, V.M., Kyrpides, N.C. The Genomes On Line Database (GOLD) in 2009: status of genomic and metagenomic projects and their associated metadata. Nucleic Acids Res. **38**: D346–354 (2010).
55. Tatusov, R.L., Koonin, E.V., Lipman, D.J. A genomic perspective on protein families. Science **278**: 631–637 (1997).
56. Tatusov, R.L., Galperin, M.Y., Natale, D.A., Koonin, E.V. The COG database: a tool for genome-scale analysis of protein functions and evolution. Nucleic Acids Res. **28**: 33–36 (2000).
57. Tatusov, R.L., Fedorova, N.D., Jackson, J.D., Jacobs, A.R., Kiryutin, B., Koonin, E.V., Krylov, D.M., Mazumder, R., Mekhedov, S.L., Nikolskaya, A.N., et al. The COG database: an updated version includes eukaryotes. BMC Bioinformatics **4**: 41 (2003).
58. Haft, D.H., Loftus, B.J., Richardson, D.L., Yang, F., Eisen, J.A., Paulsen, I.T., White, O. TIGRFAMs: a protein family resource for the functional identification of proteins. Nucleic Acids Res. **29**: 41–43 (2001).
59. Kanehisa, M., Goto, S., Kawashima, S., Okuno, Y., Hattori, M. The KEGG resource for deciphering the genome. Nucleic Acids Res. **32**: D277–280 (2004).
60. Kanehisa, M., Araki, M., Goto, S., Hattori, M., Hirakawa, M., Itoh, M., Katayama, T., Kawashima, S., Okuda, S., Tokimatsu, T., et al. KEGG for linking genomes to life and the environment. Nucleic Acids Res. **36**: D480–484 (2008).

61. Gattiker, A., Michoud, K., Rivoire, C., Auchincloss, A.H., Coudert, E., Lima, T., Kersey, P., Pagni, M., Sigrist, C.J., Lachaize, C., et al. Automated annotation of microbial proteomes in SWISS-PROT. Comput. Biol. Chem. **27**: 49–58 (2003).
62. Meyer, F., Overbeek, R., Rodriguez, A. FIGfams: yet another set of protein families. Nucleic Acids Res. **37**: 6643–6654 (2009).
63. O'Brien, K.P., Remm, M., Sonnhammer, E.L. Inparanoid: a comprehensive database of eukaryotic orthologs. Nucleic Acids Res. **33**: D476–480 (2005).
64. Chen, F., Mackey, A.J., Stoeckert, C.J., Jr., Roos, D.S. OrthoMCL-DB: querying a comprehensive multi-species collection of ortholog groups. Nucleic Acids Res. **34**: D363–368 (2006).
65. Alexeyenko, A., Tamas, I., Liu, G., Sonnhammer, E.L. Automatic clustering of orthologs and inparalogs shared by multiple proteomes. Bioinformatics **22**: e9–15 (2006).
66. Muller, J., Szklarczyk, D., Julien, P., Letunic, I., Roth, A., Kuhn, M., Powell, S., von Mering, C., Doerks, T., Jensen, L.J., et al. eggNOG v2.0: extending the evolutionary genealogy of genes with enhanced non-supervised orthologous groups, species and functional annotations. Nucleic Acids Res. **38**: D190–195 (2010).
67. Schneider, A., Dessimoz, C., Gonnet, G.H. OMA Browser – exploring orthologous relations across 352 complete genomes. Bioinformatics **23**: 2180–2182 (2007).
68. Davidsen, T., Beck, E., Ganapathy, A., Montgomery, R., Zafar, N., Yang, Q., Madupu, R., Goetz, P., Galinsky, K., White, O., et al. The comprehensive microbial resource. Nucleic Acids Res. **38**: D340–345 (2010).
69. Markowitz, V.M., Chen, I.M., Palaniappan, K., Chu, K., Szeto, E., Grechkin, Y., Ratner, A., Anderson, I., Lykidis, A., Mavromatis, K., et al. The integrated microbial genomes system: an expanding comparative analysis resource. Nucleic Acids Res. **38**: D382–390 (2010).
70. Dehal, P.S., Joachimiak, M.P., Price, M.N., Bates, J.T., Baumohl, J.K., Chivian, D., Friedland, G.D., Huang, K.H., Keller, K., Novichkov, P.S., et al. MicrobesOnline: an integrated portal for comparative and functional genomics. Nucleic Acids Res. **38**: D396–400 (2010).
71. Caspi, R., Foerster, H., Fulcher, C.A., Kaipa, P., Krummenacker, M., Latendresse, M., Paley, S., Rhee, S.Y., Shearer, A.G., Tissier, C., et al. The MetaCyc Database of metabolic pathways and enzymes and the BioCyc collection of Pathway/Genome Databases. Nucleic Acids Res. **36**: D623–631 (2008).
72. Overbeek, R., Begley, T., Butler, R.M., Choudhuri, J.V., Chuang, H.Y., Cohoon, M., de Crecy-Lagard, V., Diaz, N., Disz, T., Edwards, R., et al. The subsystems approach to genome annotation and its use in the project to annotate 1000 genomes. Nucleic Acids Res. **33**: 5691–5702 (2005).
73. Enault, F., Suhre, K., Poirot, O., Abergel, C., Claverie, J.M. Phydbac2: improved inference of gene function using interactive phylogenomic profiling and chromosomal location analysis. Nucleic Acids Res. **32**: W336–339 (2004).
74. Mellor, J.C., Yanai, I., Clodfelter, K.H., Mintseris, J., DeLisi, C. Predictome: a database of putative functional links between proteins. Nucleic Acids Res. **30**: 306–309 (2002).
75. Sneath, P.H.A., Sokal, R.R. *Numerical taxonomy*. San Francisco, CA: Freeman (1973).
76. Page, R.D.M. Maps between trees and cladistic analysis of historical associations among genes, organisms, and areas. Syst. Biol. **43**: 58–77 (1994).
77. Hulsen, T., Huynen, M.A., de Vlieg, J., Groenen, P.M. Benchmarking ortholog identification methods using functional genomics data. Genome Biol. **7**: R31 (2006).
78. Chen, F., Mackey, A.J., Vermunt, J.K., Roos, D.S. Assessing performance of orthology detection strategies applied to eukaryotic genomes. PLoS One **2**: e383 (2007).
79. Altenhoff, A.M., Dessimoz, C. Phylogenetic and functional assessment of orthologs inference projects and methods. PLoS Comput. Biol. **5**: e1000262 (2009).
80. Uchiyama, I. MBGD: a platform for microbial comparative genomics based on the automated construction of orthologous groups. Nucleic Acids Res. **35**: D343–346 (2007).

81. Besemer, J., Lomsadze, A., Borodovsky, M. GeneMarkS: a self-training method for prediction of gene starts in microbial genomes. Implications for finding sequence motifs in regulatory regions. Nucleic Acids Res. **29**: 2607–2618 (2001).
82. Delcher, A.L., Harmon, D., Kasif, S., White, O., Salzberg, S.L. Improved microbial gene identification with GLIMMER. Nucleic Acids Res. **27**: 4636–4641 (1999).
83. Haft, D.H., Selengut, J.D., Brinkac, L.M., Zafar, N., White, O. Genome Properties: a system for the investigation of prokaryotic genetic content for microbiology, genome annotation and comparative genomics. Bioinformatics **21**: 293–306 (2005).
84. Harris, M.A., Clark, J., Ireland, A., Lomax, J., Ashburner, M., Foulger, R., Eilbeck, K., Lewis, S., Marshall, B., Mungall, C., et al. The Gene Ontology (GO) database and informatics resource. Nucleic Acids Res. **32**: D258–261 (2004).
85. Wu, J., Kasif, S., DeLisi, C. Identification of functional links between genes using phylogenetic profiles. Bioinformatics **19**: 1524–1530 (2003).
86. Enault, F., Suhre, K., Claverie, J.M. Phydbac "Gene Function Predictor": a gene annotation tool based on genomic context analysis. BMC Bioinformatics **6**: 247 (2005).
87. Forterre, P. A hot story from comparative genomics: reverse gyrase is the only hyperthermophile-specific protein. Trends Genet. **18**: 236–237 (2002).
88. Makarova, K.S., Wolf, Y.I., Koonin, E.V. Potential genomic determinants of hyperthermophily. Trends Genet. **19**: 172–176 (2003).
89. Koonin, E.V. How many genes can make a cell: the minimal-gene-set concept. Annu. Rev. Genom. Hum. Genet. **1**: 99–116 (2000).
90. Koonin, E.V. Comparative genomics, minimal gene-sets and the last universal common ancestor. Nat. Rev. Microbiol. **1**: 127–136 (2003).
91. Kobayashi, K., Ehrlich, S.D., Albertini, A., Amati, G., Andersen, K.K., Arnaud, M., Asai, K., Ashikaga, S., Aymerich, S., Bessieres, P., et al. Essential *Bacillus subtilis* genes. Proc. Natl. Acad. Sci. USA **100**: 4678–4683 (2003).
92. Baba, T., Ara, T., Hasegawa, M., Takai, Y., Okumura, Y., Baba, M., Datsenko, K.A., Tomita, M., Wanner, B.L., Mori, H. Construction of *Escherichia coli* K-12 in-frame, single-gene knockout mutants: the Keio collection. Mol. Syst. Biol. **2**: 2006 0008 (2006).
93. Charlebois, R.L., Doolittle, W.F. Computing prokaryotic gene ubiquity: rescuing the core from extinction. Genome Res. **14**: 2469–2477 (2004).
94. Lan, R., Reeves, P.R. Intraspecies variation in bacterial genomes: the need for a species genome concept. Trends Microbiol. **8**: 396–401 (2000).
95. Tettelin, H., Masignani, V., Cieslewicz, M.J., Donati, C., Medini, D., Ward, N.L., Angiuoli, S.V., Crabtree, J., Jones, A.L., Durkin, A.S., et al. Genome analysis of multiple pathogenic isolates of *Streptococcus agalactiae*: implications for the microbial "pan-genome". Proc. Natl. Acad. Sci. USA **102**: 13950–13955 (2005).
96. Medini, D., Donati, C., Tettelin, H., Masignani, V., Rappuoli, R. The microbial pan-genome. Curr. Opin. Genet. Dev. **15**: 589–594 (2005).
97. Lefébure, T., Stanhope, M.J. Evolution of the core and pan-genome of *Streptococcus*: positive selection, recombination, and genome composition. Genome Biol. **8**: R71 (2007).
98. Makarova, K.S., Aravind, L., Galperin, M.Y., Grishin, N.V., Tatusov, R.L., Wolf, Y.I., Koonin, E.V. Comparative genomics of the Archaea (Euryarchaeota): evolution of conserved protein families, the stable core, and the variable shell. Genome Res. **9**: 608–628 (1999).
99. Jain, R., Rivera, M.C., Lake, J.A. Horizontal gene transfer among genomes: the complexity hypothesis. Proc. Natl. Acad. Sci. USA **96**: 3801–3806 (1999).
100. Nelson, K.E., Clayton, R.A., Gill, S.R., Gwinn, M.L., Dodson, R.J., Haft, D.H., Hickey, E.K., Peterson, J.D., Nelson, W.C., Ketchum, K.A., et al. Evidence for lateral gene transfer between Archaea and bacteria from genome sequence of Thermotoga maritima. Nature **399**: 323–329 (1999).
101. Ochman, H., Lawrence, J.G., Groisman, E.A. Lateral gene transfer and the nature of bacterial innovation. Nature **405**: 299–304 (2000).
102. Brown, J.R. Ancient horizontal gene transfer. Nat. Rev. Genet. **4**: 121–132 (2003).

103. Boucher, Y., Douady, C.J., Papke, R.T., Walsh, D.A., Boudreau, M.E., Nesbo, C.L., Case, R.J., Doolittle, W.F. Lateral gene transfer and the origins of prokaryotic groups. Annu. Rev. Genet. **37**: 283–328 (2003).
104. Doolittle, W.F. Phylogenetic classification and the universal tree. Science **284**: 2124–2129 (1999).
105. Gogarten, J.P., Doolittle, W.F., Lawrence, J.G. Prokaryotic evolution in light of gene transfer. Mol. Biol. Evol. **19**: 2226–2238 (2002).
106. de la Cruz, F., Davies, J. Horizontal gene transfer and the origin of species: lessons from bacteria. Trends Microbiol. **8**: 128–133 (2000).
107. Harris, J.K., Kelley, S.T., Spiegelman, G.B., Pace, N.R. The genetic core of the universal ancestor. Genome Res. **13**: 407–412 (2003).
108. Philippe, H., Douady, C.J. Horizontal gene transfer and phylogenetics. Curr. Opin. Microbiol. **6**: 498–505 (2003).
109. Bapteste, E., Boucher, Y., Leigh, J., Doolittle, W.F. Phylogenetic reconstruction and lateral gene transfer. Trends Microbiol. **12**: 406–411 (2004).
110. Hacker, J., Carniel, E. Ecological fitness, genomic islands and bacterial pathogenicity. A Darwinian view of the evolution of microbes. EMBO Rep. **2**: 376–381 (2001).
111. Uchiyama, I. Multiple genome alignment for identifying the core structure among moderately related microbial genomes. BMC Genomics **9**: 515 (2008).
112. Rogozin, I.B., Makarova, K.S., Murvai, J., Czabarka, E., Wolf, Y.I., Tatusov, R.L., Szekely, L.A., Koonin, E.V. Connected gene neighborhoods in prokaryotic genomes. Nucleic Acids Res. **30**: 2212–2223 (2002).
113. Shimodaira, H., Hasegawa, M. Multiple comparisons of log-likelihoods with applications to phylogenetic inference. Mol. Biol. Evol. **16**: 1114–1116 (1999).

Predicting Protein Functional Sites with Phylogenetic Motifs: Past, Present and Beyond

Dennis R. Livesay, Dukka Bahadur KC, and David La

Abstract More than sequence or structure, function imposes very tight constraints on the evolutionary variability within a protein family. As such, numerous functional site prediction methods are based on algorithms to uncover conserved regions that lead to conserved function. Nevertheless, evolution does allow for some systematic variability within functional regions. Based on this tenet, we have introduced the MINER algorithm to predict functional regions from phylogenetic motifs. Specifically, our approach identifies alignment fragments that parallel the overall phylogeny of the family, which are more likely to be functional due to increased evolutionary signature. In this chapter, we provide an overview of the method, summarize recent developments, and comment on future work.

Introduction

Due to the rapid increased in the number of solved sequences from next-generation sequencing technologies, accurate prediction of protein function and functional sites from sequence-derived data is now more important than ever. There are many different functional site prediction algorithms in the literature [1], most of which attempt to identify some sort of evolutionary feature within the input alignment. Meaning, they are primarily based on the simple and common dogma that conservation of function is the ultimate evolutionary driving force.

The evolutionary constraints imposed by function severely limits sequence variability at certain sites, which has led to myriad algorithms to predict function from conservation [2–4]. However, the constraints imposed by function need not completely limit variability within a given site. Rather, functional sites frequently vary somewhat dependent upon exact functional criteria (i.e., substrate specificity, catalytic efficiency, etc.) and the context of the rest of the protein, thus yielding systematic variations between subfamilies [5–6]. Unfortunately, prediction algorithms

D.R. Livesay (✉)
Department of Bioinformatics and Genomics, University of North Carolina at Charlotte,
9201 University City Blvd., Charlotte, NC 28223, USA
e-mail: drlivesa@uncc.edu

D. Kihara (ed.), *Protein Function Prediction for Omics Era*,
DOI 10.1007/978-94-007-0881-5_5, © Springer Science+Business Media B.V. 2011

based solely on these "evolutionary trace" positions result in an unsatisfactory number of false positives [7–9]. MINER is based on a similar notion; however, it attempts to identify phylogenetic motifs (PMs), which are contiguous alignment fragments, not alignment positions, that have co-evolved to satisfy the functional evolutionary constraints. Along with some judicious algorithmic implementation details discussed below, it is this distinction that leads to improved prediction accuracy of our approach.

The Past

Based on work published between 2005 and 2007, this section describes the original MINER algorithm. In addition, we present a summary of application of the approach to the NSS protein family, which highlights MINER's utility and limitations.

The MINER Algorithm

The MINER algorithm, originally introduced in La et al. [7], is inspired by our earlier observation that motifs taken from regions known to be functionally important a priori conserve the overall phylogeny of the family [10]. Meaning, MINER reverses this scenario to look for regions that reproduce the phylogenetic clustering, and then presents them as putative functional sites. The algorithm, which is summarized in Fig. 1, begins with a sliding sequence window that generates all possible alignment fragments of fixed width from an input alignment. Subsequently, a tree is constructed on each fragment using standard phylogenetic reconstruction algorithms, which is compared to the phylogenetic tree of the whole family using a bipartition metric algorithm that counts topological differences between the pair [11]. In the original implementation of the algorithm, all overlapping fragments that score pass some threshold are grouped into a single PM. Based on the competition between site specificity and evolutionary signal, we have determined that a window width of five is ideal in most situations [7].

Tree similarity is quantified using the ubiquitous bipartition metric algorithm [13], which is also commonly referred to as the symmetric difference or the Robinson-Foulds distance. The bipartition metric simply counts the number of partitions, defined by tree branch points, varying across the pair. To improve prediction accuracy, the bipartition metric employed by MINER is actually a slightly modified algorithm, but the details of the modification and its rationale are beyond the scope of this discussion (cf. Roshan et al. [11] for a full discussion).

Figure 1 includes a typical phylogenetic similarity spectrum for the glycolytic enzyme triosephosphate isomerase, which plots tree similarity (recast as statistical z-scores) versus fragment number. We call the MINER output values phylogenetic similarity z-scores (PSZs). Because the bipartition metric provides a distance,

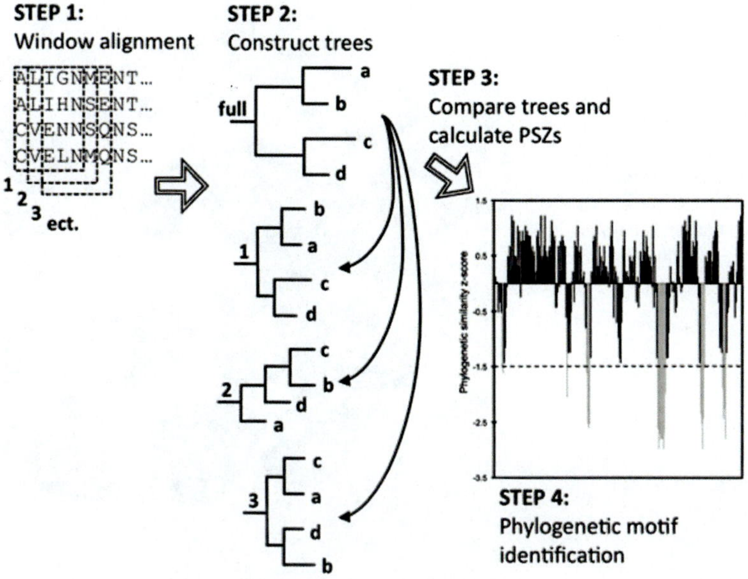

Fig. 1 A cartoon describing the original MINER algorithm. First, MINER starts with a multiple sequence alignment, from which all possible $L - W + 1$ windows are generated, where L is the alignment length and W is the window width. Second, using standard phylogenetic reconstruction techniques, a tree is generated for each window and the complete alignment. Third, the topological similarity of each window tree is compared to the overall phylogeny using a bipartition metric algorithm. The raw bipartition metric scores are then converted to statistical z-scores, called PSZs for phylogenetic similarity z-scores. Note that more negative values indicate greater tree similarity since the raw bipartition metric values represent distances. Finally, all overlapping windows scoring past some threshold are grouped into a phylogenetic motif (PM). As applied to the *triosephosphate isomerase* enzyme family, six PMs are identified. The smallest PM is composed of only a single window, whereas the largest is composed of eight contiguous windows. Note that the largest PM overlaps the PROSITE definition of the family [12]. The algorithm is more fully described in La et al. [7]

smaller values indicate greater phylogenetic similarity. Using a PSZ threshold of 1.5, Fig. 2 clearly demonstrates that the predicted functional sites map to the enzyme's active site region. In fact, with the exception of the one PM in the upper-right corner, all identified PMs (dark grey) clearly cluster around the enzyme's active site (a co-crystalized substrate analog is shown in light grey). However, this PM is actually interacting with the substrate at the active site of its homodimer partner, meaning all six PMs overlap the enzyme's active site. In a later follow-up study [14], we demonstrated the functional roles of PM residues are commonly explained in a rational way by sophisticated continuum electrostatics calculations. Therein, the biophysical calculations demonstrated that the PM residues were interacting with the strictly conserved catalytic residues to fine-tune their chemical properties. This result highlights the power of synergistically combining empirical and first principles viewpoints to understand protein function. However, biophysical calculations

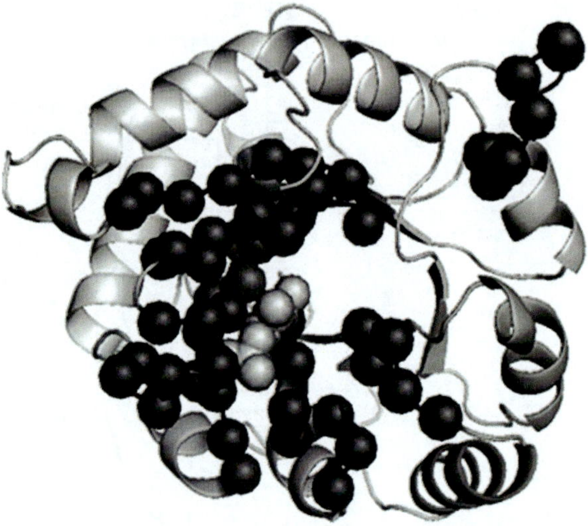

Fig. 2 Triosephosphate isomerase. The PMs identified in Fig. 1 are mapped to the structure of an example structure. *Dark grey colored spheres* represent α-carbon atoms of the predicted sites. The enzyme's substrate analog is colored *light grey* and shown in *spacefill*

are generally expensive and require structural input, thus limiting their utility for high-throughput investigations.

The PSZ threshold used can be predefined by the user or automatically determined. Threshold values of $\sim 1.5 \pm 0.5$ standard deviations are generally ideal; however, large prediction differences can occur within this range. While there are myriad signal-to-noise methods, we have developed the EXTREME algorithm to be specifically appropriate to the problem at hand [15]. The approach is based on three primary features. The first is that we pre-process the MINER output to highlight the evolutionary signal. Specifically, because they are associated with a single PM, contiguous stretches of scores within the above range are represented by a single data point, which we call sharpening. Second, the sharpened scores are then clustered into $k = 2$ groups using partition around medoids clustering. We use k-medoids clustering because it is less sensitive to outliers compared to the more common k-means clustering. The threshold is defined as the largest score within the second (more negative) cluster. Finally, there are number of algorithmic overrides that have been developed to ensure that the resultant threshold has the desired properties, such as not predicting too many PMs. A quantitative assessment of prediction accuracy on a small dataset of 32 protein families demonstrates that EXTEME leads to 69% correct predictions and 23% useful predictions. Only 11% were deemed wrong. As previously done with evolutionary trace [16], the assessment of *correct*, *useful*, and *incorrect* is determined from whether the predicted sites are, respectively, *within*, *overlapping*, or *distinct* from the known functional site. However, as we have discussed previously, this assessment is overly strict because it completely ignores functional roles outside the active site.

Prediction of Functional Sites Within the NSS Protein Family

The accuracy and utility of the MINER functional site predictions has been born out many times. As an example, we focus here on our application of the method to the neurotransmitter/sodium symport (NSS) family, which is a large and functionally diverse family of transporter proteins. In the NSS family, free energy provided by the flux of sodium and chloride ions with their electrochemical gradients across a membrane barrier is used to move chemical substrates against theirs. The chemical substrates recognized by members of the family are extremely chemically diverse, and include amino acids, biogenic amines and osmolytes. Application of MINER, along with a number of other common functional site prediction methods, identified a large number of putative functional sites, which were compared to residues identified as important from the leucine transporter transporter solved by Yamashita et al. [17] and an exhaustive survey of the experimental mutagenesis data.

MINER had the best prediction coverage of the six methods considered, predicting an impressive 62% of the benchmark sites. Moreover, MINER's overall performance was among the best considered. Interestingly, the others with similar performance were primarily conservation measures. Yet, MINER performed much better than evolutionary trace and SDPpred [18], which is another common prediction technique based on subfamily differences. To provide a balanced description of coverage and accuracy, overall performance is calculated as the Cartesian distance between (coverage, accuracy) of each method to a hypothetical perfect method (coverage = 1.00, accuracy = 1.00). The distances are normalized such that a method with 0.00 coverage and accuracy would have a value of unity. The reason that MINER's overall performance is slightly below the conservation measures is that it tends to over-predict sites. This is simply due to each prediction within MINER actually corresponding to five residues. As such, we also evaluated a relative accuracy, which is normalized by the number of predicted windows (not residues). The relative accuracy of MINER is very good, but should not necessarily be compared to the site specific methods since they are fundamentally different quantities. These results are presented in Table 1.

A very interesting result from this work was that the set of predictions from each of the six methods are generally orthogonal to each other. As such, we demonstrated that predictions based on simple intersections of the various methods significantly improve prediction accuracy. Meaning, only positions that are simultaneously predicted by multiple methods are put forth as a prediction. Impressively, prediction by any three methods (except for SDPpred that was excluded due to poor overall performance), the coverage and accuracy reached 0.56 and 0.44, respectively, which is much higher than any of the individual methods. Another interesting result from this work is based on consensus predictions. We demonstrated that predictions with better support, meaning they are predicted by multiple methods, are more likely to cluster around the leucine-binding site and the proposed transport route (cf. Fig. 3). Taken together, these two sets of results highlight the synergy and complementarity across various functional site prediction methods.

Table 1 Coverage and accuracy of the various functional site prediction schemes across all the NSS functional site benchmark

Method[a]	Coverage (%)	Accuracy[b] (%)	Overall performance[c]
Phylogenetic motif	62	24 (55%)	0.40
Motif conservation	53	35 (90%)	0.43
Position conservation	59	35	0.45
Rate4Site	50	37	0.43
Evolutionary trace	44	27	0.34
SDPpred	12	27	0.19
Intersect 2[d]	71	29	0.46
Intersect 3	56	44	0.50
Intersect 4	32	50	0.40
Intersect 5	18	67	0.37

[a]These results are reproduced from Livesay et al. [8], which provides details of the methods employed.
[b]Accuracies are reported as the ratio of correct to total alignment positions predicted. For methods that are based on alignment fragments, the relative accuracy that describes the ratio of correct predictions to the total number of alignment windows is provided in parentheses.
[c]Overall performance is calculated as the Cartesian distance between (coverage, accuracy) of each method and that of a perfect method (coverage = 1.00, accuracy = 1.00). The distance is normalized such that a method with 0.00 coverage and accuracy would have a value of unity.
[d]The Intersect predictions describe a hybrid approach composed of the unique prediction strategies. Whenever the number of predictions for a particular site are greater than the intersect value, that site is put forth as a prediction

The Present

Based on work published between 2008 and 2010, this section describes our recent attempts to improve the MINER algorithm and to explain its predictive power. Specifically, we demonstrate that the accuracy of MINER is improved by translating it into a site-specific model. Moreover, development of more rigorous hybrid methods that combine PMs and conservation provide very good predictions. Finally, we have also demonstrated, not unexpectedly, that the bulk of the predictive power of MINER comes from its topological description of evolutionary variability.

Residue Specific Predictions

As discussed above, MINER does a very good job of identifying known functional sites; however, its accuracy is somewhat tempered by its window-centric view. Moreover, the NSS family results above are just a single example, which may or may not be representative of average performance. To determine how well MINER performs relative to conservation measures, we have constructed a large well-curated and nonredundant benchmark dataset based on the catalytic site atlas

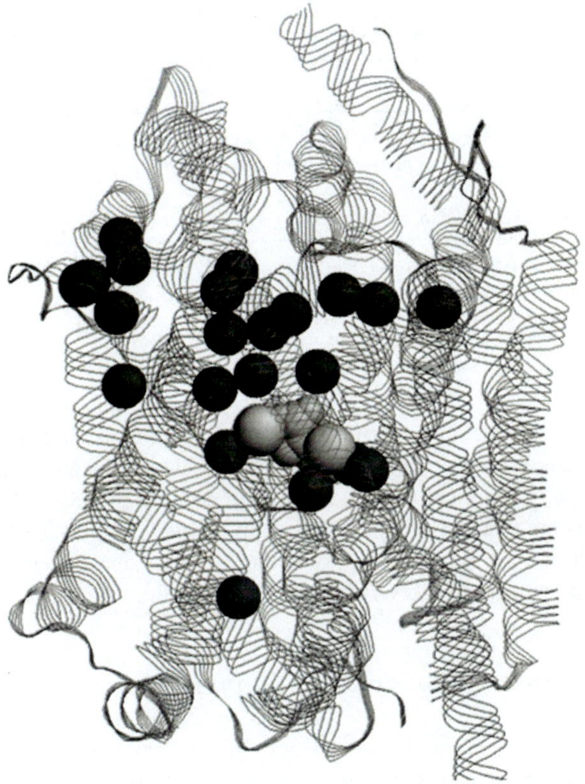

Fig. 3 The NSS family. Structural superposition of all functional site predictions within the neutrostransmitter/sodium symporter family that are predicted by at least four methods are highlighted in *dark grey* within the leucine transporter structure, which do a good job of covering the extracellular/periplasmic gate and ligand-binding site residues. The leucine, sodium ions and chloride ion are also shown in spacefill (*light grey*) at the center of the structure. Reproduced from Livesay et al. [8]

[19]. Specifically, we defined *active sites* from the catalytic residues plus all residues interacting with them [9]. To make MINER position-specific, a given alignment position is simply assigned the phylogenetic similarity score of the window centered on it.

As such, there are two key differences between this approach and what we have done prior. First, of course, we have removed MINER's inherent window-centric view. Second, we have also removed the threshold needed to group windows into a PM. Rather, like other site-specific measures, we now just have a list of scores rank-ordered from best to worst predictions. And like all methods along these lines, the appropriate cut-off to balance sensitivity and specificity is a degree of freedom to be optimized. To eliminate the arbitrariness of defining such a threshold, we apply receiver operator characteristic (ROC) analysis to quantify the balance between the two over a systematic range of cut-offs. Table 2 provides the area under curve (AUC) at a false positive rate of 0.1, which is a standard measure of the predictive power

Table 2 Receiver operator characteristic analysis for position specific predictions of active site residues across a large nonredundant dataset[a]

Method	$AUC^{b}_{0.10}$
MINER (based on phylogenetic similarity of window centered on target position)	2.13
SCORECONS[c] (which is a sum of pairs conservation score)	1.95
psMINER (based on SCORECONS)	2.38
hMINER (based on SCORECONS and $\alpha = 0.6$)	2.48

[a]These results are reproduced from KC and Livesay [9]. While not provided here, statistical significance of the improvements is discussed in the cited paper.
[b]All reported values are $\times\ 10^{-2}$.
[c] The citation for SCORECONS is Valdar [2]

of functional site prediction algorithms. (Note that AUCs at larger false positive rates are generally not considered because they would produce too many spurious predictions to be of practical usefulness as a guide for experimental studies.) Our results demonstrate that the modified PM approach is very powerful. Specifically, the PSZs result in a 9% improvement over the common sum-of-pairs conservation metric. Similar results are observed for other conservation scores.

Integrating Conservation and Evolutionary Viewpoints

Based on the complementarity between variability and conservation viewpoints discussed above, we have recently developed more rigorous algorithms to integrate both approaches. Specifically, we have developed two hybrid site-specific versions of MINER [9]. Both incorporate conservation information, but do so in distinct ways. The first approach, called psMINER for position-specific MINER, starts by rank ordering each alignment position with respect to a calculated conservation score. Next, each position is interrogated about whether or not it is found within a PM. If so, then its ranking is unaffected. However, if not, the position is shuffled to the bottom of the list and is never considered to be a possible functional site. Again, based on ROC AUC values, Table 2 demonstrates that psMINER leads to large improvements over MINER itself and the underlying conservation scores in the prediction of active site residues. In fact, the improvement over sum-of-pairs is 22% and the improvement over the PSZs is nearly 12%.

One drawback of the psMINER algorithm is that it only uses PM information as a binary, meaning a residue is either part of a PM or not. To incorporate a quantitative aspect, we have also developed a hybrid MINER (hMINER) approach that averages the phylogenetic similarity and conservation scores using optimized statistical weight α. The hMINER score for alignment position i is given by: $H_i = \alpha M_i + (1 - \alpha)C_i$, where M_i is the MINER similarity score and C_i is the conservation score. Again, Table 2 demonstrates that hMINER does an excellent job of improving predictive power. The improvement over sum-of-pairs is 27% and the improvement over the PSZs is 16%. Nevertheless, an equally interesting aspect of the hMINER approach is that the value of α dissects the relative importance of the evolutionary variability and conservation aspects of the hybrid approach. For example, depending upon the conservation score used, typical values range from $\alpha \sim 0.5 - 0.7$, indicating that descriptions of the phylogenetic variability are generally slightly more important than conservation.

The Importance of Topology

In prior work, we demonstrated that improving phylogenetic descriptions is another straightforward way of improving the predictive power of MINER [11]. Specifically, we demonstrated that we could improve prediction accuracy by focusing on phylogenetic trees reconstructed using parsimony, rather than neighbor-joining methods. However, an interesting report recently demonstrated that an algorithm similar to MINER, but instead focused on distance matrix comparisons rather than phylogenetic trees, could also provide acceptable prediction accuracies [20]. To test their assertions more rigorously, we performed an exhaustive assessment of 39 different variants of their approach over a range of window widths [21]. We considered three different types of distance matrices [22–24] and thirteen different matrix-to-matrix comparison metrics. Figure 4 summarizes our results. Specifically, it plots the ROC

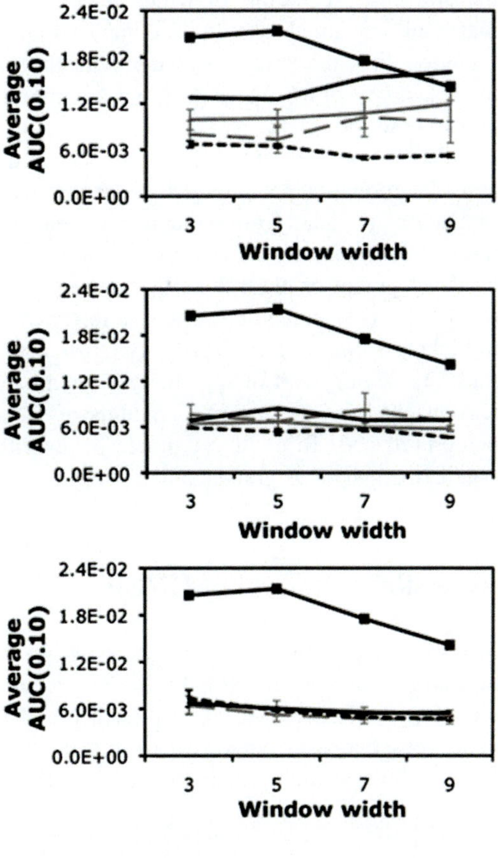

Fig. 4 The importance of topology. Average $AUC_{0.10}$ values for active site prediction are plotted for each matrix similarity metric class (*black squares* = MINER, *solid black line* = Tanimoto coefficient, *solid grey line* = distance-based, *long dashed grey line* = information theory, and *short dashed black line* = correlation coefficient). The *error bars* correspond to one standard deviation. There are no error bars for MINER and the Tanimoto coefficient because each is only a single metric. Each row corresponds to, respectively, ClustalDist, TREE-PUZZLE, and ProtDist distance matrices

$AUC_{0.10}$ values for the prediction of active sites for the distance-matrix variants and the original MINER approach. The results clearly demonstrate the superior predictive power of MINER. The three panels correspond to the three different distance matrices and each curve corresponds to the four matrix comparison metric classes + MINER over a range of window widths. Taken as a whole, these results establish that the improved predictive power arises from the added evolutionary insight provided by phylogenetic trees. Meaning, tree topologies represent a simple, yet powerful way to improve the accuracy of PM functional site predictions.

The Future

The sum of our work to date in the realm of protein functional site prediction clearly indicates that strategies based on strict conservation scores and alternate strategies based on evolutionary variability both have merit. Moreover, we have clearly demonstrated that integrating viewpoints is a convenient way to improve predictive power. However, while not specifically discussed, another general conclusion from our work to date is that *all* current functional site prediction algorithms (MINER included) lack prediction specificity. While all published methods produce better than random predictions of which positions within an alignment are important, that is all they are able to do. The methods indiscriminately identify evolutionarily important sites and/or regions, but provide little additional insight. Meaning they fail to explain *how* or *why* these positions are important. In order to provide such mechanistic descriptions, we continually attempt to layer the results from biophysical calculations on representative structures from the family onto the predicted sites to assist interpretation and provided added value. Alternately, to provide the same sorts of mechanistic detail from sequence-derived data alone, the bioinformatics community needs to identify new ways to improve functional site prediction specificity by development of algorithms that are able to distinguish between various functional roles (i.e., catalytic residues, allosteric/regulatory sites, ligand-binding sites, trafficking signals, etc.). *Second generation* functional site prediction algorithms must provide this sort of specificity if we are ever going to fully extract biochemical insight from the massive amounts of sequence information that is currently being produced. To us, development of algorithms that include such mechanistic specificity is the next grand challenge for the functional site prediction community.

Accessibility and miniMINER

MINER is accessible in three ways. First, we have developed a web-based implementation called webMINER [25]. The implementation contains full functionality, including the EXTREME algorithm. It uses ClustalW [22] to construct phylogenetic trees, and uses our own partition metric implementation to compare them. Input is either a multiple alignment or a set of unaligned sequences that we will

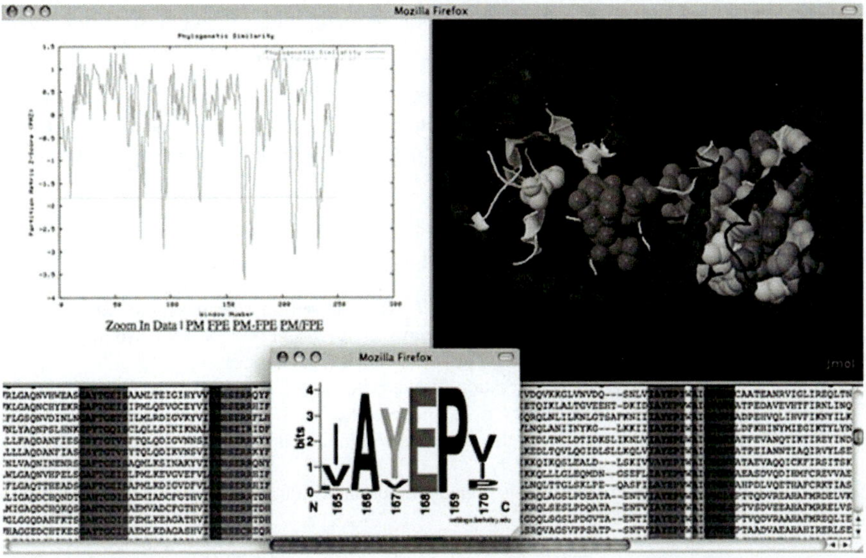

Fig. 5 Screenshot of the webMINER output. In addition to PM identification, webMINER includes a variety of additional functionality, including the option to map the PM predictions to a representative structure (as shown here). The multiple sequence alignment, which highlights the identified PMs, is hyperlinked to the structure viewer such that structural context of one or all of the PMs can be interactively analyzed. Additionally, webMINER provides sequence logo descriptions of the PMs so that the user can quickly evaluate the evolutionary variability within the identified region. Finally, all of the raw data is available for download so that user can port the data to other analysis programs

align for you. The basic ouput is a phylogenetic similarity z-score for each window, but depending upon user options, a number of additional analysis tools are also provided. The webMINER is currently accessible at http://coit-apple01.uncc.edu/MINER/. A screenshot from a typical output is provided in Fig. 5.

The second option is that, upon request, we will provide a standalone PERL program that integrates all the relevant software used by the webMINER. Meaning, it includes all of the visualization options, which can be either used or not. Unfortunately, the standalone version is rather difficult to compile and integrate. As such, if you have only a few families to analyze, we recommend that you use our web-version. Conversely, if you want to apply MINER in a large-scale way, we have recently developed a third option.

Our most recent work has focused on improving the utility of MINER by providing a streamlined version of MINER that has no dependencies upon other installed software (but Java). Specifically, this miniMINER has been programmed to ease the high-throughput use of MINER. The program simply outputs the PSZs for a given input alignment. To ease installation, we have re-implemented all of the underlying phylogenetic reconstruction and tree similarity functionalities within a self-contained Java jar file that should work seamlessly on any computer with Java installed. This miniMINER is available upon request, and a paper describing these results is currently being prepared.

Acknowledgements The development and application of MINER is based on the dedicated work of a number of people. Specifically, we wish to thank Dr. Usman Roshan, Ehsan Tabari, Brian Sutch, Patrick Kidd, and Dr. Sepehr Eskandari for contributions to the results discussed herein.

References

1. K.C., D., Livesay, D.R. A spectrum of phylogenetic-based approaches for predicting protein functional sites. Bioinformatics for systems biology Krawetz, S. (ed.) Humana Press, New York, NY: pp. 315–337 (2009).
2. Valdar, W.S. Scoring residue conservation. Proteins **48**: 227–241 (2002).
3. Pupko, T. Rate4Site: an algorithmic tool for the identification of functional regions in proteins by surface mapping of evolutionary determinants within their homologues. Bioinformatics **18**: S71–S77 (2002).
4. Capra, J.A., Singh, M. Predicting functionally important residues from sequence conservation. Bioinformatics **23**: 1875–1882 (2007).
5. Lichtarge, O., et al. An evolutionary trace method defines binding surfaces common to protein families. J. Mol. Biol. **257**: 342–358 (1996).
6. del Sol, A., et al. Automatic methods for predicting functionally important residues. Ann. Mat. Pura. Appl. **326**: 1289–1302 (2003).
7. La, D., et al. Predicting protein functional sites with phylogenetic motifs. Proteins **58**: 309–320 (2005).
8. Livesay, D.R., et al. Assessing the ability of sequence-based methods to provide functional insight within membrane integral proteins: a case study analyzing the neurotransmitter/Na+ symporter family. BMC Bioinform. **8**: 397 (2007).
9. K.C., D., Livesay, D.R. Improving position-specific predictions of protein functional sites using phylogenetic motifs. Bioinformatics **24**: 2308–2316 (2008).
10. Livesay, D.R., et al. Conservation of electrostatic properties within enzyme families and superfamilies. Biochemistry **42**: 3464–3473 (2003).
11. Roshan, U., et al. Improved phylogenetic motif identification using parsimony. Proc. IEEE Syms. Bioinform. Bioeng. **BIBE05**: 19–26 (2005).
12. Hulo, N., et al. Recent improvements to the PROSITE database. Nucleic Acids Res. **32**: D134–D137 (2004).
13. Penny, D., Hendy, M. The use of tree comparison metrics. Sys. Zoo. **34**: 75–82 (1985).
14. Livesay, D.R., La, D. Probing the evolutionary origins and catalytic importance of conserved electrostatic networks in TIM-barrel proteins. Protein Sci. **14**: 1158–1170 (2005).
15. La, D., Livesay, D.R. Accurate functional site prediction using an automated algorithm suitable for heterogeneous datasets. BMC Bioinform. **6**: 116 (2005).
16. Aloy, P., Querol, E., Aviles, F.X., Sternberg, M.J.E. Automated structure-based prediction of functional sites in proteins. J. Mol. Biol. **311**: 395–408 (2001).
17. Yamashita, A. Crystal structure of a bacterial homologue of Na+/Cl– dependent neurotransmitter transporters. Nature **437**: 215–223 (2005).
18. Kalinina, O.V. SDPpred: a tool for prediction of amino acid residues that determine differences in functional specificity of homologous proteins. Nucleic Acids Res. **32**: W424–W428 (2004).
19. Porter, C.T. The Catalytic Site Atlas: a resource of catalytic sites and residues identified in enzymes using structural data. Nucleic Acids Res. **32**: D129–D133 (2004).
20. Manning, J.R., et al. The contrasting properties of conservation and correlated phylogeny in protein functional residue prediction. BMC Bioinform. **9**: 51 (2008).

21. K.C., D., Livesay, D.R. Topology improves phylogenetic motif functional site predictions. Trans. Comp. Biol. Bioinf. (2010) In press.
22. Thompson, J.D., et al. CLUSTAL W: improving the sensitivity of progressive multiple sequence alignment through sequence weighting, position-specific gap penalties and weight matrix choice. Nucleic Acids Res. **22**: 4673–4680 (1994).
23. Felsenstein, J. PHYLIP (Phylogeny Inference Package) version 3.6. Distributed by the Author, Department of Genome Sciences, University of Washington (2004).
24. Schmidt, H. A., et al. TREE-PUZZLE: Maximum Likelihood Phylogenetic Analysis Using Quartets and Parallel Computing. Bioinformatics **18**: 502–504 (2002).
25. La, D., Livesay, D.R. MINER: software for phylogenetic motif identification. Nucleic Acids Res. **33**: W267–W270 (2005).

Exploiting Protein Structures to Predict Protein Functions

Alison Cuff, Oliver Redfern, Benoit Dessailly, and Christine Orengo

Abstract The exponential growth of experimentally determined protein structures in the Protein Data Bank (PDB) has provided structural data for an ever increasing proportion of genomic sequences. In combination with enhanced functional annotation from sequence, it has become possible to predict protein function from structure. In this chapter we discuss a range of methods which aim to recognise enzyme active sites and predict protein-ligand interactions. We then focus on algorithms developed as part of the CATH database of structural domains, where an evolutionary approach is used to recognise proteins with similar functions. While protein domains that exhibit the same structural fold tend to display related functional activities, there are a several large domain structure superfamilies that show a high degree of functional diversity. In these cases, we have built novel tools (FLORA and GeMMA) which are able to effectively identify sub-families of functionally linked domains, where standard methods of homologue detection (e.g. sequence profile and global structure alignment) fail.

Introduction

Many approaches for assigning protein functions attempt to exploit the 3D structure of the proteins, either to recognise putative active site regions and binding sites (e.g. for known ligands such as ATP), or to identify structural homologues likely to possess similar functions. The prediction of protein function from structure has become increasingly valuable as a significant proportion [1] of structures solved by the structural genomics initiatives (SGI) lack functional annotation [2]. In addition, structure-based approaches are particularly important for predicting binding sites and/or catalytic sites for the purposes of protein engineering and targeting drugs (for reviews see [1, 2]).

C. Orengo (✉)
Department of Structural and Molecular Biology, University College London, London, UK
e-mail: orengo@biochem.ucl.ac.uk

D. Kihara (ed.), *Protein Function Prediction for Omics Era*,
DOI 10.1007/978-94-007-0881-5_6, © Springer Science+Business Media B.V. 2011

Protein structures are more likely to be conserved during evolution than their sequences and structural data has been exploited to classify protein domains into evolutionary superfamilies. Nearly 40 years after the launch of the Protein Databank (PDB), established as a repository of solved 3D structures, the two major structural classifications, SCOP [3] and CATH [4] currently comprise more than 100,000 domain structures from the PDB classified into less than 3000 superfamilies. Furthermore, recent analyses have shown that nearly 70% of domain sequences in completed genomes can be predicted to belong to these families using HMM-HMM and threading protocols [5].

Both SCOP and CATH also further classify homologous structures according to their folds or topologies where structures are assigned to the same fold group if they have equivalent secondary structures, connected in the same way and oriented similarly in 3D space. Domains sharing the same fold are not necessarily evolutionary related and both classifications consider other evidence from sequence similarities or shared functional properties before classifying homologues [6]. Currently less than 1500 folds are recognised in SCOP and CATH. However, the definition of fold is somewhat subjective as no quantitative definitions exist and different protocols, employing manual inspection, are used to capture related folds by the two classifications.

There is no strong tendency for functional conservation across fold groups. Martin and Thornton explored the relationship between fold and function [7] and observed that whilst many small fold groups, comprising single evolutionary superfamilies exhibited only one molecular function, the highly populated fold groups could encompass a wide range of different functions. For example, the TIM barrel fold contains domains with more than 400 GO molecular function terms. However, there is often a tendency for particular surface features to be associated with the domain function. For example, Rossmann folds tend to bind substrates in the cleft created by the chain crossover at the C-terminal ends of the strands in the central β-sheet. Whilst structures adopting TIM barrel folds typically bind substrates in the large pocket at the base of the β-barrel. Russell and co-workers described these common sites as supersites [8]. These supersites may hint at remote homologies but whatever the cause of the similarity, fold recognition can help in identifying residues that are likely to be functionally important.

Whilst grouping protein domains into evolutionary families is important for studying their evolution, it is also valuable for predicting the functions of uncharacterised proteins since many analyses have revealed conservation of functional properties, particularly molecular function, within protein superfamilies [9]. However, it is clear that the degree of functional conservation varies with the domain superfamily as some superfamilies have diverged considerably in their structures and functions during evolution.

In this chapter we review the challenges faced when exploiting protein structures to predict function and describe some of the approaches that have been developed to cope with these challenges. We focus in particular on global methods of structure comparison and methods, developed within our group, which perform structure comparisons across a superfamily to identify specific structural features that are highly conserved within functional subfamilies in the superfamily.

Divergence of Protein Structures and Functions During Evolution

Analyses of structural superfamilies have revealed that many superfamilies are structurally very highly conserved during evolution and that this is accompanied by considerable conservation of function [10, 11]. The CATH classification captures structural divergence by determining the number of structurally distinct subgroups within a superfamily. A structural subgroup clusters together domains whose structures can be superposed with a normalised RMSD of 5Å. This is defined as:

$$\text{Normalised RMSD} = \frac{(\text{max length}) \times \text{RMSD}}{N} \tag{1}$$

where maxlength = number of residues in the largest structure, and N = total number of aligned residues.

Many CATH superfamilies (45%) comprise a single structural subgroup. Appendix A (http://www.biochem.ucl.ac.uk/~cuff/appendixA.html) lists the structurally diverse superfamilies containing more than one structural subgroup and shows the number of distinct GO and EC terms that can be identified for each of these superfamilies.

Figure 1 shows that there is a correlation between the number of structural subgroups and the number of distinct functional categories identified within the superfamily. Previous studies have shown that 75 superfamilies (<4% of CATH superfamilies) have diverged highly in their structures and functions [9]. These superfamilies tend to be highly recurrent in the genomes accounting for nearly 40%

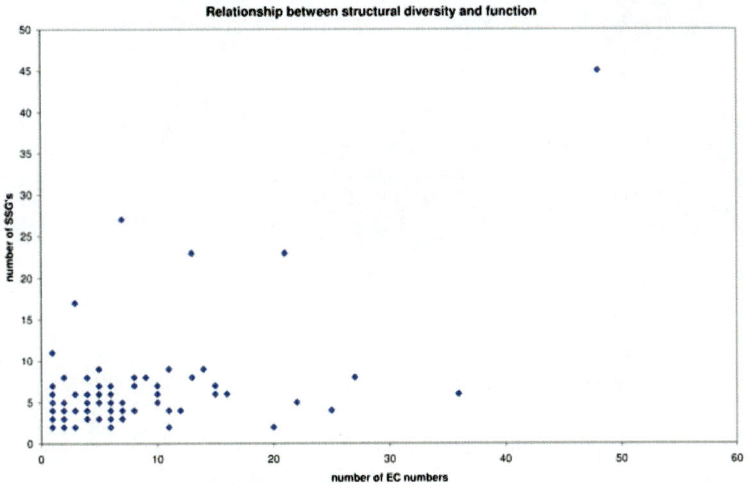

Fig. 1 Graph showing the correlation between number of structural subgroups and number of EC numbers identified within the superfamily

of predicted structural domains in completed genomes. They are also the oldest superfamilies, with the majority found in all three kingdoms of life and therefore probably present in the last common ancestor.

It is likely that the extensive duplication of these superfamilies within genomes and the divergence of structure and function in the duplicated or paralogous domains is accompanied by recruitment of the paralogues to different metabolic pathways or biological processes. Several studies have shown evidence for this in highly duplicated enzyme families [12] where homologues are frequently recruited to different pathways where perhaps they bring a chemical activity characteristic of their superfamily [13, 14]. Other large, diverse, superfamilies display conservation of parts of their ligands [15], possibly as the result of metabolic pathway retrograde evolution where the duplicated copy of an enzyme is recruited to catalyse the previous reaction in the same metabolic pathway [13, 16].

Extensive analyses of structural variation across these superfamilies has characterised the extent to which secondary structures are inserted and/or deleted during evolution. Whilst the secondary structures in the core of the domain tend to be very highly conserved, there can be considerable embellishment of additional secondary structures to this conserved core. Figure 2 illustrates structural divergence across some relatives from the large HAD domain superfamily, showing the conserved core and secondary structure embellishments.

Studies of the 31 most structurally and functionally divergent superfamilies showed that secondary structure insertions are generally distributed along the whole

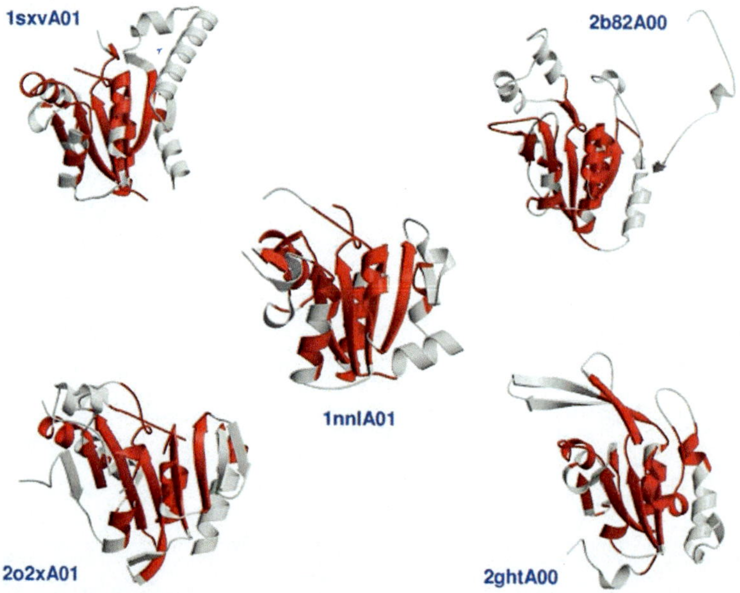

Fig. 2 Structural divergence among members of the mechanistically diverse haloacid dehalogenase (HAD) superfamily. The common structural core is coloured *red* and the structural embellishments are *grey*

length of the polypeptide chain with typically less than 3 being adjacent together in the sequence [11]. However, they accumulate in relatively few locations in 3D to give larger structural features. They were found to be modifying active site geometry or providing alternative protein interaction surfaces in relatives with different embellishments. Superfamilies adopting layered domain architectures such as αβα, αβ and β sandwiches appear more able to accommodate structural embellishments to the domain core [11].

Structural changes in domain relatives can also bring about changes in the domain partners and changes in the protein partners and oligomerisation states which can further modify functional sites or provide additional functional sites. Examples of these phenomena are given in Todd et al. [12], Reeves et al. [11] and Dessailly et al. [17]. Other evolutionary mechanisms causing structural change include circular permutations [18, 19], segment-swapping [18], addition of major structural embellishments to a conserved structural core [11], or more dramatic fold changes [20].

Despite the considerable divergence in structure observed in some superfamilies, some aspect of the function is generally conserved. Early studies by Todd et al. [12] revealed conservation of one or more chemical intermediates along the reaction pathway occurring in many highly diverse superfamilies. Such superfamilies, which are mechanistically diverse but share some common functional feature are being increasingly studied. The SFLD established by Babbitt and her group [21] now describes 6 such superfamilies and sequence diverse relatives within these superfamilies have been deliberately targeted by associated structural genomics initiatives to provide structures for characterising the diverse functional subfamilies. This work has been accompanied by extensive experimental characterisation of relatives within the superfamilies. Similarly the Structural Genomics Consortium (SGC), headed by the Edwards group in Canada, is targeting relatives from large superfamilies, highly expanded in human, to characterise relatives having different ligand specificities. These initiatives, which combine structural characterisation with biochemical studies, will be very useful in expanding the repertoire of diverse structural relatives within superfamilies with known functions which can be used to validate structure function prediction algorithms.

To What Extent Can Function Be Predicted from the Structure of the Domain

Global Structure Comparison

Since most structural domain superfamilies (>70% of superfamilies in CATH) are rather homogeneous in function [13], classifying a new domain in one of these superfamilies generally allows inheritance of function from one of the other experimentally characterised superfamily members [22, 23]. Over the last 20 years a plethora of structure comparison algorithms have been developed which attempt to handle the diverse structural changes that can occur during evolution. For very

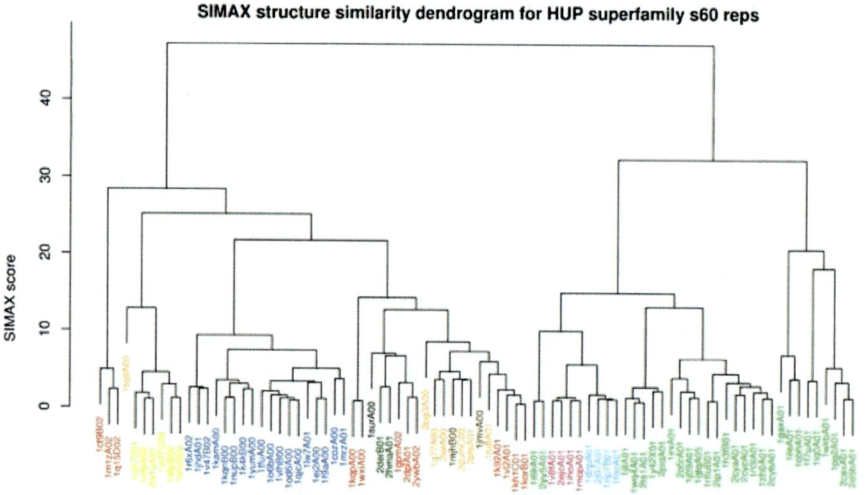

Fig. 3 Dendrogram showing non-redundant relatives of the HUP superfamily clustered by a normalised RMSD score (SIMAX). Domains that share similar functions are highlighted in the same colour

remote homologues in these superfamilies, function can often be assigned using reliable structure comparison methods (e.g. CE [24], DALI [25], CATHEDRAL [26], Structal [27], FatCat [28]; see also [2, 29] for reviews).

Whilst a number of fast structure comparison methods exist [26, 28, 30] most of which compare secondary structures between proteins and can be used to search the PDB for putative fold matches, the most accurate methods compare residue positions between proteins [2, 29]. Some of these algorithms exploit the dynamic programming algorithms or other sophisticated optimisation protocols like simulated annealing to handle residue insertions and deletions. However, whilst global structural similarity is quite a good indication of functional similarity and can be used to cluster together relatives sharing common functions within structural superfamilies (see Fig. 3), rather high thresholds on similarity are required to ensure significant conservation of function (see Fig. 4).

Assigning Functions Based on Local Structural Similarity

Various studies suggest that domains that seem unrelated as a whole may contain evolutionarily-conserved subparts [31, 32] such as their active sites [33].

As structure is more conserved across protein families than sequence [10], structure comparison methods are able to detect far more distant relationships than the most powerful profile methods. However, as discussed already, even domains in the same superfamily can exhibit large amounts of structural variation [11]. This may be due to different protein or domain interactions, or requirements to attach to distinct cellular environments, or might simply be due to random evolutionary drift.

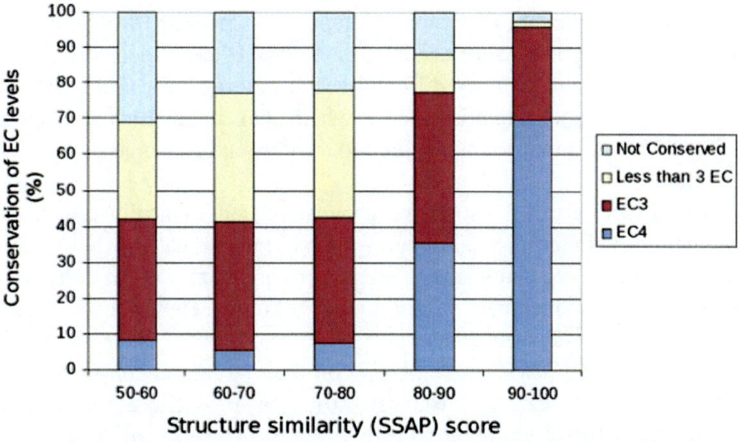

Fig. 4 Plot showing relationship between structural similarity and functional conservation (measured as conservation of EC levels). The SSAP score ranges from 0 to 100 (for identical structures)

Consequently, these structural deviations can mean that even an accurate alignment of two structures can produce a global similarity score that falls below reliable thresholds for transferring a specific function.

In a similar vein to the way PRINTS [34] and PROSITE [35] focus on smaller conserved sequence patterns, there are several approaches to identifying local structure motifs that are associated with specific functions. For example, the Catalytic Site Atlas [36] concentrates on building 3D motifs of residues that are directly involved in ligand binding or the catalytic mechanism in an enzyme. As ab initio prediction of functional residues is a complex problem in itself, the Thornton group at the European Bioinformatic Institute (EBI) have focussed on mining the primary literature to obtain the information on which to build templates. Torrance et al. [37] analysed the performance of this approach for enzymes with more than 2 catalytic residues. They were able to discriminate related proteins from random with 85% accuracy and found that it was important to focus on C-alpha/C-beta residues as their position is better conserved than side chain atoms. However, even by capturing the correct functionally active residues – for example, the catalytic triad in the serine proteases – the flexibility of active sites significantly impacts on the ability of these templates to detect mobile residues in X-ray crystal structures with different bound ligands.

Methods That Search for Patterns of Conservation Without Having Functional Groups or Motifs Defined

In contrast to exploiting information on known functional residues, the DRESPAT method [38] uses graph theory to extract recurring structural patterns across superfamilies in the SCOP database [3]. DRESPAT makes no assumptions about the location or nature of the motif positions, except by excluding hydrophobic residues. A statistical model is built to assess the significance of each recurring pattern and the authors were able to identify different metal binding sites in distantly related proteins. However, as with many methods which seek small structural motifs, distinguishing between genuine similarities and background is hampered by high false positive rates.

The PINTS methods [39] also shows promise for automatically detecting structural motifs in protein families, although is not able to annotate novel proteins with high accuracy. Again, recurring side chain patterns are identified through a pair-wise comparison of diverse members within a protein family. These motifs can then be used to scan against a novel structure.

Instead of detecting 3D templates based on their structural conservation across an enzyme family, Polacco and Babbitt [40] used a genetic algorithm (GASP) to generate a functional template from a given structure based on its ability to identify members of the same enzyme superfamily against a background of unrelated proteins in the SCOP database. An initial PSI-BLAST step builds a multiple sequence alignment for each enzyme structure that is used to create a set of conserved residues, from which a small number ($\sim$10) are selected at random to build a template. The performance of each template is then evaluated by using a geometric matching algorithm, SPASM, to score matches to the functional relatives and the SCOP library. Interestingly, the best template generally contains known functional amino acids, although there are also a few additional residues with no known functional role. This method is a promising development, although each template takes up to 18 h to generate and the performance was only evaluated for five superfamilies.

Methods That Search for Structural Differences Between Defined Functional Groups to Identify Functional Determinants

The FLORA Algorithm

A novel approach was [41], developed recently in our group to provide structural templates for assigning uncharacterised structures to functional subfamilies in the CATH classification, performs global structural comparisons between relatives within a superfamily to identify structural features that are highly conserved within a functional subfamily but less conserved across the complete superfamily.

FLORA does not exploit information on known functional residues such as catalytic residues from the Catalytic Site Atlas (CSA) to characterise functionally important positions in the protein. Functionally relevant positions are identified from structural comparisons within and between the functional subfamilies within a superfamily.

Benchmark Dataset

The method was originally benchmarked by deriving a dataset of functional subfamilies in 29 large, enzyme superfamilies. Only functionally diverse superfamilies were included, with relatives accounting for at least 3 different Enzyme Classification (EC) codes. A non-redundant set of structures were used for each superfamily, generated by clustering relatives sharing 60% or more sequence identity. This threshold was used as it has been shown to be associated with a high likelihood of functional similarity in the EC classification [41]. Subsequently, structures were clustered into functional subfamilies if they shared at least the first 3 EC numbers. A CATH superfamily was then included in the dataset only if it contained at least 3 functional subfamilies, where each subfamily contained at least 4 structures. These criteria were chosen to create a sufficiently diverse data set, which could be effectively assessed using leave-one-out benchmarking. The final dataset contained 82 functional subfamilies from 29 diverse CATH superfamilies (900 domains in total) and constitutes one of the largest datasets available for evaluating structure to function prediction algorithms. Furthermore, although these superfamilies account for <2% of the CATH superfamilies (currently 2600), they are very large comprising nearly 50% of sequences in functionally diverse CATH superfamilies.

Overview of Method

Figure 5 shows a flowchart of the FLORA method. FLORA does not rely on initial seeds of known functional residues but explores the whole structure of the domains in order to find discriminating positions. This information is then captured by generating vectors between these positions which can be compared against query structures to recognise functional homologues.

Structural comparisons within and between functional subfamilies are performed using the CATHEDRAL algorithm, another in-house method [26]. This is a relatively fast comparison method which exploits graph theory and double dynamic

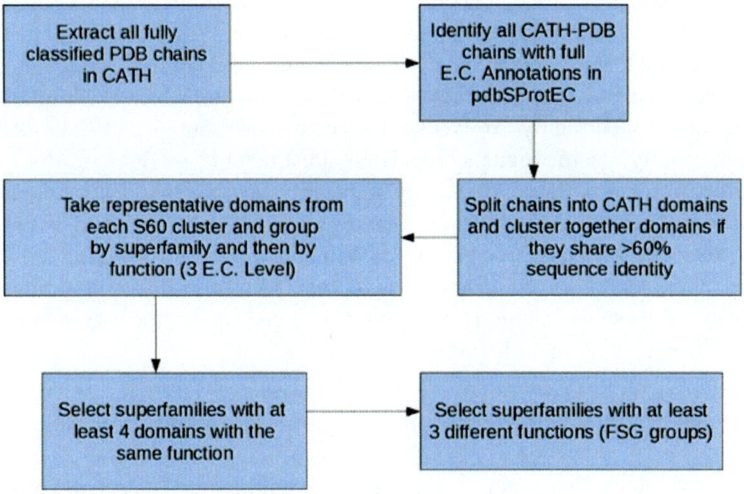

Fig. 5 Flowchart of the FLORA method

programming and had been shown to perform well against other established methods (e.g. DALI, Structal, CE) and which aligns the largest proportion of equivalent residues with respect to manually curated alignments [26].

Step 1: Identify Structurally Conserved Residues

CATHEDRAL is used to perform pair-wise structural comparisons between all representatives in the given CATH functional subfamily. Subsequently, for each domain, residues are only considered if they can be aligned against residues in at least 75% of other relatives from the subfamily (equivresidues). For each domain, vectors were calculated between the equivresidues.

Vectors were calculated between the C_β atoms of the equivalent residues (A→B) and then multiplied by a co-ordinate frame calculated from the tetrahedral geometry of the bonds of the C_α of residue A as described in [42]. As the C_α geometry of residues A and B are not identical, vectors were calculated in both the A→B and B→A direction. However, we found that taking only one of these vectors forward to the next steps in the algorithm gave the same performance as using both, but increased the speed of FLORA.

Vectors for each domain in the superfamily were then compared against equivalent vectors in all other domain representatives from the superfamily. Equivalent vectors were determined from the structural alignment of the two domains being compared. Vectors were scored using the formula given in Eq. (2) below, where the values for a and b were determined from trials. The optimal values were a=b=2.

$$\text{score} = \frac{a}{|v1 - v2| + b} \tag{2}$$

The next step is to identify those vectors for a given domain that are structurally more conserved between members of the same functional subfamily than compared to members of different functional subfamilies. The aim of this step is to eliminate any vectors that are conserved across the whole superfamily. These vectors are likely to be associated with the core of the domain structure which is common to all members of the superfamily. Any remaining vectors are more likely to be associated with functionally specific regions on the domain structure.

In order to identify these "functionally specific" vectors, two distributions were calculated for each vector considered. One captures the scores obtained by comparing the vector to equivalent vectors in domains in the same functional subfamily and the other, scores for comparisons involving vectors in different subfamilies. The means of these distributions were calculated and the vector was identified as functionally specific if the following condition was met:

$$\text{mean (functional subfamily distribution)} - \text{mean (superfamily distribution)} > 1$$

The set of selected vectors is reduced by jack-knifing the data set and repeating the calculation above. That is, each domain is removed in turn and a vector is only selected as specific if the inequality is always satisfied. At the end of this process, each domain is associated with a *template* set of functionally specific vectors.

Scoring Query Structures Against FLORA Template Sets for Individual Domains

In order to determine whether a query structure can be assigned to a specific functional subfamily within a CATH superfamily, the query is structurally aligned to all representatives in the superfamily, using the CATHEDRAL algorithm again, and a score calculated for each comparison.

When scoring the alignment of the query structure against a given member of functional subfamily, the algorithm only scores the similarity over the set of functionally specific vectors for the subfamily domain. Thus the algorithm is effectively calculating a local score using the correspondences determined by a global structure comparison. Each vector in the template set is scored against the equivalent vector in the query domain using the following formula:

$$\text{florascore} = \frac{\sum_{i=0}^{N} \text{score}(v1, v2)}{N} \tag{3}$$

where N = number of template vectors; $v1$ = template vector; $v2$ = equivalent vector in query domain.

Any vectors that are not aligned (i.e. gapped positions) are given a score of zero. The total similarity of the query domain against enzyme domain (the *florascore*) is simply the sum of these similarities, normalised by the total number of vectors in the template (Eq. (3)).

In order to take account of the different degrees of structural-functional diversity in different superfamilies this score is converted to a Z-score which could be applied regardless of the superfamily being considered.

Assessing the Performance of FLORA

FLORA was benchmarked using the dataset of 29 functionally diverse CATH enzyme superfamilies described above. In order to assess the performance in an unbiased manner we used a standard leave-one-out approach. That is, for a given superfamily being evaluated, one domain member is removed from the set which is then used as a training set for the algorithm. The selected test domain is then scored against FLORA templates for all superfamilies.

We compared the performance of FLORA against global structure comparison algorithms CE [24], CATHEDRAL [26] and against another publicly available structure–function prediction method, Reverse Templates [43]. Unfortunately few structure–function prediction algorithms are available but Reverse Templates is one of the leading methods. We plotted sensitivity (i.e. tp/(tp + fn)) versus

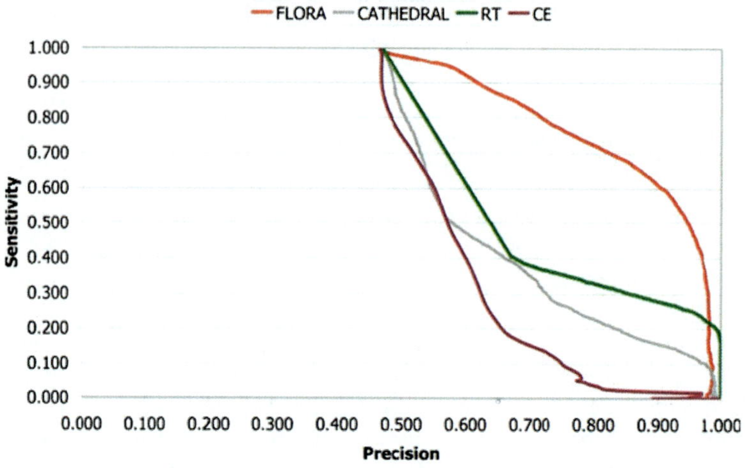

Fig. 6 Graph of sensitivity versus precision to show the performance of CE, CATHEDRAL, RT and FLORA for the prediction of enzyme family

precision (tp/(tp + fp)) and assessed the performance on individual superfamilies by calculating AUC value (area under ROC curve).

It can be seen from Fig. 6 that both global structure comparison methods, CE and CATHEDRAL, are poor at recognising the correct functional subfamily to which a query domain should be assigned. CATHEDRAL outperforms CE, most likely because it is able to align more equivalent positions, as identified in previous studies [26]. However, neither method was specifically designed for recognising functional homologues.

Even at high precision (>95%) FLORA significantly outperforms CE, CATHEDRAL and Reverse Templates. At 90% precision it captures twice the number of functional homologues than Reverse Templates. The sensitivity of the algorithm derives from the fact that although FLORA uses an alignment derived by CATHEDRAL, it only scores positions deemed to be functionally specific (i.e. in the FLORA template set). By exploiting multiple structures from a functional subfamily it can more easily identify these specific positions.

We have also examined the effect on the FLORA performance of using whole protein chains rather than protein domains. There was negligible impact on performance which suggests that there is enough signal in the domain structure to recognise the specific function of the protein containing the domain. This is encouraging if we wish to exploit FLORA as a general function prediction method since the majority of proteins differ between organisms [44] whilst the domain components within them are related and can therefore, from these results, be used to suggest functions for the whole proteins.

Visualisation of Functionally Specific Positions Detected by FLORA

The power of FLORA lies in its ability to identify residues beyond the common structural core of the domain subfamily. Our previous analyses observed that nearly 70% of residue positions identified by FLORA were located close to functional sites [41]. Other FLORA positions were found to be close to interface surfaces involved in protein interactions. To manually assess the ability of FLORA to recognise functionally relevant sites in the domain structures, FLORA positions were mapped onto representative structures from the HUP domain superfamily, which is one of the largest and most structurally and functionally diverse superfamilies in CATH, comprising more than 9 different functional subfamilies.

Domains in this superfamily adopt a Rossmann-like fold with a central parallel β-sheet surrounded on both sides by α-helices. The main active site is always located in the C-terminal half of the central β-sheet and is generally involved in nucleotide-binding.

Figure 7 illustrates residue locations identified by FLORA templates for a subfamily from the HUP domain superfamily. A representative structure for this functional subfamily was chosen as the structure with the highest cumulative structural similarity score to all other non-redundant members (at 100% sequence identity) of the subfamily. Residue positions are highlighted if at least 30% of

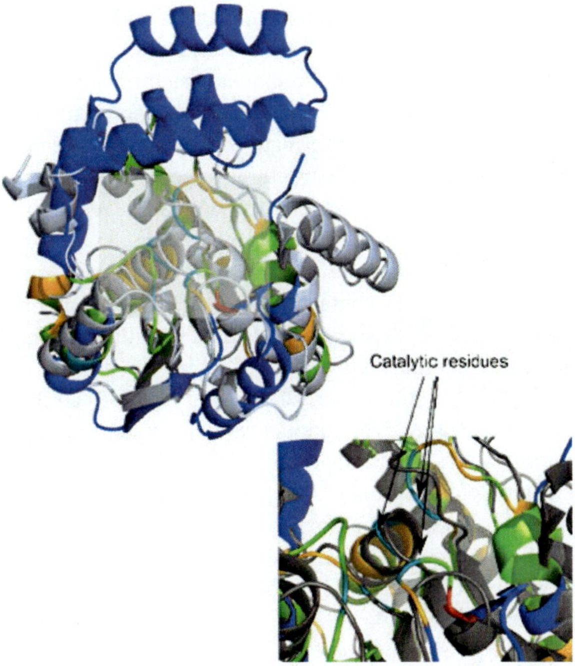

Fig. 7 Superposition of PSI structure 2pbl (*dark grey*) with 1tqh (superfamily 3.40.50.1820, EC 3.1.1.-). Flora residues are coloured *green*, or *gold* if they are conserved across the whole superfamily, and catalytic residues are shown in *light blue*. It can be seen that there is reasonable agreement in the region of the active site

FLORA templates for this subfamily include these positions. Any positions conserved across a majority of the superfamily (i.e. 75% or more of the relatives) are coloured gold.

Incorporating Sequence Based Protocols with FLORA to Identify Functionally Specific Residues

We explored the effects of including sequence matching within the FLORA algorithm. That is including a contribution to the score reflecting identical or similar residues between the query and the template structure. However, this tended to degrade the performance and was not included in the final version of the algorithm.

Instead we have developed a separate sequence based protocol (GeMMA [45]) for identifying residue positions likely to be associated with the function. This allowed us to annotate structural domains within each functional subfamily with residue positions identified as functionally specific from both structural data (FLORA) and sequence data (GeMMA).

More importantly, GeMMA allows to identify functional subgroups amongst all the sequences assigned to a superfamily, even those without known structures. Since the number of sequence relatives can be up to 100-fold greater than the number of structures for some superfamilies, this gives a more accurate representation of functional divergence across the superfamily. Functional subfamilies identified by GeMMA can be used as sets for training the FLORA algorithm, provided they contain three or more non-redundant structures and can therefore be used to identify positions associated with function which are structurally conserved.

GeMMA exploits information from all the predicted sequence domains assigned to a particular CATH structural superfamily and contained within our Gene3D resource (http://gene3d.biochem.ucl.ac.uk/Gene3D). For example in the HUP superfamily mentioned above, there are 85 non-redundant structures (at 60% sequence identity) and 9484 non-redundant sequences stored within CATH-Gene3D. Gene3D contains all the predicted domain sequences for CATH superfamilies identified using HHM models built from the sequences of non-redundant structural domains in CATH [23].

GeMMA initially compares (using BLAST) all the sequences against each other and then progressively merges similar sequences into functional subgroups or subfamilies. This is initially done on the basis of pairwise sequence similarity but as the clusters grow and there are enough sequences to make a sequence profile, profile–profile comparisons are performed between clusters. Clusters are merged provided the E-value returned from the comparison is below a threshold obtained by benchmarking with superfamilies for which there are extensive experimental functional characterisations [45].

Since profile–profile comparisons can be very computationally expensive, we have developed a strategy for reducing the number of comparisons that need to be performed and for running a modified version of the protocol on multiple compute nodes. Alternative sequence based strategies for identifying functional subfamilies within superfamilies tend to exploit tree based approaches that rely on a multiple sequence alignment of all the sequences to build the tree. However, the most functionally diverse superfamilies in CATH, which account for more than half the sequences in the genomes, contain more than 10,000 sequences. This number of sequences is beyond the scope of most multiple sequence alignment methods. Even when non-redundant datasets are generated at 60% sequence identity to ensure functional coherence, there are still large numbers of sequences in these very large superfamilies (i.e. > 5000). Therefore, the iterative clustering protocol used by GeMMA (also described as agglomerative clustering), is the most tractable approach for these very large and functionally diverse superfamilies.

FLORA templates can be derived for GeMMA functional subfamilies which contain 3 or more non-redundant structures (at 30% sequence identity). As mentioned above FLORA analyses can exploit the structural data in these subfamilies to identify structurally conserved positions associated with functional sites (e.g. active sites and protein–protein interaction surfaces). GeMMA identifies >100 functional subfamilies in the diverse HUP superfamily. Figure 8 shows a representative from one of these subfamilies with residue positions highlighted according to whether they are identified as sequence conserved by GeMMA or structurally conserved by FLORA or both sequence and structure conserved. Mapping these conserved residues onto the structure is clearly useful in suggesting the location of functional sites on the protein domain. In the future CATH-Gene3D will be providing information on GeMMA functional subfamilies for selected CATH domain superfamilies being targeted for structural genomics by the protein structure initiative (PSI) in the United States.

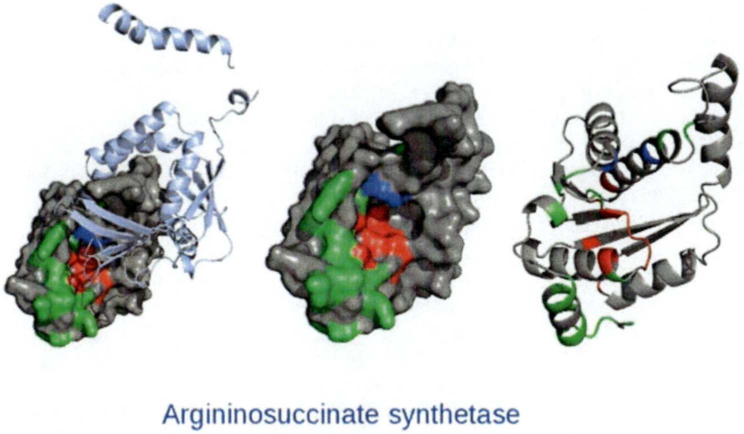

Fig. 8 Representative structure from one of the HUP protein subfamilies. Residues that are conserved by structure are coloured *green*, those conserved by sequence are coloured *blue* and those conserved by both sequence and structure are coloured *red*

References

1. Watson, J.D., Laskowski, R.A., et al. Predicting protein function from sequence and structural data. Curr. Opin. Struct. Biol. **15**(3): 275–284 (2005).
2. Lee, D., Redfern, O., et al. Predicting protein function from sequence and structure. Nat. Rev. Mol. Cell Biol. **8**(12): 995–1005 (2007).
3. Murzin, A.G., Brenner, S.E., et al. SCOP: a structural classification of proteins database for the investigation of sequences and structures. J. Mol. Biol. **247**(4): 536–540 (1995).
4. Orengo, C.A., Michie, A.D., et al. CATH--a hierarchic classification of protein domain structures. Structure **5**(8): 1093–1108 (1997).
5. Sillitoe, I., Dibley, M., et al. Assessing strategies for improved superfamily recognition. Protein Sci. **14**(7): 1800–1810 (2005).
6. Cuff, A., Redfern, O., et al. Classification of protein structures. *Computational structural biology, methods and applications*. Schwede, T., Peitsch, M.C. (eds.). Singapore: World Scientific, pp. 153–188 (2008).
7. Martin, A.C., Orengo, C.A., et al. Protein folds and function. Structure **15**;6(7): 875–884 (Jul 1998).
8. Russell, R.B., Sasieni, P.D., et al. Supersites within superfolds. Binding site similarity in the absence of homology. J. Mol. Biol. **282**(4): 903–918 (1998).
9. Cuff, A., Redfern, O.C., et al. The CATH hierarchy revisited-structural divergence in domain superfamilies and the continuity of fold space. Structure **17**(8): 1051–1062 (2009).
10. Chothia, C., Lesk, A.M. The relation between the divergence of sequence and structure in proteins. EMBO J. **5**(4): 823–826 (1986).
11. Reeves, G.A., Dallman, T.J., et al. Structural diversity of domain superfamilies in the CATH database. J. Mol. Biol. **360**(3): 725–741 (2006).
12. Todd, A.E., Orengo, C.A., et al. Plasticity of enzyme active sites. Trends Biochem. Sci. **27**(8): 419–426 (2002).
13. Todd, A.E., Orengo, C.A., et al. Evolution of function in protein superfamilies, from a structural perspective. J. Mol. Biol. **307**(4): 1113–1143 (2001).
14. Gerlt, J.A. A protein structure (or function?) initiative. Structure **15**(11): 1353–1356 (2007).
15. Chiang, R.A., Sali, A., et al. Evolutionarily conserved substrate substructures for automated annotation of enzyme superfamilies. PLoS Comput. Biol. **4**(8): e1000142 (2008).
16. Rison, S.C., Thornton, J.M. Pathway evolution, structurally speaking. Curr. Opin. Struct. Biol. **12**(3): 374–382 (2002).
17. Dessailly, B.H., Redfern, O.C., Cuff, A.L., Orengo, C.A. Detailed analysis of function divergence in a large and diverse domain superfamily: toward a refined protocol of function classification. Structure. **18**(11): 1522–1535 (2010 Nov 10).
18. Andreeva, A., Murzin, A.G. Evolution of protein fold in the presence of functional constraints. Curr. Opin. Struct. Biol. **16**(3): 399–408 (2006).
19. Taylor, W.R. Evolutionary transitions in protein fold space. Curr. Opin. Struct. Biol. **17**(3): 354–361 (2007).
20. Grishin, N.V. Fold change in evolution of protein structures. J. Struct. Biol. **134**(2–3): 167–185 (2001).
21. Pegg, S.C., Brown, S.D., et al. Leveraging enzyme structure-function relationships for functional inference and experimental design: the structure-function linkage database. Biochemistry **45**(8): 2545–2555 (2006).
22. Wilson, D., Madera, M., et al. The SUPERFAMILY database in 2007: families and functions. Nucleic Acids Res. **35**(Database issue): D308–D313 (2007).
23. Yeats, C., Lees, J., et al. Gene3D: comprehensive structural and functional annotation of genomes. Nucleic Acids Res. **36**(Database issue): D414–D418 (2008).
24. Shindyalov, I.N., Bourne, P.E. Protein structure alignment by incremental combinatorial extension (CE) of the optimal path. Protein Eng. **11**(9): 739–747 (1998).
25. Holm, L., Sander, C. Protein structure comparison by alignment of distance matrices. J. Mol. Biol. **233**(1): 123–138 (1993).

26. Redfern, O.C., Harrison, A., et al. CATHEDRAL: a fast and effective algorithm to predict folds and domain boundaries from multidomain protein structures. PLoS Comput. Biol. **3**(11): e232 (2007).
27. Kolodny, R., Koehl, P., et al. Comprehensive evaluation of protein structure alignment methods: scoring by geometric measures. J. Mol. Biol. **346**(4): 1173–1188 (2005).
28. Ye, Y., Godzik, A. Flexible structure alignment by chaining aligned fragment pairs allowing twists. Bioinformatics **19**(Suppl 2): ii246–ii255 (2003).
29. Redfern, O.C., Dessailly, B., et al. Exploring the structure and function paradigm. Curr. Opin. Struct. Biol. **18**(3): 394–402 (2008).
30. Zhang, Y., Skolnick, J. TM-align: a protein structure alignment algorithm based on the TM-score. Nucleic Acids Res. **33**(7): 2302–2309 (2005).
31. Soding, J., Lupas, A.N. More than the sum of their parts: on the evolution of proteins from peptides. Bioessays **25**(9): 837–846 (2003).
32. Manikandan, K., Pal, D., et al. Functionally important segments in proteins dissected using gene ontology and geometric clustering of peptide fragments. Genome Biol. **9**(3): R52 (2008).
33. Xie, L., Bourne, P.E. Detecting evolutionary relationships across existing fold space, using sequence order-independent profile-profile alignments. Proc. Natl. Acad. Sci. USA **105**(14): 5441–5446 (2008).
34. Attwood, T.K., Bradley, P., et al. PRINTS and its automatic supplement, prePRINTS. Nucleic Acids Res. **31**(1): 400–402 (2003).
35. Hulo, N., Bairoch, A., et al. The PROSITE database. Nucleic Acids Res. **34**(Database issue): D227–D230 (2006).
36. Porter, C.T., Bartlett, G.J., et al. The catalytic site atlas: a resource of catalytic sites and residues identified in enzymes using structural data. Nucleic Acids Res. **32**(Database issue): D129–D133 (2004).
37. Torrance, J.W., Bartlett, G.J., et al. Using a library of structural templates to recognise catalytic sites and explore their evolution in homologous families. J. Mol. Biol. **347**(3): 565–581 (2005).
38. Wangikar, P.P., Tendulkar, A.V., et al. Functional sites in protein families uncovered via an objective and automated graph theoretic approach. J. Mol. Biol. **326**(3): 955–978 (2003).
39. Stark, A., Russell, R.B. Annotation in three dimensions. PINTS: patterns in non-homologous tertiary structures. Nucleic Acids Res. **31**(13): 3341–3344 (2003).
40. Polacco, B.J., Babbitt, P.C. Automated discovery of 3D motifs for protein function annotation. Bioinformatics **22**(6): 723–730 (2006).
41. Redfern, O.C., Dessailly, B.H., et al. FLORA: a novel method to predict protein function from structure in diverse superfamilies. PLoS Comput. Biol. **5**(8): e1000485 (2009).
42. Aravind, L., Anantharaman, V., et al. Monophyly of class I aminoacyl tRNA synthetase, USPA, ETFP, photolyase, and PP-ATPase nucleotide-binding domains: implications for protein evolution in the RNA. Proteins **48**(1): 1–14 (2002).
43. Laskowski, R.A., Watson, J.D., et al. Protein function prediction using local 3D templates. J. Mol. Biol. **351**(3): 614–626 (2005).
44. Grant, A., Lee, D., et al. Progress towards mapping the universe of protein folds. Genome Biol. **5**(5): 107 (2004).
45. Lee, D.A., Rentzsch, R., et al. GeMMA: functional subfamily classification within superfamilies of predicted protein structural domains. Nucleic Acids Res. **38**(3): 720–737.

Sequence Order Independent Comparison of Protein Global Backbone Structures and Local Binding Surfaces for Evolutionary and Functional Inference

Joe Dundas, Bhaskar DasGupta, and Jie Liang

Abstract Alignment of protein structures can help to infer protein functions and can reveal ancient evolutionary relationship. We discuss computational methods we developed for structural alignment of both global backbones and local surfaces of proteins that do not depend on the ordering of residues in the primary sequences. The algorithm for global structural alignment is based on fragment assembly, and takes advantage of an approximation algorithm for solving the maximum weight independent set problem. We show how this algorithm can be applied to discover proteins related by complex topological rearrangement, including circularly permuted proteins as well as proteins related by complex higher order permutations. The algorithm for local surface alignment is based on solving the bi-partite graph matching problem through comparison of surface pockets and voids, such as those computed from the underlying alpha complex of the protein structure. We also describe how multiple matched surfaces can be used to automatically generate signature pockets and a basis set that represents the ensemble of conformations of protein binding surfaces with a specific biological function of binding activity. This is followed by illustrative examples of signature pockets and a basis set computed for NAD binding proteins, along with a discussion on how they can be used for discriminating NAD-binding enzymes from other enzymes.

Introduction

To understand the molecular basis of cellular processes, it is important to gain a comprehensive understanding of the biological functions of protein molecules. Although an increasing number of sequences and structures of proteins are now available, there are many proteins whose biological functions are not known, or knowledge of their biological roles is incomplete. This is evidenced by the existence

J. Liang (✉)
Bioinformatics Program, Department of Bioengineering, University of Illinois at Chicago, Chicago, IL 60607, USA
e-mail: jliang@uic.edu

D. Kihara (ed.), *Protein Function Prediction for Omics Era*,
DOI 10.1007/978-94-007-0881-5_7,

of a large number of partially annotated proteins, as well as the accumulation of a large number of protein structures from structural genomics whose biological functions are not well characterized [1, 2]. Researchers have turned to in silico methods to gain biological insight into the functional roles of these uncharacterized proteins, and there has been a number of studies addressing the problem of computationally predicting the biological function of proteins [3–8].

A relatively straightforward method for inferring protein function is to transfer annotation based on homology analysis of shared characteristics between proteins. If a protein shares a high level of sequence similarity to a well characterized family of proteins, frequently the biological functions of the family can be accurately transferred onto that protein [9–11]. At lower levels of sequence similarity, probabilistic models such as profiles can be constructed using local regions of high sequence similarity [11–13]. The large amount of information of protein such as those deposited in the SWISS-PROT database [14] provides rich information for constructing such probabilistic models.

However, limitations to sequence-based homology transfer for function prediction arise when the sequence identity between a pair of proteins is less than 60% [15]. An alternative to sequence analysis is to infer protein function based on structural similarity. It is now well known that protein structures are much more conserved than protein sequences, as proteins with little sequence identity often fold into similar three-dimensional structures [16].

Protein structure and protein function are strongly correlated [17]. Conceptually, knowledge of three-dimensional structures of proteins should enable inference of protein function. Computational tools and databases for structural analysis are indispensable for establishing the relationship between protein function and structure. Among databases of protein structures, the SCOP [18] and CATH [19] databases organize protein structures hierarchically into different classes and folds based on their overall similarity in topology and fold. Such classification of protein structures generally depends on a reliable structural comparison method. Although there are several widely used methods, including Dali [20] and CE [21], current structural alignment methods cannot guarantee to give optimal results and structural alignment methods do not have the reliability and interpretability comparable to that of sequence alignment methods.

Comparing protein structures is challenging. First, it is difficult to obtain a quantitative measure of structural similarity that is generally applicable to different types of problems. Similar to sequence alignment methods, one can search for global structural similarity between overall folds or focus on local similarity between surface regions of interest. Defining a quantitative measure of similarity is not straightforward as illustrated by the variety of proposed structural alignment scoring methods [22]. Unlike sequence alignment, in which the scoring systems are largely based on evolutionary models of how protein sequence evolve [23, 24], scoring systems of structural alignment must take into account both the three-dimensional positional deviations between the aligned residues or atoms, as well as other characteristics that are biologically important. Second, many alignment methods assume the ordering of the residues follow that of the primary sequence when seeking to optimize structure similarity [21, 25]. This assumption can be problematic, as similar three-dimensional placement of residues may arise from residues with different sequential ordering. This problem is frequently encountered when comparing local regions on proteins structures. When comparing global structures of proteins, the existence of circular and higher ordered permutations [26, 27] also poses significant problems. Third, proteins may undergo minor residue side chain structural fluctuations as well as large backbone conformational changes in vivo. These structural fluctuations are not represented in a static snapshot of a crystallized structures in the Protein Data Bank (PDB) [28]. Many structural alignment methods, which assume rigid bodies and cannot account for structural changes that may occur.

In this chapter, we will first discuss several overall issues important for protein structural alignment. We then discuss a method we have developed for sequence order independent structural alignment at both the global and local level of protein structure. This is followed by discussion on how this method can be used to detect protein pairs that appear to be related by simple and complex backbone permutations. We will then describe the use of local structural alignment in automatic construction of *signature pockets* of binding surfaces, which can be used to construct *basis set* for a specific biological function. These constructs can detect structurally conserved surface regions and can be used to improve the accuracy of protein function prediction.

Structural Alignment

Protein structural alignment is an important problem [22]. It is particularly useful when comparing two proteins with low sequence identity between them. A widely used measure of protein structural similarity is the root mean squared distance (RMSD) between the equivalent atoms or residues of the two proteins. When the equivalence relationship between structural elements are known, a superposition described by a rotation matrix R and a translation vector T that minimizes the root mean squared distances (RMSD) between the two proteins can be found by solving the minimization problem:

$$\min \sum_{i=1}^{N_B} \sum_{j=1}^{N_A} |T + RB_i - A_j|^2, \tag{1}$$

where N_A is the number of points in structure A and N_B is the number of points in structure B and it is assumed that $N_A = N_B$. The least-squares estimation of the transformation parameters R and T in Eq. (1) can be found using the technique of singular value decomposition [29].

However, it is often the case that the equivalences between the structural elements are not known a priori. For example, when two proteins have diverged significantly. In this case, one must use heuristics to determine the equivalence relationship, and the problem of protein structural alignment becomes a multi-objective problem. That is, we are interested in finding the maximum number of equivalent elements as well as in minimizing the RMSD upon superposition of the equivalent elements of the two proteins.

A number of methods that are heuristic in nature have been developed for aligning protein structures [30–37]. These methods can be divided into two categories. *Global* structural alignment methods, which are suited for detecting similarities between the overall backbones of two proteins, while *local* structural alignment methods are suited for detecting similarities between local regions or sub-structures within the two proteins. As discussed earlier, many structural alignment algorithms are constrained to find only structural similarities where the order of the structural elements follows their order in the primary sequence. Sequence order independent methods ignore the sequential ordering of the structural elements and are better suited to find more complex global structural similarities. They are also very effective for all atom comparison of protein sub-structures, as in the case of binding surface alignment. Below we discuss methods for both global and local sequence order independent structural alignment.

Global Sequence Order Independent Structural Alignment

Global sequence order independent structural alignment is a powerful tool that can be used to detect similarities between two proteins that have complex topological rearrangements, including permuted structures. Permuted proteins can be described as two proteins with similar three-dimensional spatial arrangement of secondary structures, but with a different backbone connection topology. An example of permuted proteins are proteins with circular permutations, which can be thought of as ligation of the N- and C-termini of a protein, and cleavage somewhere else on the protein. Circular permutations are interesting not only because they tend to have similar three-dimensional structure but also because they often maintain the same biological function [26]. Circularly permuted proteins may provide a generic mechanism for introducing protein diversity that is widely used in evolution. Detecting circular permutations is also important for homology modeling, for studying protein folding, and for designing protein.

A Fragment Assembly Based Approach to Sequence Order Independent Structural Alignment

We have developed a sequence order independent structural alignment method that is well-suited for detecting circular permutation as well as more complex topological rearrangement relationships among proteins [27]. Our algorithm is capable of aligning two protein backbone structures independent of the secondary structure element connectivity. Briefly, the two proteins to be aligned are first separately and exhaustively fragmented. Each fragment $\lambda^A_{i,k}$ from protein structure S_A is then pair-wise superimposed onto each fragment $\lambda^B_{j,k}$ from protein structure S_B, forming a set of fragment pairs $\chi_{i,j,k}$, where $i \in S_A$ and $j \in S_B$ are the indices in the primary sequence of the first residue of the two fragment, respectively. Here

$k \in \{5, 6, 7\}$ is the length of the fragment. For each fragment, we assign a similarity score,

$$\sigma(\chi_{i,j,k}) = \alpha \left[C - s(\chi_{i,j,k}) \cdot \frac{cRMSD}{k^2} \right] + SCS, \tag{2}$$

where *cRMSD* is the measured RMSD value after optimal superposition of the two fragments, α and C are two constants, $s(\chi_{i,j,k})$ is a scaling factor to the measured RMSD values that depends on the secondary structure of this fragment, and *SCS* is a BLOSSUM-like measure of similarity in sequence of the matched fragments [24]. Details of the similarity score and the parameters α and C can be found in [27].

The goal of structural alignment for the moment seeks to find a consistent set of fragment pairs $\Delta = \{\chi_{i_1,j_1,k_1}, \chi_{i_2,j_2,k_2}, \ldots, \chi_{i_t,j_t,k_t}\}$ that minimize the global RMSD. Finding the optimal combination of fragment pairs is a special case of the well known maximum weight independent set problem in graph theory. This problem is MAX-SNP-hard. We employ an approximation algorithm that was originally described for scheduling split-interval graphs [38] and is itself based on a fractional version of the local-ratio approach.

Our method begins by creating a conflict graph $G = (V, E)$, where a vertex is defined for each aligned fragment pair. Two vertices are connected by an edge if any of the fragments $\left(\lambda^A_{i,k}, \lambda^A_{i',k'}\right)$ or $\left(\lambda^B_{j,k}, \lambda^B_{j',k'}\right)$ from the aligned pair is not disjoint, that is, if both fragments from the same protein share one or more residues. For each vertex representing aligned fragment pair, we assign three indicator variables x_χ, $y_{\chi_{\lambda_A}}$, and $y_{\chi_{\lambda_B}} \in \{0, 1\}$ and a closed neighborhood Nbr$[\chi]$. x_χ indicates whether the fragment pair should be used ($x_\chi = 1$) or not ($x_\chi = 0$) in the final alignment. $y_{\chi_{\lambda_A}}$, and $y_{\chi_{\lambda_B}}$ are artificial indicator values for λ_A and λ_B, which allow us to encode consistency in the selected fragments. The closed neighborhood of a vertex χ of G is $\{\chi' | \{\chi, \chi'\} \in E\} \cup \{\chi\}$, which is simply χ and all vertices that are connected to χ by and edge.

Our algorithm for sequence order independent structural alignment can now be described as follows. To begin, we initialize the structural alignment Δ equal to the entire set of aligned fragment pairs. We then:

1. Solve a linear programming (LP) formulation of the problem:
   maximize

$$\sum_{\chi \in \Delta} \sigma(\chi) \cdot x_\chi \tag{3}$$

subject to

$$\sum_{a_t \in \lambda^A} y_{\chi_{\lambda_A}} \quad \leq 1 \quad \forall a_t \in S_A \tag{4}$$

$$\sum_{b_t \in \lambda^B} y_{\chi_{\lambda_B}} \quad \leq 1 \quad \forall b_t \in S_B \tag{5}$$

$$y_{\chi\lambda_A} - x_\chi \geq 0 \quad \forall \chi \in \Delta \tag{6}$$

$$y_{\chi\lambda_B} - x_\chi \geq 0 \quad \forall \chi \in \Delta \tag{7}$$

$$x_\chi, y_{\chi\lambda_A}, y_{\chi\lambda_B} \geq 0 \quad \forall \chi \in \Delta \tag{8}$$

2. For every vertex $\chi \in V_\Delta$ of G_Δ, compute its *local conflict number* $\alpha_\chi = \sum_{\chi' \in \mathrm{Nbr}_\Delta[\chi]} x_{\chi'}$. Let $\chi_{\min}$ be the vertex with the *minimum* local conflict number. Define a new similarity function σ_{new} from σ as follows:

$$\sigma_{\mathrm{new}}(\chi) = \begin{cases} \sigma(\chi), & \text{if} \quad \chi \notin \mathrm{Nbr}_\Delta[\chi_{\min}] \\ \sigma(\chi) - \sigma(\chi_{\min}), & \text{otherwise} \end{cases}$$

3. Create $\Delta_{\mathrm{new}} \subseteq \Delta$ by removing from Δ every substructure pair χ such that $\sigma_{\mathrm{new}}(\chi) \leq 0$. Push each removed substructure on to a stack in arbitrary order.
4. If $\Delta_{\mathrm{new}} \neq \emptyset$ then repeat from step 1, setting $\Delta = \Delta_{\mathrm{new}}$ and $\sigma = \sigma_{\mathrm{new}}$. Otherwise, continue to step 5.
5. Repeatedly pop the stack, adding the substructure pair to the alignment as long as the following conditions are met:

   a. The substructure pair is consistent with all other substructure pairs that already exist in the selection.
   b. The *cRMSD* of the alignment does not change beyond a threshold. This condition bridges the gap between optimizing a local similarity between substructures and optimizing the tertiary similarity of the alignment. It guarantes that each substructure from a substructure pair is in the same spatial arrangement in the global alignment.

Detecting Permuted Proteins

This algorithm is used in a large scale study, where a subset with 3,336 protein structures taken from the PDBSELECT 90 data set % [39] are structurally aligned in a pair-wise fashion. Our goal is to determine if we could detect structural similarities with complex topological rearrangements such as circular permutations. From this subset of 3,336 proteins, we aligned two proteins if they met the following conditions: the difference in their lengths was no more than 75 residues, and they had approximately the same secondary structure content (see [27] for details). Within the approximately 200,000 alignments, we found many known circular permutations, and three novel circular permutations previously unknown, as well as a pair of non-cyclic complex permuted proteins. Below we describe in some details the circular permutations we found between a neucleoplasmin-core and an auxin binding protein, as well as details of the more complex non-cyclic permutation.

Nucleoplasmin-Core and Auxin Binding Protein

A novel circular permutation was detected between the nucleoplasmin-core protein in *Xenopu laevis* (PDB ID `1k5j`, chain E) [40] and the auxin binding protein in maize (PDB ID `1lrh`, chain A, residues 37 through 127) [41]. The structural alignment between `1k5jE` (Fig. 1a, top) and `1lrhA` (Fig. 1a, bottom) consisted of 68 equivalent residues superimposed with an RMSD of 1.36 Å. This alignment is statistically significant with a p-value of 2.7×10^{-5} after Bonferroni correction. Details of p-value calculation can be found in reference [27]. The short loop connecting two antiparallel strands in nucleoplasmin-core protein (in circle, top of Fig. 1b) becomes disconnected in auxin binding protein 1 (in circle, bottom of Fig. 1b), and the N- and C- termini of the nucleoplasmin-core protein (in square, top of Fig. 1b) are connected in auxin binding protein 1 (square, bottom of Fig. 1b). For details of other circular permutations we discovered, including permutations between aspartate racemase and type II 3-dehydrogenase and between microphage migration inhibition factor and the C-terminal domain of arginine repressor, please see [27].

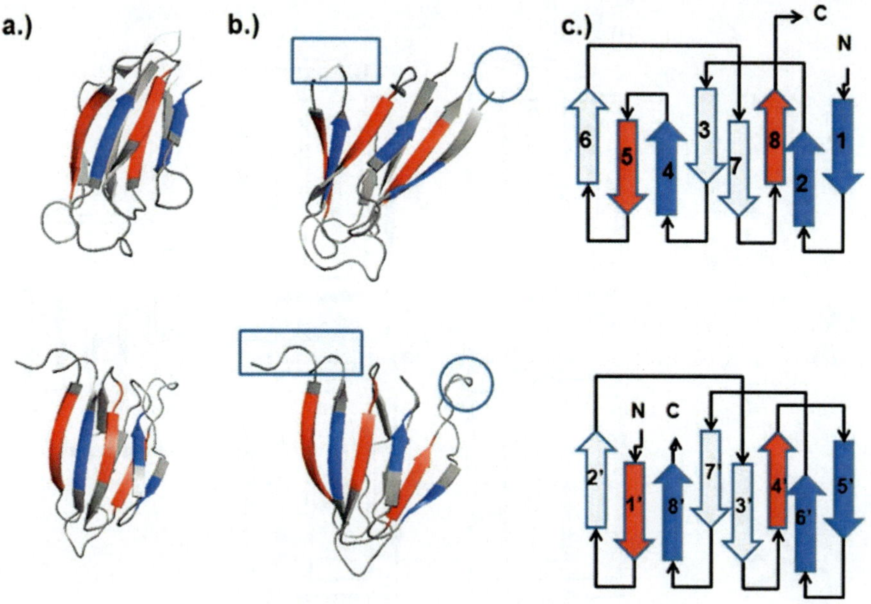

Fig. 1 A newly discovered circular permutation between nucleoplasmin-core (`1k5j`, chain E, *top panel*), and a fragment of auxin binding protein 1 (residues 37–127) (`1lrh`, chain A, *bottom panel*). **a** These two proteins align well with a RMSD value of 1.36 Å over 68 residues, with a significant p-value of 2.7×10^{-5} after Bonferroni correction. **b** The loop connecting strand 4 and strand 5 of nucleoplasmin-core (in *rectangle*, *top*) becomes disconnected in auxin binding protein 1. The N- and C- termini of nucleoplasmin-core (in *rectangle*, *top*) become connected in auxin binding protein 1 (in *rectangle*, *bottom*). To aide in visualization of the circular permutation, residues in the N-to-C direction before the cut in the nucleoplasmin-core protein are colored *red*, and residues after the cut are colored *blue*. **c** The topology diagram of these two proteins. In the original structure of nucleoplasmin-core, the electron density of the loop connecting strand 4 and strand 5 is missing in the PDB structure file. This figure is modified from [27]

Beyond Circular Permutation

Because of its relevance in understanding the functional and folding mechanism of proteins, circular permutations have received much attention [28, 42]. A more challenging class of permuted proteins is that of the non-cyclic permutation with more complex topological changes. Very little is known about this class of permuted proteins, and the detection of non-cyclic permutations is challenging task [43–46].

Non-cyclic permutations of the Arc repressor were created artificially were found to be thermodynamically stable. It can refold on the sub-millisecond time scale, and can bind operator DNA with nanomolar affinity [47], indicating that naturally occurring non-cyclic permutations may be as rich as the cyclic permutations. Our database search uncovered a naturally occurring non-cyclic permutation between chain F of AML1/Core Binding Factor (AML1/CBF, PDB ID `1e50`, Fig. 2a, top) and chain A of riboflavin synthase (PDB ID `1pkv`, Fig. 2a, bottom) [48, 49]. The

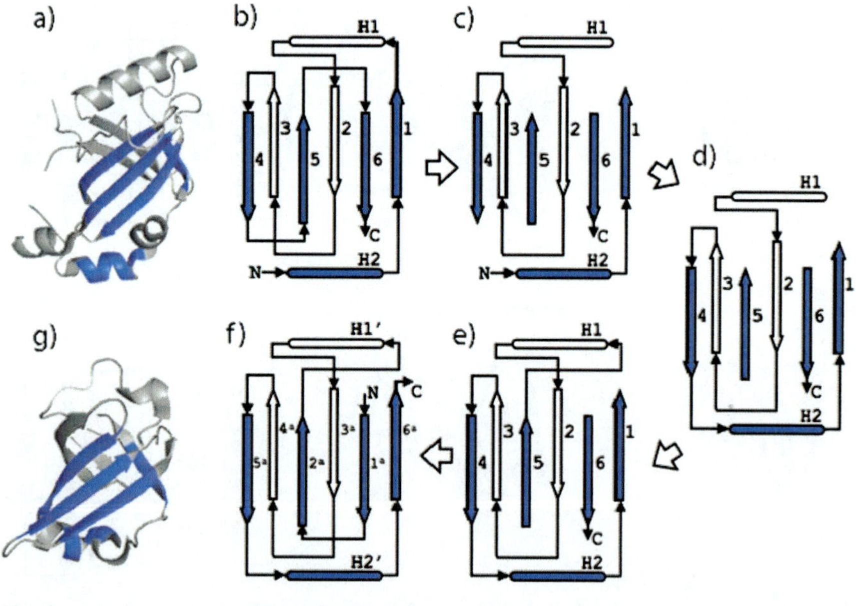

Fig. 2 A non-cyclic permutation discovered between AML1/Core Binding Factor (AML1/CBF, PDB ID `1e50`, Chain F, *top*) and riboflavin synthase (PDBID `1pkv`, chain A, *bottom*) **a** These two proteins structurally align with an RMSD of 1.23 Å over 42 residues , and has a significant *p*-value of 2.8×10^{-4} after Bonferroni correction. The residues that were assigned equivalences from the structural alignment are colored blue. **b** These proteins are related by a complex permutation. The steps to transform the topology of AML1/CBF (*top*) to riboflavin (*bottom*) are as follows: **c** Remove the loops connecting strand 1 to helix 2, strand 4 to strand 5, and strand 5 to helix 6; **d** Connect the C-terminal end of strand 4 to the original N-termini; **e** Connect the C-terminal end of strand 5 to the N-terminal end of helix 2; **f** Connect the original C-termini to the N-terminal end of strand 5. The N-terminal end of strand 6 becomes the new N-termini and the C-terminal end of strand 1 becomes the new C-termini. We now have the topology diagram of riboflavin synthase. This figure was modified from [27]

two structures align well with an RMSD of 1.23 Å, at an alignment length of 42 residues, with a significant p-value of 2.8×10^{-4} after Bonferroni correction.

The topology diagram of AML1/CBF (Fig. 2b) can be transformed into that of riboflavin synthase (Fig. 2f) by the following steps: Remove the loops connecting strand 1 to helix 2, strand 4 to strand 5, and strand 5 to strand 6 (Fig. 2c). Connect the C-terminal end of strand 4 to the original N-termini (Fig. 2d). Connect the C-terminal end of strand 5 to the N-terminal end of helix 2 (Fig. 2e). Connect the original C-termini to the N-terminal end of strand 5. The N-terminal end of strand 6 becomes the new N-termini and the C-terminal end of strand 1 becomes the new C-termini (Fig. 2f).

Local Sequence Order Independent Structural Alignment

The comparison of overall structural folds regardless of topological reconnections can lead to insight into distant evolutionary relationship. However, similarity in overall fold is not a reliable indicator of similar function [50–52]. Several studies suggest that structural similarities between local surface regions where biological function occurs, such as substrate binding sites, are a better predictor of shared biological function [8, 53–57].

Substrate binding usually occurs at concave surface regions, commonly referred to as *surface pockets* [55, 58–60]. A typical protein has many surface pockets, but only a few of them present a specific three-dimensional arrangement of chemical properties conducive to the binding of a substrate. This protein must maintain this physiochemical environment throughout evolution in order to maintain its biological function. For this reason, shared structural similarities between *functional surfaces* among proteins may be a strong indicator of shared biological function. This has lead to a number of promising studies, in which protein functions can be inferred by similarity comparison of local binding surfaces [55, 61–64].

A challenging problem with the structural comparison of protein pockets lies in the inherent flexibility of the protein structure. A protein is not a static structure represented by a Protein Data Bank entry. The whole protein as well as the local functional surface may undergo large structural fluctuations. The use of a single surface pocket structure as a representative template for a specific protein function will often result in many false negatives. This is due to the inability of a single representative to capture the full functional characteristics across all conformations of the protein.

To address this problem, we have developed a method that can automatically identify the structurally preserved atoms across a family of protein structures that are functionally related. Based on sequence-order independent surface alignments across the functional pockets of a family of protein structure, our method creates *signature pockets* by identifying structurally conserved atoms and measuring their fluctuations. As more than one signature pocket may result for a single functional class, the signature pockets can be organized into a *basis set* of signature pockets for that functional family. These signature pockets of the binding surfaces then can be used for scanning a protein structure database for function inference.

Bi-partite Graph Matching Approach to Structural Alignment

Our method for surface alignment is sequence order independent. It is based on a maximum weight bi-partite graph matching formulation of [65] with further modifications. This alignment method is a two step iterative process. First, an optimal set of equivalent atoms under the current superposition are found using a bi-partite graph representation. Second, a new superposition of the two proteins is determined using the new equivalent atoms from the previous step. The two steps are repeated until a stopping condition has been met.

To establish the equivalence relationship, two protein functional pocket surfaces S_A and S_B are represented as a graph, in which a node on the graph represent an atom from one of the two functional pockets. The graph is bi-partite if edges only connect nodes from protein S_A to nodes from protein S_B. In our implementation, directed edges are only drawn from nodes of S_A to nodes of S_B if a similarity threshold is met. The similarity threshold used in our implementation is a function of spatial distances and chemical differences between the corresponding atoms (see [66] for details). Each edge $e_{i,j}$ connecting node i to node j is assigned a weight $w(i,j)$ equal to the similarity score between the two corresponding atoms. A set of equivalence relations between atoms of S_A and atoms of S_B can be found by selecting a subset of the edges connecting nodes of S_A to S_B, with maximized total edge weight, where at most one edge can be selected for each atom [67]. A solution to the maximum weight bi-partite graph matching problem can be found using the Hungarian algorithm [68].

The Hungarian method works as follows. To begin, an overall score $F_{\text{all}} = 0$ is initialized, and an artificial source node s and an artificial destination node d are added to the bi-partite graph. Directed edges with 0-weight from the source node s to each node of S_A and from each node of S_B to the destination node d are also added. The algorithm then proceeds as follows:

1. Find the shortest distance $F(i)$ from the source node s to every other node i using the Bellman-Ford [69] algorithm.
2. Assign a new weight $w'(i,j)$ to each edge that does not originate from the source node s as follows,

$$w'(i,j) = w(i,j) + [F(i) - F(j)]. \tag{9}$$

3. Update F_{all} as $F_{\text{all}}' = F_{\text{all}} - F(d)$
4. Reverse the direction of the edges along the shortest path from s to d.
5. If $F_{\text{all}} > F(d)$ and a path exists between s and d then start again at step 1.

The Hungarian algorithm terminates when either there is no path from s to d or when the shortest distance from the source node to the destination node $F(d)$ is greater than the current overall score F_{all}. The bi-partite graph will now consist of directed edges that have been reversed (point from nodes of S_B to nodes of S_A). These flipped edges represent the current equivalence relationships between atoms of S_A and atoms of S_B.

The equivalence relations can then be used to superimpose the two proteins. After superposition, a new bi-partite graph is created and the maximum weight bi-partite matching algorithm is called again. This process is repeated iteratively until the change in RMSD upon superposition falls below a threshold.

Signature Pockets and Basis Set of Binding Surface for a Functional Family of Proteins

Based on the pocket surface alignment algorithm, we have developed a method that automatically generate structural templates of local surfaces, called *signature pockets*, which can be used to represent an enzyme function or a binding activity. These signature pockets contain broad structural information as well as discriminating ability.

A signature pocket is derived from an optimal alignment of precomputed surface pockets in a sequence-order-independent fashion, in which atoms and residues are aligned based on their spatial correspondence when maximal similarity is obtained, regardless how they are ordered in the underlying primary sequences. Our method does not require the atoms of the signature pocket to be present in all member structures. Instead, signature pockets can be created at varying degrees of partial structural similarity, and can be organized hierarchically at different level of binding surface similarity.

The input to the signature pocket algorithm is a set of functional pockets from a pre-calculated database of surface pockets and voids on proteins, such as those contained in the CASTp database [60]. The algorithms begins by performing all vs all pair-wise sequence order independent structural alignment on the input functional surface pockets. A distance score, which is a function of the RMSD and the chemistry of the paired atoms from the structural alignment, is recorded for each aligned pair of functional pockets (see [66] for details). The resulting distance matrix is then used by an agglomerative clustering method, which generates a hierarchical tree. The signature of the functional pockets can then be computed using a recursive process following the hierarchical tree.

The process begins by finding the two closest siblings (pockets S_A and S_B), and combining them into a single surface pocket structure S_{AB}. Because of the recursive nature of this algorithm, either of the two structures being combined may themselves already be a combination of several structures. When combining the two structures, we follow the criteria listed below:

1. If two atoms were considered equivalent in a structural alignment, a single coordinate is created in the new structure to represent both atoms. The new coordinate is calculated by averaging the coordinates of all underlying atoms that are currently represented by the two coordinates to be averaged.
2. If no equivalence was found for an atom during the structural alignment, the coordinates of that atom are transferred directly into the new pocket structure.

During each step in combining two surface pockets, a count of the number of times that an atom at the position i was present in the underlying set of pockets is recorded, which is then divided by the number of the constituent pockets. This is the *preservation ratio* $\rho(i)$. In addition, the mean distance of the coordinates of the aligned atoms to their geometric center is recorded as the *location variation* v. At the end of each step, the new structure S_{AB} replaces the two structures S_A and S_B in the hierarchical tree, and the process is repeated on the updated hierarchical tree. At a specific height of the hierarchical tree, different signature pockets can be created with different extents of structural preservation by selecting a similarity threshold value.

The signature pocket algorithm can be terminated at any point during its traversal of the hierarchical tree. Figure 3 illustrates this point by showing three different stopping thresholds (horizontal dashed lines). Depending on the choice of the threshold, one or multiple signature pockets may result. Figure 3a shows a low threshold which results in a set of 3 signature pockets. Raising the threshold can produce fewer signature pockets (Fig. 3b). A single signature pocket that represents all surface pockets in the data set can be generated by raising the threshold even further (Fig. 3c). Since clusters from the hierarchical tree represent a set of surface pockets that are similar within certain threshold, if a stopping threshold is chosen such that there exist multiple clusters in the hierarchical tree, a signature pocket will be created for each cluster. The set of signature pockets from different clusters collectively form a *basis set* of signature pockets, which represent the ensemble of differently sampled conformations for a functional family of proteins. As a basis set of signatures can represent many possible variations in shapes and chemical textures, it can represent structural features of an enzyme function with complex binding activities, and can also be used to accurately predict enzymes function.

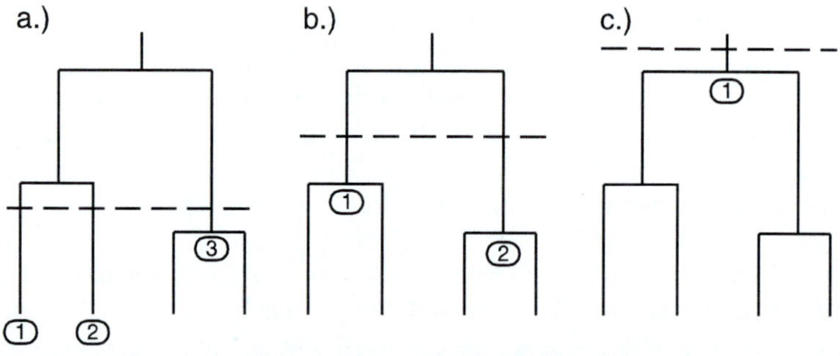

Fig. 3 Different basis sets of signature pockets can be produced at different levels of structural similarity by raising or lowering the similarity threshold (*vertical dashed line*). **a** A low threshold will produce more signature pockets. **b** As the threshold is raised, fewer signature pockets will be created. **c** A single signature pocket can in principle be created to represent the full surface pocket data set by raising the threshold

Signature Pockets of NAD Binding Proteins

To illustrate how signature pockets and basis set help to identify key structural elements important for binding and how they can facilitate function inference, we discuss a study of the nicotinamide adenine dinucucleotide (NAD) binding proteins. NAD consists of two nucleotides, nicotinamide and adenine, which are joined by two phosphate groups. NAD plays essential roles in metabolism where it acts as a coenzyme in redox reactions, including glycolysis and the citric acid cycle.

Using a set of 457 NAD binding proteins of diverse fold structures and diverse evolutionary origin, we first extracted the NAD binding surfaces from precomputed CASTp database of protein pockets and voids [60]. Based on similarity values from a comprehensive all-against-all sequence order independent surface alignment, we obtain a hierarchical tree of NAD binding surfaces. The resulting 9 signature pockets of the NAD binding pocket form a basis set, which are shown in Fig. 4.

These signature pockets contain rich biological information. Among the NAD-binding oxioreductase, three signature pockets (Fig. 4e, h, and i) are for clusters of oxioreductases that act on the CH-OH group of donors (alcohol oxioreductases), one signature pocket (Fig. 4j) is for a cluster that act on the aldehyde group of donors, and the remaining two signature pockets (Fig. 4f and g) are for oxioreductases that act on the CH-CH group of donors. For NAD-binding lyase, one of the two signature pockets (Fig. 4d) represent lyase that cleave both C–O and P–O bonds. The other signature pocket (Fig. 4b) represent lyases that cleave both C–O and C–C bonds. These two signatures come from two clusters of lyase conformations, each with a very different class of conformations of the bound NAD cofactor.

We found that the structural fold and the conformation of the bound NAD cofactor are the two major determinants of the formation of the clusters of the NAD binding pockets (Fig. 4a). It can be seen in Fig. 4b–j that there are two general conformations of the NAD coenzyme. The NAD coenzymes labeled C (Fig. 4b, c, f, g, h, and j) have a closed conformation, while the coenzymes labeled X (Fig. 4d, e, and i) have an extended conformation. This indicates that the binding pocket may take multiple conformations yet bind the same substrate in the same general structure. For example, the two structurally distinct signature pockets shown in Fig. 4f, g are derived from proteins that have the same biological function and SCOP fold. All of these proteins bind to the same NAD conformation.

We have further evaluated the effectiveness of the NAD binding site basis set by determining its accuracy in correctly classifying enzymes as either NAD-binding or non-NAD-binding. We constructed a test data set of 576 surface pockets from the CASTp database [60] independent of the training set of 457 NAD binding proteins. These 576 surface pockets were selected by taking the top 3 largest pockets in volume from 142 randomly chosen proteins and 50 proteins that have NAD bound in the PDB structure, with the further constraint that they were not in our training data set. We then structurally aligned all 576 pockets in our test data set against each of the nine NAD signature pockets in the resulting basis set. The testing pocket was assigned to be an NAD binding pocket if it structurally aligned to one of the nine NAD signature pockets, with its distance under a predefined threshold. Otherwise it

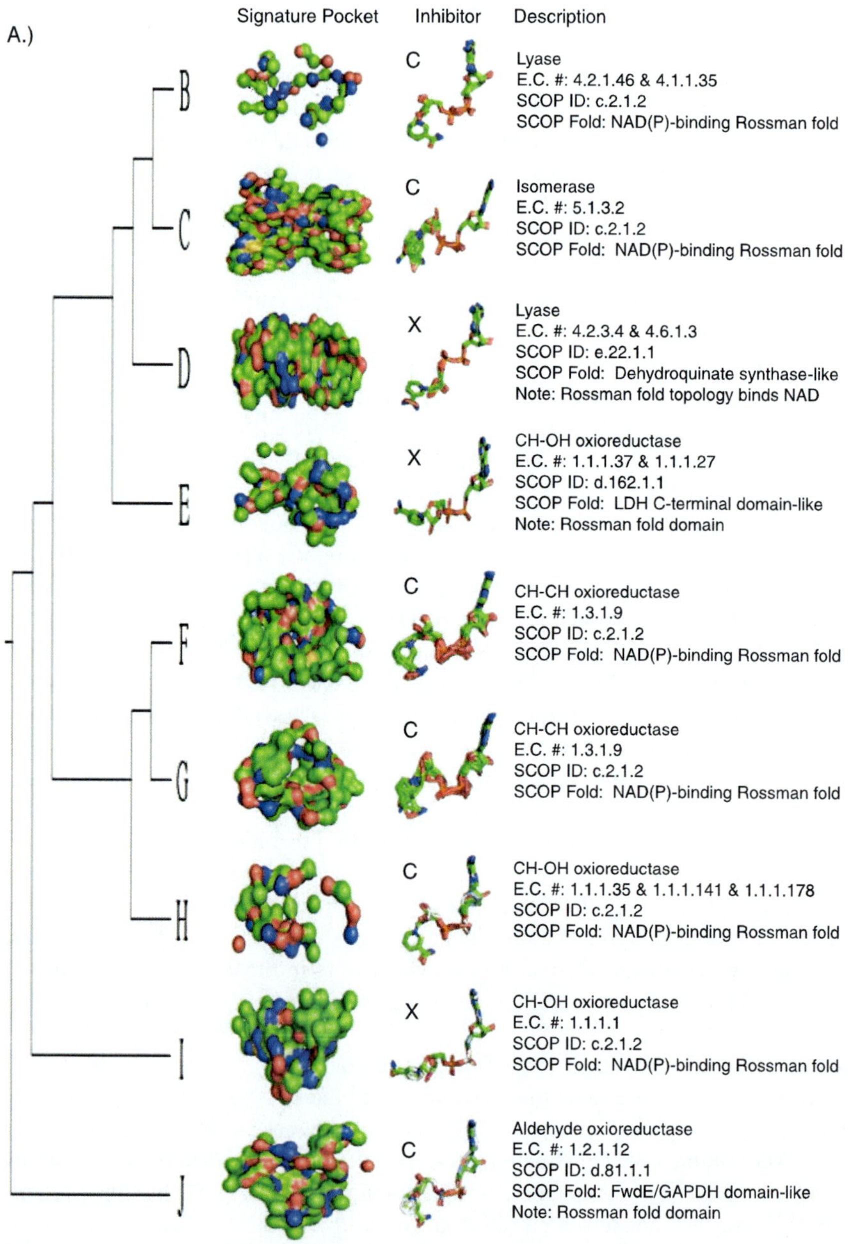

Fig. 4 The topology of the hierarchical tree and signature pockets of the NAD binding pockets. **a** The resulting hierarchical tree topology. **b–j** The resulting signature pockets of the NAD binding proteins, along with the superimposed NAD molecules that were bound in the pockets of the member proteins of the respective clusters. The NAD coenzymes have two distinct conformations. Those in an extended conformation are marked with an X and those in a compact conformation are marked with a C

was classified as non-NAD binding. The results show that the basis set of 9 signature pockets can classify the correct NAD binding pocket with sensitivity and specificity of 0.91 and 0.89, respectively. We performed further testing to determine whether a single representative NAD binding pocket, as opposed to a basis set, is sufficient for identifying NAD-binding enzymes. We chose a pocket representative pocket from one of the 9 clusters that were used to construct the 9 signature pockets. Here, a testing pockets was classified as NAD-binding if its structural similarity to the single representative pocket was above the same pre-defined threshold used in the basis set study. We repeat this exercise nine times, each time using a different representative from a different cluster. We found that the results deteriorated significantly, with an average sensitivity and specificity of only 0.36 and 0.23, respectively. This study strongly indicates that the construction of a basis set of signatures as a structural template provides significant improvement for a set of proteins binding the same co-factor but with diverse evolutionary origin. Further details of the NAD-binding protein study can be found in [66], along with an in-depth study of the metalloendopeptidase, including the construction of its signatures and basis set, as well as their utility in function prediction.

Conclusion

In this chapter, we have discussed methods that provide solutions to the problem of aligning protein global structures as well as aligning protein local surface pockets. Both methods disregard the ordering of residues in the protein primary sequences. For global alignment of protein structures, such a method can be used to address the challenging problem of identifying proteins that are topologically permuted but are spatially similar. The approach of fragment assembly based on the formulation of a relaxed integer programming problem and an algorithm based on scheduling split-interval graphs works well, and is characterized by a guaranteed approximation ratio. In a scaled up study, we showed that this method enables in discovery of circularly permuted proteins, including several previously unrecognized protein pairs. It also uncovered a case of two proteins related by higher order permutations.

We also described a method for order-independent alignment of local spatial surfaces that is based on bi-partite graph matching. By assessing surface similarity for a group of protein structures of the same function, this method can be used to automatically construct signatures and basis set of binding surfaces characteristic of a specific biological function. We showed that such signatures can reveal useful mechanistic insight on enzyme function, and can correlate well with substrate binding specificity.

In this chapter, we neglected an important issue in our discussion of comparing protein local surfaces for inferring biochemical functions, namely, how to detect evolutionary signals and how to employ such information for protein function prediction. Instead of going into details, we first point readers to the general approach of constructing continuous time Markovian models to study protein evolution [70, 71]. In addition, a Bayesian Monte Carlo method that can separate selection pressure due to biological function from selection pressure due to the constraints of protein folding stability and folding dynamics can be found in [57] and in [72]. The Bayesian Monte Carlo approach can be used to construct customized scoring matrices that are specific to a particular class of proteins of the same function. Details of how such method works and how it can be used to accurately predict enzyme functions from structure with good sensitivity and specificity for 100 enzyme families can be found in a recent review [72] and original publications [8, 57]. The task of computing surface pockets and voids using alpha shape is discussed in a recent review [73].

Acknowledgements This work was supported by NIH grants GM079804, GM081682, GM086145, NSF grants DBI-0646035 and DMS-0800257, ONR grant N00014-09-1-0028.

References

1. Binkowski, A., Joachimiak, A., Liang, J. Protein surface analysis for function annotation in high-throughput structural genomics pipeline. Protein Sci. **14**: 2972–2981 (2005).
2. Pazos, F., Sternberg, M.J.E. Automated prediction of protein function and detection of functional sites from structure. PNAS **101**:14, 14754–14759 (2004).
3. Andrade, M.A., Brown, N.P., Leroy, C., Hoersch, S., de Daruvar, A., Reich, C., Franchini, A., Tamames, J., Valencia, A., Ouzounis, C., Sander, C. Automated genome sequence anlysis and annotation. Bioinformatics **15**: 391–412 (1999).
4. Jensen, L.J., Gupta, R., Blom, N., Devos, D., Tamames, J., Kesmir, C., Nielsen, H., Staerfeldt, H.H., Rapacki, K., Workman, C., Andersen, C.A.F., Knudsen, S., Krogh, A., Valencia, A., Brunak, S. Prediction of human protein function from post-translational modifications and localization features. J. Mol. Biol. **319**: 1257–1265 (2002).
5. Pal, D., Eisenberg, D. Inference of protein function from protein structure. Structure **13**: 121–130 (2005).
6. Laskowski, R.A., Watson, J.D., Thornton, J.M. ProFunc: a server for predicting protein function from 3D structure. Nucleic Acids Res. **33**: W89–93 (2005).
7. Deng, M., Zhang, K., Mehta, S., Chen, T., Sun, F. Prediction of protein function using protein-protein interaction data. J. Comput. Biol. **10**(6): 947–960 (2003).
8. Tseng, Y.Y., Dundas, J., Liang, J. Predicting protein function and binding profile via matching of local evolutionary and geometric surface patterns. J. Mol. Biol. **387**(2): 451–464 (2009).
9. Shah, I., Hunterm, L. Predicting enzyme function from sequence: a systematic appraisal. ISMB **5**: 276–283 (1997).
10. Altschul, S.F., Warren, G., Miller, W., Myers, E.W., Lipman, D.J. Basic local alignment search tool. J. Mol. Biol. **215**: 403–410 (1990).
11. Altschul, S.F., Madden, T.L., Schaffer, A.A., Zhang, J., Zhang, Z., Miller, W., Lipman, D.J. Gapped BLAST and PSI-BLAST: a new generation of protein database search programs. Nucleic Acids Res. **25**(17): 3389–3402 (1997).
12. Karplus, K., Barret, C., Hughey, R. Hidden Markov Models for detecting remote protein homologues. Bioinformatics **14**: 846–856 (1998).
13. Hulo, N., Sigrist, C.J.A., Le Saux, V. Recent improvements to the PROSITE database. Nucleic Acids Res. **32**: D134–D137 (2004).
14. Boeckmann, B., Bairoch, A., Apweiler, R., Blatter, M.C., Estreicher, A., Gasteiger, E., Martin, M.J., Michoud, K., O'Donovan, C., Phan, I., Pilbout, S., Schneider, M. The SWISS-PROT protein knowledgebase and its supplement TrEMBL in 2003. Nucleic Acids Res. **31**: 365–370 (2003).

15. Weidong, T., Skolnick, J. How well is enzyme function conserved as a function of pairwise sequence identity. J. Mol. Biol. **333**: 863–882 (2003).
16. Rost, B. Twilight zone of protein sequence alignments. Protein Eng. **12**: 85–94 (1999).
17. Hegyi, H., Gerstein, M. The relationship between protein structure and function: a comprehensive survey with application to the yeast genome. J. Mol. Biol. **288**: 147–164 (1999).
18. Murzin, A.G., Brenner, S.E., Hubbard, T., Chothia, C. SCOP: a structural classification of proteins database for the investigation of sequences and structures. J. Mol. Biol. **247**: 536–540 (1995).
19. Orengo, C.A., Michie, A.D., Jones, D.T., Swindells, M.B., Thornton, J.M. CATH: a hierarchical classification of protein domain structures. Structure **5**: 1093–1108 (1997).
20. Holm, L., Sander, C. Protein structure comparison by alignment of distance matrices. J. Mol. Biol. **233**: 123–138 (1993).
21. Shindyalov, I.N., Bourne, P.E. Protein structure alignment by incremental combinatorial extension (CE) of the optimal path. Protein Eng. **11**(9): 739–747 (1998).
22. Hasegawa, H., Holm, L. Advances and pitfalls of protein structural alignment. Curr. Opin. Struct. Biol. **19**: 341–348 (2009).
23. Dayhoff, M.O., Schwartz, R.M., Orcutt, B.C. A model of evolutionary change in proteins. Atlas Protein Seq. Struct. **5**(3): 345–352 (1978).
24. Henikoff, S., Henikoff, J.G. Amino acid substitution matrices from protein blocks. PNAS **89**(22): 10915–10919 (1992).
25. Teichert, F., Bastolla, U., Porto, M. SABERTOOTH: protein structure comparison based on vectorial structure representation. BMC Bioinformatics **8**: 425 (2007).
26. Lindqvist, Y., Schneider, G. Circular permutations of natural protein sequences: structural evidence. Curr. Opin. Struct. Biol. **7**: 422–427 (1997).
27. Dundas, J., Binkowski, T.A., DasGupta, B., Liang, J. Topology independent protein structural alignment. BMC Bioinformatics **8**(388) doi:10.1186/1471-2105-8-388 (2007).
28. Berman, H.M., Westbrook, J., Feng, Z., Gilliland, G., Bhat, T.N., Weissig, H., Shindyalov, I.N., Bourne, P.E. The protein data bank. Nucleic Acids Res. **28**: 235–242 (2000).
29. Umeyama, S. Least-squares estimation of transformation parameters between two point patterns. IEEE Trans. Pattern Anal. Mach. Intell. **13**(4): 376–380 (1991).
30. Veeramalai, M., Gilbert, D. A novel method for comparing topological models of protein structures enhanced with ligand information. Bioinformatics **24**(23): 2698–2705 (2008).
31. Aghili, S.A., Agrawal, D., El Abbadi, A. PADS: protein structure alignment using directional shape signatures. In DASFFA (2004).
32. Szustakowski, J.D., Weng, Z. Protein structure alignment using a genetic algorithm. Proteins: Struct. Funct. Genet. **38**: 428–440 (2000).
33. Standley, D.M., Toh, H., Nakamura, H. Detecting local structural similarity in proteins by maximizing number of equivalent residues. Proteins: Struct. Funct. Genet. **57**: 381–391 (2004).
34. Roach, J., Sharma, S., Kapustina, M., Cater Jr., C.W. Structure alignment via delaunay tetrahedralization. Proteins: Struct. Funct. Genet. **60**: 66–81 (2005).
35. Teyra, J., Paszkowski-Rogacz, M., Anders, G., Pisabarro, M.T. SCOWLP classification: structural comparison and analysis of protein binding regions. BMC Bioinformatics doi:10.1186/1471-2105-9-9 (2008).
36. Gold, N.D., Jackson, R.M. Fold independent structural comparisons of protein-ligand binding sites for exploring functional relationships. J. Mol. Biol. **355**: 1112–1124 (2006).
37. Zhu, J., Weng, Z. A novel protein structure alignment algorithm. Proteins: Struct. Funct. Bioinform. **58**: 618–627 (2005).
38. Bar-Yehuda, R., Halldorsson, M.M., Naor, J., Shacknai, H., Shapira, I. Scheduling split intervals. 14th ACM-SIAM Symposium on Discrete Algorithms, Baltimore, MD, pp. 732–741 (2002).

39. Hobohm, U., Sander, C. Enlarged representative set of protein structures. Protein Sci. **33**: 522 (1994).
40. Dutta, S., Akey, I.V., Dingwall, C., Hartman, K.L., Laue, T., Nolte, R.T., Head, J.F., Akey, C.W. The crystal structure of nucleoplasmin-core implication for histone binding and nucleosome assembly. Mol. Cell **8**: 841–853 (2001).
41. Woo, E.J., Marshall, J., Bauly, J., Chen, J.G., Venis, M., Napier, R.M., Pickersgill, R.W. Crystal structure of the auxin-binding protein 1 in complex with auxin. EMBO J. **21**: 2877–2885 (2002).
42. Uliel, S., Fliess, A., Amir, A., Unger, R. A simple algorithm for detecting circular permutations in proteins. Bioinformatics **15**(11): 930–936 (1999).
43. Alexandrov, N.N., Fischer, D. Analysis of topological and nontopological structural similarities in the PDB: new examples with old structures. Proteins **25**: 354–365 (1996).
44. Dror, O., Benyamini, H., Nussinov, R., Wolfson, H.J. MASS: multiple structural alignment by secondary structures. Bioinformatics **19**: i95–i104 (2003).
45. Shih, E.S., Hwang, M.J. Alternative alignments from comparison of protein structures. Proteins **56**: 519–527 (2004).
46. Ilyin, V.A., Abyzov, A., Leslin, C.M. Structural alignment of proteins by a novel TOPOFIT method, as a superimposition of common volumes at a topomax point. Protein Sci. **13**: 1865–1874 (2004).
47. Tabtiang, R.K., Cezairliyan, B.O., Grant, R.A., Cochrane, J.C., Sauer, R.T. Consolidating critical binding determinants by noncyclic rearrangement of protein secondary structure. PNAS **7**: 2305–2309 (2004).
48. Warren, A.J., Bravo, J., Williams, R.L., Rabbitts, T.H. Structural basis for the heterodimeric interaction between the acute leukemia-associated transcription factors AML1 and CBFbeta. EMBO J. **19**: 3004–3015 (2000).
49. Meining, W., Eberhardt, S., Bacher, A., Ladenstein, R. The structure of the N-terminal domain of riboflavin synthase in complex with riboflavin at 2.6A resolution. J. Mol. Biol. **331**: 1053–1063 (2003).
50. Lichtarge, O., Bourne, H.R., Cohen, F.E. An evolutionary trace method defines binding surfaces common to protein families. J. Mol. Biol. **7**: 39–46 (1994).
51. Norel, R., Fischer, H., Wolfson, H., Nussinov, R. Molecular surface recognition by computer vision-based technique. Protein Eng. **7**(1): 39–46 (1994).
52. Fischer, D., Norel, R., Wolfson, H., Nussinov, R. Surface motifs by a computer vision-technique: searches, detection, and implications for protein-ligand recognition. Proteins **16**: 278–292 (1993).
53. Meng, E., Polacco, B., Babbitt, P. Superfamily active site templates. Proteins **55**: 962–967 (2004).
54. Orengo, C., Todd, A., Thornton, J. From protein structure to function. Curr. Opin. Struct. Biol. **9**: 374–382 (1999).
55. Binkowski, A., Adamian, L., Liang, J. Inferring functional relationship of proteins from local sequence and spatial surface patterns. J Mol Biol. **332**: 505–526 (2003).
56. Jeffery, C. Molecular mechanisms for multi-tasking: recent crystal structures of moon-lighting proteins. Curr. Opin. Struct. Biol. **14**: 663–668 (2004).
57. Tseng, Y.Y., Liang, J. Estimation of amino acid residue substitution rates at local spatial regions and application in protein function inference: a Bayesian Monte Carlo approach. Mol. Biol. Evol. **23**: 421–436 (2006).
58. Liang, J., Edelsbrunner, H., Woodward, C. Anatomy of protein pockets and cavities: measurement of binding site geometry and implications for ligand design. Protein Sci. **7**: 1884–1897 (1998).
59. Edelsbrunner, H., Facello, M., Liang, J. On the definition and the construction of pockets in macromolecules. Disc Appl. Math. **88**: 83–102 (1998).
60. Dundas, J., Ouyang, Z., Tseng, J., Binkowski, A., Turpaz, Y., Liang, J. CASTp: computed atlas of surface topography of proteins with structural and topographical mapping of functionally annotated residues. Nucleic Acids Res. **34**: W116–W118 (2006).

61. Lee, S., Li, B., La, D., Fang, Y., Ramani, K., Rustamov, R., Kihara, D. Fast protein tertiary structure retrieval based on global surface shape similarity. Proteins **72**: 1259–1273 (2008).
62. Binkowski, T.A., Joachimiak, A. Protein functional surfaces: global shape matching and local spatial alignments of ligand binding sites. BMC Struct. Biol. **8**: 45 (2008).
63. Bandyopadhyay, D., Huan, J., Liu, J., Prins, J., Snoeyink, J., Wang, W., Tropsha, A. Functional Neighbors: Inferring relationships between non-homologous protein families using family-specific packing motifs. Proc. IEEE Int. Conf. Bioinform. Biomed. **14**(5): 1137–1143 (2008).
64. Mol, M., Kavraki, L.E. LabelHash: A flexible and extensible method for matching structural motifs. Automated Function Prediction Meetings, Toronto, Canada (2008).
65. Chen, L., Wu, L.Y., Wang, R., Wang, Y., Zhang, S., Zhang, X.S. Comparison of protein structures by multi-objective optimization. Genome Inform. **16**(2): 114–124 (2005).
66. Dundas, J. Adamian, L. Liang, J. Structural signatures of enzyme binding pockets from order-independent surface alignment: a study of metalloendopeptidase and nad binding proteins. J. Mol. Biol. **406**(5): 713–729 (2011 Mar).
67. Corment, T.H., Leiserson, C.E., Rivest, R.L., Stein, C. *Introduction to algorithms*, 2nd edn. Cambridge, MA: MIT Press (2001).
68. Kuhn, H.W. The hungarian method for the assignment problem. Nav. Res. Logist. Q. **2**: 83–97 (1995).
69. Bellman, R. On a routing problem. Q. Apply Math. **16**(1): 87–90 (1958).
70. Yang, Z., Nielsen, R., Hasegawa, M. Models of amino acid substitution and applications to mitochondrial protein structures. Mol. Biol. Evol. **15**: 1600–1611 (1998).
71. Huelsenbeck, J.B., Ronquist, R., Nielsen, R., Bollback, J. Bayesian inference of phylogeny and its impact on evolutionary biology. Science **294**: 2310–2314 (2001).
72. Liang, J., Tseng, Y.Y., Dundas J., Binkowski, A., Joachimiak, A., Ouyang, Z., Adamian, L. Chapter 4: predicting and characterizing protein functions through matching geometric and evolutionary patterns of binding surfaces. Adv. Protein Chem. **75**: 107–141 (2008).
73. Liang, J., Kachalo, S., Li, X., Ouyang, Z., Tseng, Y.Y., Zhang, J. Geometric structures of proteins for understanding folding, discriminating natives and predicting biochemical functions. *The World is a Jigsaw*. van de Weygaert R. (ed.). Springer (2009).

Protein Binding Ligand Prediction Using Moments-Based Methods

Rayan Chikhi, Lee Sael, and Daisuke Kihara

Abstract Structural genomics initiatives have started to accumulate protein structures of unknown function in an increasing pace. Conventional sequence-based function prediction methods are not able to provide useful function information to most of such structures. Thus, structure-based approaches have been developed, which predict function of proteins by capturing structural characteristics of functional sites. Particularly, several approaches have been proposed to identify potential ligand binding sites in a query protein structure and to compare them with known ligand binding sites. In this chapter, we introduce computational methods for describing and comparing ligand binding sites using two dimensional and three dimensional moments. An advantage of moment-based methods is that the tertiary structure of pocket shapes is described compactly as a vector of coefficients of series expansion. Thus a search against an entire PDB-scale database can be performed in real-time. We evaluate two binding pocket representations, one based on two-dimensional pseudo-Zernike moments and the other based on three-dimensional Zernike moments. A new development of pocket comparison method is also mentioned, which allows partial matching of pockets by using local patch descriptors.

Introduction

Functional assignment of proteins is a fundamental and challenging problem in biology and bioinformatics [1]. In recent years structural genomics projects have been solving an increasing number of protein structures which were not able to be characterized by traditional sequence based methods [2, 3]. Therefore, much effort has been devoted recently to the development of function prediction methods based on structural information. Structure-based function prediction methods aim either to

D. Kihara (✉)
Department of Biological Sciences; Department of Computer Science; Markey Center for Structural Biology, College of Science, Purdue University, West Lafayette, IN 47907, USA
e-mail: dkihara@purdue.edu

D. Kihara (ed.), *Protein Function Prediction for Omics Era*,
DOI 10.1007/978-94-007-0881-5_8,

capture global structure or local structure similarity to proteins of known function in the structure database (PDB [4]). Global approaches are motivated by the observation that protein folds are better conserved than primary sequences. Alternatively, local methods aim to capture properties of functional sites, where interactions with ligand molecules or other proteins take place. Ligand binding sites are intrinsically unrelated to global folds, as two sequentially and structurally dissimilar proteins may bind the same ligand molecule [5]. Because two proteins with similar folds often have different functions [6] ligand binding sites are of particular interest in structure-based function prediction. In most cases ligand molecules bind to a protein at its surface pocket regions [7], hence detecting pockets enables identification of binding sites [8–10]. Binding ligand prediction approaches have two logical steps: (i) detection of the pocket region in a given protein surface and (ii) comparison of a pocket against a database of known sites.

Several methods have been developed to predict the location of ligand binding sites in a protein surface. These methods are based on the detection of specific geometric properties on the protein surface. For instance, gaps can be detected on a protein surface using probe spheres [11–13]. Grid-based methods [7, 14, 15] scan protein surface points for various properties, e.g. voids, the visibility, or the depth. Voronoi diagrams have also been applied to identify pockets by recognizing depressed regions [16]. Recent methods combine geometrical criteria with evolutionary information [17–20] and energetics [21–23].

Comparison of binding sites relies on how pockets are represented. These representations are either based on coordinates of residues/atoms or shape of pocket surfaces. In the former representation, protein binding pockets are described as sets of three dimensional coordinates of key residues [24–26], for which pair-wise similarity is computed, for example, with the root mean square deviation (RMSD). The geometric hashing [27] technique defines a distance between two binding sites by the number of spatially matching atoms. Alternatively, in a type of fingerprinting methods, a site is represented by all the distances between residues, which are then grouped by types for fast matching [28, 29]. Similar fingerprinting approaches have also been applied to atoms on the solvent accessible surface [5, 30].

Surface-based representations of binding sites are based on a wide spectrum of computational techniques. Moments-based methods belong to this category and are thoroughly discussed in the next section. Graph-based representation is an alternative choice for representing protein surfaces. Klebe et al. employed subgraph matching algorithm to describe the surface geometry and the electrostatic potential of binding pockets [31]. Kinoshita et al. used a clique detection algorithm [32] for local surface similarity retrieval in their method named eF-Site. Using the eF-Site and its associated tool, eF-Seek, users can search functional sites in an unannotated query structure [32, 33]. Another approach uses the spin-image, a 2D histogram representation for protein surface points, which describes relative geometrical position of each point to the other points [34]. Generally speaking, moment-based descriptors have advantage over graph methods and 2D histograms in terms of lower time complexity (thus faster running time).

In this chapter, we review three-dimensional (3D) and two-dimensional (2D) geometric moments for the representation and comparison of protein ligand binding sites. Concretely, we describe application of the 2D pseudo-Zernike moments and the 3D Zernike descriptors. The 2D pseudo-Zernike (p-Z) moments are employed to describe the projection of the pocket surface on a 2D image. The 3D Zernike descriptors (3DZD) can directly represent 3D pocket surface properties. These moments compactly represent a binding pocket by a vector of coefficients of the series expansion. The rest of this chapter is organized as follows: First, an overview of the theoretical differences between these moments is given. In addition to the p-Z moments and the 3DZD descriptors, we also discuss the spherical harmonics in comparison with the two methods. Then, we describe our recent works on the application of the 2D p-Z moments and the 3DZD for binding ligand prediction for proteins. The methodology quantifies similarity of pockets by the Euclidean distance of the vector of p-Z/3DZD coefficients of pockets and uses a *k*-NN classifier to make final prediction of binding ligand for a query pocket. Our methods are benchmarked on two datasets. Finally, recent ongoing development in our group on a new pocket comparison method is discussed, which uses local surface patch descriptors to allow matching of flexible binding ligands.

Pocket Surface Shape Descriptors

In this section we briefly describe 3D and 2D moment-based pocket descriptors, which will be used in the subsequent sections. For the 3D descriptors of pockets, we introduce the spherical harmonics and the 3D Zernike descriptors. For the 2D descriptors, we introduce the p-Z moments.

Spherical Harmonics

Spherical harmonics are a set of mathematical moments which are applied for 3D volumetric representation of objects [35]. The object shape is approximated as a spherical function $f(\theta,\phi)$ defined on the unit sphere, which describes the distance to the outermost surface of the object from the center for the direction (θ,ϕ). The function $f(\theta,\phi)$ is then expanded as a series of spherical harmonics

$$f(\theta,\phi) \approx \sum_{l=0}^{l_{\max}} \sum_{m=-l}^{l} c_{lm} \, \mathrm{Re}\left[Y_{lm}(\theta,\phi)\right], \tag{1}$$

where $l_{\max}$ is the moments order, $\mathrm{Re}[Y_{lm}(\theta,\phi)]$ is the real part of the spherical harmonic functions, and c_{lm} are the associated coefficients. The similarity of two objects can be measured by the Euclidean distance of the vectors of coefficients c_{lm} of the two pockets. Since spherical harmonics are not rotationally invariant, in principle pose normalization of object is needed.

Kahraman and colleagues [8] defined the Interact Cleft Model for ligand binding sites by employing spherical harmonics as follows. For a ligand binding pocket, the volume of a ligand binding pocket is defined by SURFNET [11] spheres within 0.3 Å to protein atoms interacting with the bound ligand. The software HBPLUS [36] is used to determine such atoms. To achieve rotation invariance, a coordinate system is defined at the center of gravity of the pocket volume. The moment of inertia tensor for the pocket volume V is a matrix of components

$$I_{i,j} = \int_V (r^2 \delta_{i,j} - r_i r_j) dV, \tag{2}$$

where $i,j = x, y, z$ and r is the vector from the center of gravity to a point in the volume. The pocket is rotated so that its moment of inertia tensor is diagonal with maximal values in x followed by y then followed by z. The outermost surface of these spheres is then expanded as a spherical harmonics series $f(\theta,\phi)$ where the order $l_{\max}$ is set to 16.

3D Zernike Descriptors

We have applied the 3D Zernike descriptors (3DZD), which also give a series expansion of a 3D function. It allows a compact and rotationally invariant representation of a 3D object. Mathematical foundation of the 3DZD was laid by Canterakis [37], then Novotni and Klein [38] have applied it to 3D shape retrieval. Here we provide a brief mathematical derivation of the 3DZD. Refer to the two papers [37, 38] for more technical details.

The surface of a ligand binding pocket is extracted using the Connolly surface [39] of protein heavy atoms within 8 Å to any heavy atom of the bound ligand, then placed on a 3D grid. To represent a surface shape, each grid cell (voxel) is assigned the value of 1 if it contains the protein surface and the value of 0 otherwise. For representing other physicochemical properties, such as the electrostatic potentials and hydrophobicity values, values are also assigned only to the surface voxels. The resulting voxels-values mapping is considered as a 3D function, $f(x)$, which is expanded into a series in terms of Zernike-Canterakis basis [38] defined by the following collection of functions:

$$Z_{nl}^{m}(r, \vartheta, \varphi) = R_{nl}(r) Y_l^m(\vartheta, \varphi), \tag{3}$$

with $-l < m < l$, $0 \leq l \leq n$, and $(n - l)$ even. The function $Y_l^m(\vartheta, \varphi)$ are the spherical harmonics [40] and $R_{nl}(r)$ are radial functions defined by Canterakis, constructed so that $Z_{nl}^m(r, \vartheta, \varphi)$ can be converted to polynomials, $Z_{nl}^m(\mathbf{x})$, in Cartesian coordinates. Now 3D Zernike moments of $f(\mathbf{x})$ are defined as the coefficients of the expansion in this orthonormal basis, i.e. by the formula

$$\Omega_{nl}^{m} = \frac{3}{4\pi} \int_{|\mathbf{x}| \leq 1} f(\mathbf{x}) \bar{Z}_{nl}^{m}(\mathbf{x}) d\mathbf{x}. \tag{4}$$

Finally, the rotational invariance is obtained by defining the 3DZD series, F_{nl}, as norms of vectors Ω_{nl}:

$$F_{nl} = \sqrt{\sum_{m=-l}^{m=l} (\Omega_{nl}^{m})^2} \tag{5}$$

The parameter n is called the order of 3DZD, which determines the resolution of the descriptor. As stated above, n defines the range of l and a 3DZD is a series of invariants (Eq. (5)) for each pair of n and l, where n ranges from 0 to the specified order. We use order $n = 20$ in the pocket comparison. which was shown to provide sufficient accuracy in a previous works of shape comparison [38]. The order $n = 20$ yields 121 invariant numbers (Eq. (5)).

As for the surface electrostatic potentials, 3DZD is computed separately for the pattern of positive values and for the negative values and later concatenated into a single vector. The separation of negative patterns and positive patterns is done by creating an input grid only of negative values and only of positive values and calculating 3DZD for each grid separately [46].

The obtained 3DZD is normalized to a unit vector by dividing each moment by the norm of the whole descriptor. This normalization is found to reduce dependency of 3DZD on the number of voxels used to represent a protein [46]. An example of the invariant values of the 3DZD of a ligand binding pocket (Fig. 1a) is shown in Fig. 1b.

In our previous works, we have applied the 3DZD successfully to various protein and ligand structure analyses [41–43], including rapid protein global shape analysis (http://kiharalab.org/3d-surfer) [44, 45], quantitative comparison for protein surface physicochemical property [46], small ligand molecule comparison [47], protein–protein docking prediction [48], and comparison of low-resolution electron density maps [49].

2D Pocket Model with Pseudo-Zernike Moments

We have also developed a new computational pocket model using two dimensional moments [50]. The key aspect of this method is the projection on a 2D plane of a spherical panoramic picture computed from the center of the binding pocket. The 3D to 2D dimensional reduction relies on the finding that pockets can be quite reliably pre-aligned using their opening.

Here, the shape of a pocket is extracted using the same procedure as 3DZD. A 3D Cartesian coordinate system $(\vec{x}, \vec{y}, \vec{z})$ is defined relative to a binding pocket, following the representation in Fig. 1c. The origin of the coordinate system is the center of gravity of the binding pocket, provided the latter is not inside the protein volume; otherwise, the origin is defined as any of the closest points outside. The *opening* of a binding pocket is the set of rays starting at the center of gravity which do not intersect the volume of the pocket. The unit vector of the x-axis is defined

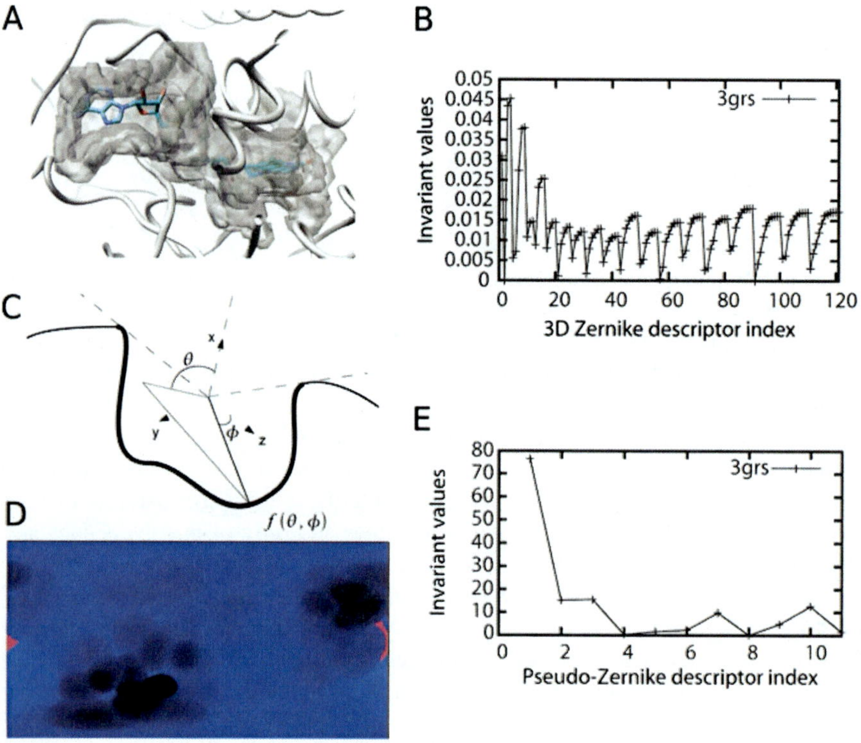

Fig. 1 Examples of the binding pocket representation with the 3DZD and 2D pseudo-Zernike descriptors. **a**, flavin adenine dinucleotide (FAD) binding pocket in PDB entry, 3grs. Pocket surface region within 8.5 Å to the ligand is shown. **b**, the 3DZD of the pocket. **c**, the coordinate system for projecting the pocket to the 2D map. **d**, the projected pocket. The distance from the center of the pocket to the pocket surface is represented in a color code from *blue* (closer) to *black* (more distant). *Pink* region shows aperture of the pocket. The x-axis is for θ and the y-axis shows ϕ. **e**, the pseudo-Zernike descriptor of the pocket

as a collinear vector to the average vector of the pocket opening. In cases where the opening is empty, the x-axis is arbitrarily defined. The remaining two axes, $\vec{y}$ and $\vec{z}$, are defined arbitrarily such that the basis $(\vec{x}, \vec{y}, \vec{z})$ is orthogonal. Optionally, an additional pre-alignment step can be applied. The z-axis is rotated such that its principal moment of inertia is maximized over all possible directions on the plane orthogonal to the x-axis. Simulations showed that this pre-alignment step is not necessary when using rotationally invariant descriptors, such as the p-Z moments used here (See fig. 3 in Chikhi et al. [50]).

A spherical function $f(\theta, \phi)$ is defined for the outermost surface of the binding pocket. Practical computation of $f(\theta, \phi)$ can be done using ray-casting. Rays are shot in every direction $(\theta,\ \phi)$ from the center of gravity of the pocket to the pocket surface, and a value for the direction (e.g. distance from the center) is taken from the surface point which first intersects the ray. If a ray never intersects the protein surface, a null value is assigned to the ray direction. Note that the function $f(\theta, \phi)$

can also describe any surface property, such as geometry or electrostatic potential [50]. Then, the function f is mapped to a 2D plane in order to be described using two dimensional moments.

Since no 3D to 2D projection preserves area, shape, and distance properties altogether, there is no solution to perfectly map function f to a 2D plane without distortion. It was found that a simple distance preserving projection, the *plate-carrée* projection, is sufficient for the purpose of pocket matching. By mapping $f(\theta, \phi)$ to a planar image using this projection, the bottom of the pocket $(\theta = \pi)$ is projected to the center of the image and the opening of the pocket $\left(\theta = 0,\ \phi = \frac{\pi}{2}\right)$, is projected to the sides (Fig. 1d). The resolution of the picture is 360 × 180, as coordinates are mapped to integer values of (θ, ϕ), resulting in 64,800 rays shot from the pocket center of gravity to each (θ, ϕ) direction. Rotations around the x-axis of the pocket correspond to rotations around the center of the image. However, since the $\vec{z}$ axis is arbitrarily defined under the 2D pocket model, the ϕ coordinate has no reference. Conveniently, the p-Z moments are mathematically invariant around the center of the image. And practically, as we will see in the results, these moments can robustly describe a projected pocket despite the lack of reference for the $\vec{z}$ axis.

2D Pseudo-Zernike Moments

The p-Z moments [51] have been employed for describing an image shape in pattern recognition applications, and they are shown to be less sensitive to noise than conventional 2D Zernike moments [52, 53]. The p-Z moments use a set of complete and orthogonal basis functions defined over the unit circle $(x^2+y^2 \leq 1)$ as follows:

$$V_{n,m}(x, y) = e^{im\theta} R_{nm}(r) = e^{im\theta} \sum_{s=0}^{n-|m|} \frac{(-1)^s (2n+1-s)!\rho^{(n-s)}}{s!(n+|m|+1-s)!(n-|m|-s)!} \tag{6}$$

where $\rho = \sqrt{x^2 + y^2}$, $\theta = \tan^{-1}(y/x)$, and $n \geq 0$, $|m| \leq n$. Using the polynomials, the p-Z moments of the order n and the repetition m for a 2D image $f(x, y)$ are defined as:

$$A_{n,m} = \frac{n+1}{\pi} \int_{x^2+y^2 \leq 1} f(x, y) V^*_{n,m}(x, y) dx dy \tag{7}$$

The asterisk $(*)$ denotes the complex conjugate. In this study, the order of moments $n = 4$ is used for most of the computation. An example of the pseudo Zernike values is shown in Fig. 1e.

Theoretical Comparison of Moments in Shape Descriptors

We briefly discuss differences between these three moments from a mathematical point of view. Obviously, the 2D p-Z moments describe a 2D function, hence a pocket 3D structure needs to be initially projected to a 2D image. On the other hand, the spherical harmonics and the 3DZD represent 3D objects, thus they can directly handle the 3D coordinates of a pocket. The coordinate system defined in Fig. 1c makes the p-Z moments rotationally invariant around the center of the image. However, a disadvantage arises from distortions caused by the projection, although in the benchmark study the 2D pocket model showed comparable performance with the 3DZD [50].

Comparing the spherical harmonics and the 3DZD, the 3DZD has a radial function $R_{nl}(r)$, (Eq. (3)), while the spherical harmonics do not. This difference results in an advantage of the 3DZD over the spherical harmonics in describing 3D pockets which intersect with a ray of a certain direction (θ, ϕ) for multiple times at different distance, r. The 3DZD can naturally handle such shapes (non star-like shapes), because it can assign a different value at each r. On the other hand, naïve use of the spherical harmonics can only take one value per direction. Therefore, usually only the outermost (or innermost) surface of an object is described by the spherical harmonics. To describe non star-like shapes, Funkhouser et al. used multiple concentric spherical shells [35].

Another advantage of the 3DZD over the spherical harmonics is that it is invariant to rotation of the object (Eq. (5)), while the direct use of the spherical harmonics is not. Thus, the 3DZD does not need pose normalization (pre-alignment) of objects for comparing and computing the similarity. This is advantageous in constructing a database of pockets since the 3DZD of pockets can be pre-computed and stored. The spherical harmonics can obtain the rotational invariance by the aforementioned use of concentric spherical shells. However, there are several drawbacks to this approach. As radial consistency of objects is not preserved (shells can be rotated with no impact on the descriptors), a certain amount of shape information is lost. Furthermore, because of polar sampling, spherical harmonics descriptors are not practically robust to rotation [54]. Also the 3DZD is more compact than the spherical harmonics by one order of magnitude [55], because adjacent spherical shells in the spherical harmonics descriptors are highly correlated.

Overall, the 3DZD is an improvement over spherical harmonics descriptors. The p-Z moments have not yet been formally studied for the description of 3D objects using a single projection, since this approach seems to be relatively specific to the description of binding pockets.

Binding Ligand Prediction Using the Pocket Descriptors

Using the 3DZD and the p-Z descriptors discussed above, we built a binding ligand prediction method for protein structures named Pocket-Surfer. Since both representations describe a pocket as a vector of coefficients, similarity of two pockets can be quantified by computing the Euclidean distance of their descriptors (vectors).

The 3DZD and the p-Z descriptors contain shape information of the pockets. However, the information of the size of the pockets is lost since the pockets are first fit to a unit sphere (for 3DZD) or a unit circle (for the pseudo Zernike descriptors) in the process of computing the moments. Therefore, we add the size information of a pocket into the vector as follows:

$$\mathrm{Descriptor}(P) = (w \cdot S_{\mathrm{P}}, A_1^{\mathrm{P}}, A_2^{\mathrm{P}}, \ldots, A_k^{\mathrm{P}}, \ldots, A_N^{\mathrm{P}}), \tag{8}$$

where S_{P} is the size of the pocket P weighted by a factor w, ${A_k}^{\mathrm{P}}$ is the kth value of the pocket moments (either 2D or 3D), and N is the total number of values of the moments. As the pocket size Sp, we used the average distance from the center of gravity G of the pocket to the pocket surface.

Equipped with the pocket descriptors and a similarity metric (i.e. Euclidean distance), pockets in a database are sorted according to the distance to a query pocket. Using the k nearest pockets to the query, the binding ligand for the query pocket (pocket type) is predicted using a k-nearest neighbors (k-NN) classifier as follows. The scoring function for a binding pocket of a ligand type F is defined as

$$\mathrm{Pocket\_score}(F) = \sum_{i=1}^{k} \left(\delta_{l(i),F} \log \left(\frac{n}{i} \right) \right) \cdot \frac{\sum_{i=1}^{k} \delta_{l(i),F}}{\sum_{i=1}^{n} \delta_{l(i),F}}, \tag{9}$$

where $l(i)$ is a function that returns the ligand type (AMP, FAD, etc.) of the ith closest pocket to the query, n is the total number of pockets in the database, and the indicator function $\delta_{X,Y}$ equals to 1 if X is of type Y, and is null otherwise. The role of the first term in this scoring function is to assign higher scores to pockets with higher ranks, within the top k results. The second term is a normalization factor of the score by considering the number of pockets of the type F in the database. The numerator is the number of pockets of the type F retrieved within top k and the denominator is the number of all the pockets of the type F in the database. Using Eq. (9), the score is computed for all the pocket types and they are sorted by the score.

To summarize the Pocket-Surfer procedure, a flow-chart is presented in Fig. 2. Given a query protein structure, ligand binding pockets on the protein surface are detected (i.e. predicted) by geometrical criteria using a method like LigSite [18] or VisGrid [7]. In the benchmark study, known ligand binding pockets are used (i.e. pockets are extracted as the surface regions which are in contact to the binding ligand molecule) to test the pocket comparison and the ranking ability of the procedure. Then, the Connolly surface [39] of pockets is constructed. Next, the pocket descriptor (Eq. (8)), either the 3DZD or the p-Z descriptor, is computed for the query pocket. Finally, the distance from the query to all pocket descriptors pre-computed and stored in a database is computed, and the ligand type for the query is predicted using Eq. (9). Pocket-Surfer has been implemented as a web server at http://kiharalab.org/pocket-surfer/. Currently the pocket database to be searched holds only a limited number of pockets used in the benchmark study of the published paper [50]. Expansion of the database is under way to make the server bear practical use of binding ligand prediction.

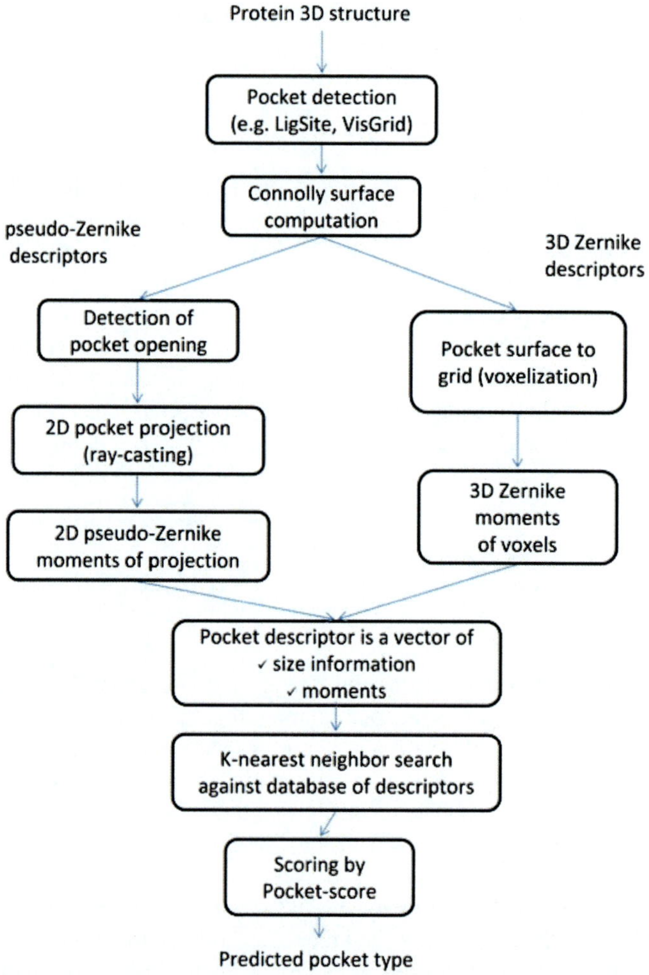

Fig. 2 Schematic flow chart of the binding ligand type prediction procedure using 2D pseudo-Zernike or 3D Zernike descriptors

Benchmark Results of Binding Ligand Prediction

In a recent paper [50], we benchmarked the performance of binding ligand with the p-Z, spherical harmonics, and 3DZD pocket models on two datasets. We briefly summarize the results in this section. The first dataset (the Kahraman set, named after the author [8] who compiled this dataset) consists of 100 evolutionary-distant proteins binding one of nine different ligand molecules (see the legends of Fig. 3). This dataset is used to train parameters and compare the performance of 3DZD and p-Z with spherical harmonics. The second dataset (the Huang set [18]) is independent from the first one in terms of proteins and ligand types. It contains 175 proteins,

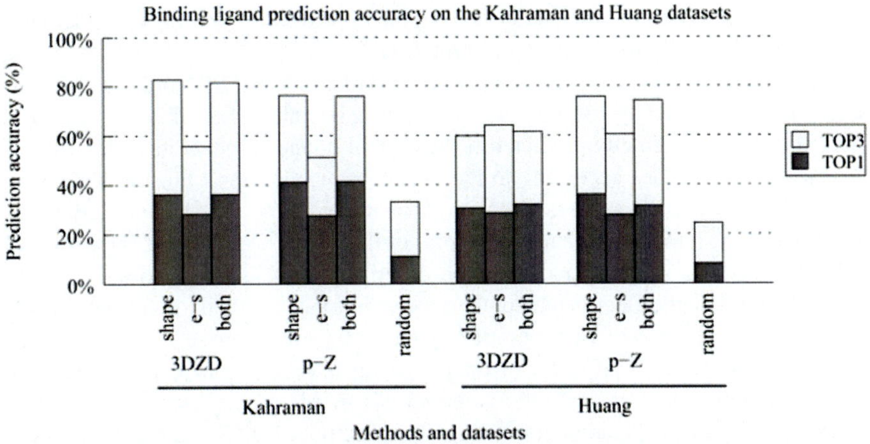

Fig. 3 The average binding ligand prediction accuracy for the Kahraman and the Huang dataset with the 3DZD and the 2D p-Z descriptors. The Kahraman dataset contains 100 proteins, each of which binds one of the following nine different ligands: adenosine-monophosphate (AMP) (9), adenosine-5′-triphosphate (ATP) (14), flavin adenine dinucleotide (FAD) (10), flavin mononucleotide (FMN) (6), glucose (GLC) (5), heme (HEM) (16), nicotinamide adenine dinucleotide (NAD) (15), phosphate (PO4) (20), and steroid (STR) (5). In the second parentheses the number of entries is shown. The Huang dataset consists of 175 proteins, which bind either of twelve ligand molecules: adenosine (ADN) (11), biotin (BTN) (12), fructose 6-phosphate (F6P) (12), fucose (FUC) (14), galactose (GAL) (36), guanine (GUN) (12), mannose (MAN) (18), O1-methyl mannose (MMA) (10), 2-phenylimidazole (PIM) (5), palmitic acid (PLM) (26), retinol (RTL) (5), and 2′-deoxyuridine 5-monophosphate (UMP) (13). The average Top-1 and Top-3 success rates of binding ligand prediction for all ligand type are reported. Results are shown for the shape descriptors, the electrostatics descriptors and both combined. For the combination of the shape and the electrostatic (e-s) potential descriptors, the average Euclidean distance by the pocket shape and the electrostatic potential descriptors are used

each of which binds one of twelve ligand molecules. Based on the performance on the Kahraman dataset, the descriptors parameters for p-Z (resp. 3DZD) descriptors were set to $w = 4.5$ (0.04) and $n = 4$ (20). The number of neighbors used in the k-NN classifier was set to $k = 24$. Using the two datasets, performance of the binding ligand prediction was examined for the pocket shape descriptors that combine the pocket and size shape information (Eq. (8)) and also for the electrostatic potential descriptors. To compute the surface electrostatic potential descriptors, the electrostatic potential on the protein surface is mapped on the 2D image for the p-Z while on the voxels of the 3D grid for the 3DZD [50].

First, we compared the performance of the shape descriptor of 3DZD and the p-Z with that of the spherical harmonics on the Kahraman dataset. The value for the spherical harmonics was taken from the paper by Kahraman et al. [8]. Both 3DZD and the p-Z performed slightly better than the spherical harmonics in terms of the Area Under the Curve (AUC) values of the receiver operating characteristic (ROC) curve [56]. The AUC values of the 3DZD, the p-Z and the spherical harmonics were 0.81, 0.79 and 0.77, respectively. We have also examined the performance of the p-Z with pocket pre-alignment (Eq. (2)) but the improvement was only 0.75%. Thus the p-Z is practically robust enough to rotation.

Figure 3 shows the Top-1 and Top-3 success rate of the 3DZD and the p-Z averaged over all ligand types. For the Top-3 success rate, a ligand for a pocket is considered to be correctly predicted if the correct ligand is included within the top 3 scoring ligand types according to the *Pocket_score* (Eq. (9)). For the Kahraman dataset, the best Top-1 success rate was achieved by the pocket shape descriptor of the p-Z (41.2%), while the shape descriptor of the 3DZD was the best for the Top-3 success rate (82.7%). The pocket shape descriptors (left bars) performed significantly better than the electrostatic potential descriptors (middle bars) for both 3DZD and the p-Z. Because of this, combining them did not improve the performance (81.5% by the 3DZD and 75.9% by the p-Z). This observation is in agreement with a previous report that electrostatic potential is variable within families of binding pockets [57]. For the Huang dataset (Fig. 3, right), both 3DZD and the p-Z showed lower success rate by the shape descriptor as compared with the Kahraman dataset. On the other hand, the electrostatic potential descriptors of both 3DZD and the p-Z showed a higher success rate on this dataset relative to the Kahraman set. As a result, for the 3DZD, the combination of the shape and the electrostatic potential descriptors showed improvement over the shape descriptor. The best Top-1 (35.9%) and the Top-3 success rate (75.6%) were achieved by the p-Z shape descriptor.

Figure 4 shows the Top-1 and Top-3 accuracy for individual pocket types on both datasets. PO_4 was predicted very well because it is distinguished by its smaller size from the other ligands. Some ligands, such as FMN, were poorly predicted. FMN is the most flexible ligand among the three smallest ligands in the dataset (GLC, FMN, and STR) with an average RMSD of 1.08 Å. The success rate largely differs from ligand to ligand and the trends are consistent for the 3DZD and the p-Z. This implies that the difference in the performance for each ligand is attributed not to the characteristics of the approaches but to the actual similarity of pockets of particular ligand types.

Performance with Ligand-Free Pockets and Predicted Pockets

In practical situations of binding ligand prediction, one of the two cases may arise: (1) the binding pocket in a query structure is known, but it is in a ligand-free conformation or (2) a binding pocket is unknown, hence it needs to be predicted. The challenge for the first case is the difference in shapes of ligand-free and ligand-bound binding pockets. To assess this difference, we searched the Huang dataset with ligand-free pockets and determined pocket retrieval accuracy with the p-Z and 3DZD pocket shape descriptors. For the p-Z (resp. 3DZD) descriptors, in 11 (resp. 7) out of 12 ligands the ligand-free pockets are retrieved with a similar or often better AUC value than the closest ligand-bound pockets [50]. The RMSD value of the ligand-bound and ligand-free proteins ranges from 0.19 to 2.48 Å with an average value of 0.86 Å. This is consistent with a recent study [58] that reports the average

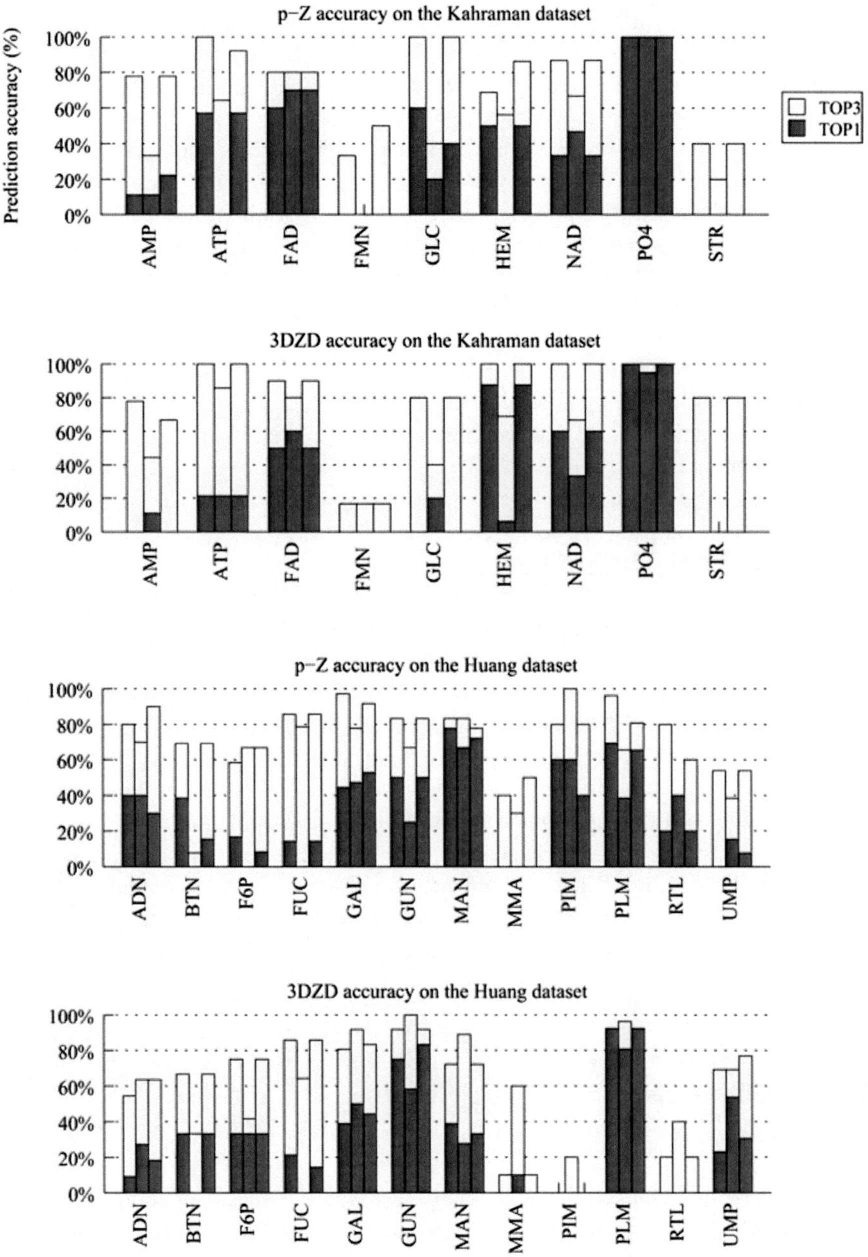

Fig. 4 The Top-1 and Top-3 success rates of binding ligand prediction for individual ligand types in the Kahraman and the Huang dataset. Results are shown for the shape descriptors, the electrostatics descriptors and both combined

RMSD between ligand-bound and ligand-free form is 0.74 Å. Our results indicate that the p-Z and 3DZD descriptors are robust enough with respect to the actual range of conformational difference between ligand-bound and ligand-free forms of binding pockets.

For simulating the second case, the situation where binding pockets are not known beforehand, we examined how well the p-Z and 3DZD descriptors perform with predicted pockets. We predicted pockets by running the LIGSITE [18] program for each protein in the Kahraman dataset and queried against the dataset (thus, dataset of known pockets) [50]. This resulted in a significant deterioration of the performance: the AUC value of the p-Z and the 3DZD dropped from 0.79 to 0.52 (p-Z) and from 0.88 to 0.53 (3DZD). The Top-3 success rates of the 3DZD dropped from 82.7 to 38.9% while for the p-Z it dropped from 77.3 to 41.0%. We note that inaccuracies in binding pocket prediction largely accounts for the unsuccessful retrieval of predicted pockets, hence more accurate prediction methods [17, 20] are likely to improve results.

Computational Time of Pocket-Surfer

We estimated the running times for computing the p-Z and the 3DZD descriptors and searching against a database of binding pockets. For computing the p-Z descriptor of a pocket, pocket projection and p-Z moments computation steps typically take about 10 s [50]. Surface voxelization and 3DZD descriptors are computed in around 40 s. Searching a query descriptor against a database of 100 descriptors takes around 12 milliseconds for the p-Z descriptors and 20 milliseconds for the 3DZD due to different moment orders [50]. By extrapolation to a PDB-scale database, searching a query pocket can be done in about a few seconds with the p-Z and 3DZD. This is significantly faster than the other methods of similar purpose. Hence, the p-Z and the 3DZD pocket descriptors realize real-time pocket database searches, where users can retrieve a search result instantly sitting in front of a computer.

Pocket Comparison with Local Surface 3D Zernike Descriptors

At the last of this chapter, we will briefly describe our recent ongoing development of binding ligand prediction method which considers similarity of local surface regions in pockets. Shape of pockets for the same ligand molecule can significantly vary due to several reasons, including the flexibility of ligand molecules and binding of solvent molecules [8]. Therefore, pockets which bind the same ligand may be better detected by scoring the local similarity of pockets. Comparing local regions of pockets can be done by segmenting the pockets into local patches and comparing the patches separately. The outline of the algorithm of local pocket surface comparison

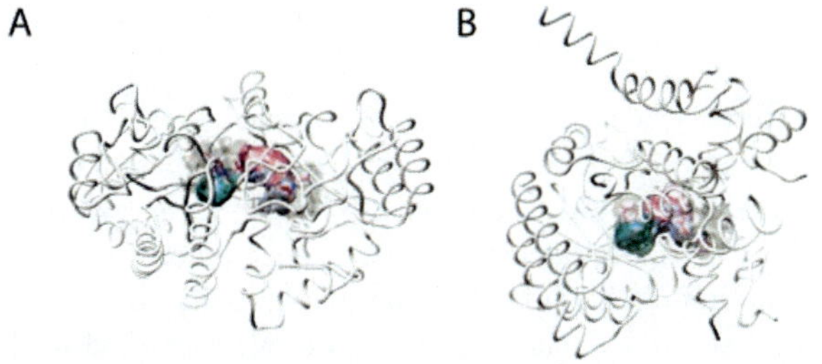

Fig. 5 Local binding site matching of FAD binding proteins. Three local binding patches from **a**, protein 1e8g; and **b**, protein 1k87. Each patch color indicates equivalent position relative to the ligand molecule

method works as follows [59]: First, seed points are evenly distributed on the pocket surface. Then, the shape of surface patch region which is within a sphere centering at each seed point is encoded by the 3DZD. Thus, a whole pocket shape is described as a set of 3DZDs each of which encodes local patch shape. For example, ATP binding pockets are represented as 29.5 overlapping local surface patches on average, while NAD binding pockets have on average 36.8 surface patches of a 5 Å radius. The surface electrostatic potentials and other properties can be also computed in the same manner. To compute the similarity of two pockets, we seek for a set of pairs of surface patches, each taken from the two pockets, which maximizes the overall score for the set. The score will consider the similarity of patches in each pair, the relative position of the patches in each pocket, and the size of the pocket.

Figure 5 shows an example of a pair of FAD binding pockets for which the local pocket surface method yields a better result. Using the global 3DZD, querying the FAD binding pocket of protein 1e8g against the Kahraman dataset retrieved the first FAD binding pocket at the 7th rank (1jqi). In contrast, the local surface comparison method retrieved a FAD binding pocket, 1k87, at the 2nd rank. It is shown in Fig. 5 that the overall pocket shape of 1e8g and 1k87 is quite different because FAD molecule is in a stretched form in 1e8g but bent in 1k87. Despite of the different overall shape, the local patch comparison method could identify the similarity between the two by detecting similarity of the patch pairs shown in the same color.

We have further applied the local protein surface representation by the 3DZD for characterization and classification of protein surface properties [60]. Here, the aim is to annotate entire protein surfaces but not only to compare pocket regions. We extracted local surface patches, which was defined as the surface within a sphere of a 6 Å radius, from 609 representative proteins. This yielded in total of 118,009 patches. A patch was characterized by two features, the shape and the electrostatic potential, and both are described by the 3DZD. We classified the patches using the emergent self-organizing map (ESOM) [61]. The classification resulted in 30–50 clusters of local surfaces of different characteristics. These clusters can be used as surface "alphabet", with which protein surface can be labeled and classified. For example, surface regions of certain biological function, e.g. DNA binding or protein-binding, can be described as a set of the surface alphabets.

Summary

In this chapter, we described moment-based approaches for representing shape of protein surfaces, which are applied for binding ligand prediction by comparing binding pockets. 2D and 3D Zernike moments are able to capture various local protein surface properties of binding pockets. While several other methods exist for binding sites representation and comparison, the moments-based methods benefit from fast computational speed for database search, as well as good retrieval accuracy. However, structure-based function prediction methods are in general vulnerable to structural variability of proteins. To accommodate this problem, we are developing the local pocket surface comparison method where two pockets are compared in terms of matching pairs of local sites.

Comparison of the tertiary structure of proteins, both global and local, is more complicated than comparison of one dimensional protein sequences. Therefore, there have not been as many structure-based methods developed as the sequence-based methods. The p-Z and the 3DZD we introduced in this chapter have potential to change this situation, as they provide very convenient, compact and rotation invariant representation of protein global and local surfaces.

Acknowledgements This work was supported by the National Institute of General Medical Sciences of the National Institutes of Health (R01 GM075004). DK also acknowledges funding from the National Science Foundation (DMS800568, IIS0915801, EF0850009).

References

1. Watson, J.D., Laskowski, R.A., Thornton, J.M. Predicting protein function from sequence and structural data. Curr. Opin. Struct. Biol. **15**: 275–284 (2005).
2. Chandonia, J.M., Brenner, S.E. The impact of structural genomics: expectations and outcomes. Science **311**: 347 (2006).
3. Hawkins, T., Kihara, D. Function prediction of uncharacterized proteins. J. Bioinform. Comput. Biol. **5**: 1–30 (2007).
4. Berman, H.M., et al. The protein data bank. Nucleic Acids Res. **28**: 235–242 (2000).
5. Minai, R., Matsuo, Y., Onuki, H., Hirota, H. Method for comparing the structures of protein ligand-binding sites and application for predicting protein-drug interactions. Proteins: Struct. Funct. Bioinform. **72**: 367–381 (2008).
6. Orengo, C.A., Jones, D.T., Thornton, J.M. Protein superfamilies and domain superfolds. Nature **372**: 631–634 (1994).
7. Li, B., et al. Characterization of local geometry of protein surfaces with the visibility criterion. Proteins **71**: 670–683 (2007).
8. Kahraman, A., Morris, R.J., Laskowski, R.A., Thornton, J.M. Shape variation in protein binding pockets and their ligands. J. Mol. Biol. **368**: 283–301 (2007).
9. Liang, J., Edelsbrunner, H., Woodward, C. Anatomy of protein pockets and cavities: measurement of binding site geometry and implications for ligand design. Protein Sci. **7**: 1884–1897 (1998).
10. Laskowski, R.A., Luscombe, N.M., Swindells, M.B., Thornton, J.M. Protein clefts in molecular recognition and function. Protein Sci. **5**: 2438–2452 (1996).
11. Laskowski, R.A. SURFNET: a program for visualizing molecular surfaces, cavities, and intermolecular interactions. J. Mol. Graph. **13**: 323–328 (1995).

12. Levitt, D.G., Banaszak, L.J. POCKET: a computer graphics method for identifying and displaying protein cavities and their surrounding amino acids. J. Mol. Graph. **10**: 229–234 (1992).
13. Kawabata, T., Go, N. Detection of pockets on protein surfaces using small and large probe spheres to find putative ligand binding sites. Proteins **68**: 516–529 (2007).
14. Weisel, M., Proschak, E., Schneider, G. PocketPicker: analysis of ligand binding-sites with shape descriptors. Chem. Cent. J. **1**: 7 (2007).
15. Hendlich, M., Rippmann, F., Barnickel, G. LIGSITE: automatic and efficient detection of potential small molecule-binding sites in proteins. J. Mol. Graph. Model. **15**: 359–363, 389 (1997).
16. Kim, D., et al. Pocket extraction on proteins via the Voronoi diagram of spheres. J. Mol. Graph. Model. **26**: 1104–1112 (2008).
17. Tseng, Y.Y., Dundas, J., Liang, J. Predicting protein function and binding profile via matching of local evolutionary and geometric surface patterns. J. Mol. Biol. **387**: 451–464 (2009).
18. Huang, B., Schroeder, M. LIGSITEcsc: predicting ligand binding sites using the Connolly surface and degree of conservation. BMC Struct. Biol. **6**: 19 (2006).
19. Ota, M., Kinoshita, K., Nishikawa, K. Prediction of catalytic residues in enzymes based on known tertiary structure, stability profile, and sequence conservation. J. Mol. Biol. **327**: 1053–1064 (2003).
20. Capra, J.A., Laskowski, R.A., Thornton, J.M., Singh, M., Funkhouser, T.A. Predicting protein ligand binding sites by combining evolutionary sequence conservation and 3D structure. PLoS Comput. Biol. **5**: e1000585 (2009).
21. Laurie, A.T., Jackson, R.M. Q-SiteFinder: an energy-based method for the prediction of protein-ligand binding sites. Bioinformatics **21**: 1908–1916 (2005).
22. Elcock, A.H. Prediction of functionally important residues based solely on the computed energetics of protein structure. J. Mol. Biol. **312**: 885–896 (2001).
23. An, J., Totrov, M., Abagyan, R. Pocketome via comprehensive identification and classification of ligand binding envelopes. Mol. Cell Proteomic. **4**: 752–761 (2005).
24. Porter, C.T., Bartlett, G.J., Thornton, J.M. The catalytic site atlas: a resource of catalytic sites and residues identified in enzymes using structural data. Nucleic Acids Res. **32**: D129–D133 (2004).
25. Arakaki, A.K., Zhang, Y., Skolnick, J. Large-scale assessment of the utility of low-resolution protein structures for biochemical function assignment. Bioinformatics **20**: 1087–1096 (2004).
26. Ferre, F., Ausiello, G., Zanzoni, A., Helmer-Citterich, M. SURFACE: a database of protein surface regions for functional annotation. Nucleic Acids Res. **32**: D240–D244 (2004).
27. Gold, N.D., Jackson, R.M. Fold independent structural comparisons of protein-ligand binding sites for exploring functional relationships. J. Mol. Biol. **355**: 1112–1124 (2006).
28. Kalidas, Y., Chandra, N. PocketMatch: a new algorithm to compare binding sites in protein structures. BMC Bioinformatics **9**: 543 (2008).
29. Xiong, B., et al. BSSF: a fingerprint based ultrafast binding site similarity search and function analysis server. BMC Bioinformatics **11**: 47 (2010).
30. Binkowski, T.A., Joachimiak, A. Protein functional surfaces: global shape matching and local spatial alignments of ligand binding sites. BMC Struct. Biol. **8**: 45 (2008).
31. Schmitt, S., Kuhn, D., Klebe, G. A new method to detect related function among proteins independent of sequence and fold homology. J. Mol. Biol. **323**: 387–406 (2002).
32. Kinoshita, K., Nakamura, H. Identification of protein biochemical functions by similarity search using the molecular surface database eF-site. Protein Sci. **12**: 1589–1595 (2003).
33. Ramensky, V., Sobol, A., Zaitseva, N., Rubinov, A., Zosimov, V. A novel approach to local similarity of protein binding sites substantially improves computational drug design results. Proteins **69**: 349–357 (2007).

34. Bock, M.E., Garutti, C., Guerra, C. Cavity detection and matching for binding site recognition. Theor. Comput. Sci. **408**: 151–162 (2008).
35. Kazhdan, M., Funkhouser, T., Rusinkiewicz, S. Rotation invariant spherical harmonic representation of 3D shape descriptors. Proc. 2003 Eurographics/ACM SIGGRAPH Symp. Geometry Process. **43**: 156–164 (2003).
36. McDonald, I.K., Thornton, J.M. Satisfying hydrogen bonding potential in proteins. J. Mol. Biol. **238**: 777–793 (1994).
37. Canterakis, N. 3D Zernike moments and Zernike affine invariants for 3D image analysis and recognition. Proceedings of the 11th Scandinavian Conference on Image Analysis, Kanger-lussuaq, Greenland, pp. 85–93 (1999).
38. Novotni, M., Klein, R. 3D Zernike descriptors for content based shape retrieval. ACM Symposium on Solid and Physical Modeling, Proceedings of the 8th ACM Symposium on Solid Modeling and Applications, pp. 216–225 (2003).
39. Connolly, M.L. Shape complementarity at the hemoglobin alpha 1 beta 1 subunit interface. Biopolymers **25**: 1229–1247 (1986).
40. Dym, H., McKean, H. *Fourier series and integrals*. New York, NY: Academic (1972).
41. Sael, L., Kihara, D. Protein surface representation and comparison: New approaches in structural proteomics. *Biological data mining*. Chen, J., Lonardi, S. (eds.). Kumar, V. (series ed.). Boca Raton, FL: Chapman & Hall/CRC Press, Chapter 3, pp. 89–109 (2009).
42. Venkatraman, V., Sael, L., Kihara, D. Potential for protein surface shape analysis using spherical harmonics and 3D Zernike descriptors. Cell Biochem. Biophys. **54**: 23–32 (2009).
43. Kihara, D., Sael, L., Chikhi, R. Molecular surface representation using 3D Zernike descriptors for protein shape comparison and docking. Curr. Protein Peptide Sci. (2011) (In Press).
44. La, D., et al. 3D-SURFER: software for high-throughput protein surface comparison and analysis. Bioinformatics **25**: 2843 (2009).
45. Sael, L., et al. Fast protein tertiary structure retrieval based on global surface shape similarity. Proteins: Struct. Funct. Bioinform. **72**: 1259–1273 (2008).
46. Sael, L., La, D., Li, B., Rustamov, R., Kihara, D. Rapid comparison of properties on protein surface. Proteins: Struct. Funct. Bioinform. **73**: 1–10 (2008).
47. Venkatraman, V., Chakravarthy, P.R., Kihara, D. Application of 3D Zernike descriptors to shape-based ligand similarity searching. J. Cheminform. **1**: 19 (2009).
48. Venkatraman, V., Yang, Y.D., Sael, L., Kihara, D. Protein-protein docking using region-based 3 D Zernike descriptors. BMC Bioinformatics **10**: 407 (2009).
49. Sael, L., Kihara, D. Protein surface representation for application to comparing low-resolution protein structure data. BMC Bioinformatics **11**:S2 (2010).
50. Chikhi, R., Sael, L., Kihara, D. Real-time ligand binding pocket database search using local surface descriptors. Proteins: Struct. Funct. Bioinform. **78**: 2007–2028 (2010).
51. Bhatia, A.B., Wolf, E. On the circle polynomials of Zernike and related orthogonal sets. Proc. Camb. Philos. Soc. **50**: 40–48 (1954).
52. Zernike, F. Beungungsthoerie des Schneiden-verfahrens und seiner verbesserten Form. Physica **1**: 689–701 (1934).
53. Teh, C.H., Chin, R.T. On image-analysis by the methods of moments. IEEE Trans. Pattern Anal. Mach. Intell. **10**: 496–513 (1988).
54. Laga, H., Takahashi, H., Nakajima, M. Spherical wavelet descriptors for content-based 3D model retrieval. IEEE International Conference on Shape Modeling and Applications (SMI2006), Sendai, Japan, pp. 75–85 (June 2006).
55. Novotni, M., Klein, R. Shape retrieval using 3D Zernike descriptors. Comput. Aided Des. **36**: 1047–1062 (2004).
56. Fawcett, T. An introduction to ROC analysis. Pattern Recognit. Lett. **27**: 861–874 (2006).
57. Kahraman, A., Morris, R.J., Laskowski, R.A., Favia, A.D., Thornton, J.M. On the diversity of physicochemical environments experienced by identical ligands in binding pockets of unrelated proteins. Proteins: Struct. Funct. Bioinform. **78**: 1120–1136 (2010).

58. Brylinski, M., Skolnick, J. What is the relationship between the global structures of apo and holo proteins? Proteins **70**: 363–377 (2008).
59. Sael, L., Kihara, D. Binding ligand prediction for proteins using partial matching of local surface patches. Int. J. Mol. Sci. **11**(12): 5009–5026 (2010).
60. Sael L., Kihara, D. Characterization and classification of local protein surfaces using self-organizing map. Int. J. Knowl. Discov. Bioinfo. **1**: 32–47 (2010).
61. Ultsch, A. Maps for the visualization of high-dimensional data spaces. Proceedings of the Workshop on Self Organizing Maps, Hibikino, Kitakyushu, Japan, pp. 225–230 (2003).

Computational Methods for Predicting DNA-Binding Sites at a Genomic Scale

Shandar Ahmad

Abstract High throughput analysis of protein-DNA interactions is required to make sense of omics-level data on protein and DNA sequences. Machine learning approaches have been successful in mapping binding information available from 3-dimensional structures of complexes to sequences. This has allowed us to develop methods to study interactions directly from sequence information. In this chapter, a motivation to such analysis is provided and most significant works in this direction has been reviewed. Primarily high-speed and coarse-grained approaches for de-novo predictions, usually derivable or predictable from sequence are discussed.

Introduction

Due to their obvious role in controlling gene expression and hence almost all natural or environment-induced biological processes in the living system, protein–DNA interactions have been studied extensively for a long time [1–17]. Although the genetic information is encoded in a simple four-letter (nucleic acid) language [18, 19] and expressed in terms of a 20-letter (amino acid) syntax, a cascade of events regulate the communication between information potential and functional performance [20]. Robust as well as flexible mechanisms of switching, regulating, repairing and catalyzing expression are rampant, creating fascinating and complex transcriptional machinery that influences all the essential, benign as well as disease-causing molecular events in the living systems. Although, one of the primary functions of gene expression is to make proteins, it is known for a long time that these very proteins, encoded by some genes control the expression of other genes [21–24]. Proteins (e.g. transcription factors) encoded by one gene control the expression of another seemingly unrelated gene on the genomic DNA, which in turn regulates another gene and so on, thus providing a very complex set of

S. Ahmad (✉)
National Institute of Biomedical Innovation, Ibaraki, Osaka, Japan
e-mail: shandar@nibio.go.jp

D. Kihara (ed.), *Protein Function Prediction for Omics Era*,
DOI 10.1007/978-94-007-0881-5_9,

connections between genes, which cannot be understood by simply reading the genomic DNA [25, 26]. Neither all proteins perform the function of gene regulation, nor do the regulatory proteins affect all genes in the genome. Hence, it becomes essential to first identify the proteins, which can interact with the genomic DNA and then, in order to pharmaceutically or otherwise intervene into these interactions, elucidate the exact mechanism of interaction, or at least identify the parts of the protein sequence that directly participate in interactions. This essentially defines the main motivation to develop methods for predicting DNA-binding proteins and binding sites in them. In principle, it is possible to perform in vitro as well as in vivo experiments to gain insights into these interactions and power of technological advancement is growing rapidly, enabling to generate direct and indirect information about protein–DNA interactions. However, accurate methods are prohibitively expensive and frustratingly time-consuming, and high throughput methods throw away an enormous amount of data, which cannot be examined manually. It is in this background that computational approaches to predict protein–DNA interactions at a large scale are needed. Since, most genomic information is encoded in sequences and high throughput experiments also yield small or large sequence-fragments, computational methods which can directly use protein and DNA sequences, even if they are less accurate, are critically important for the large-scale analysis of genome-scale behavior of interactions, as well as for providing the initial leads to perform more accurate experiments.

Computational methods to study protein–DNA interactions have focused on different aspects of the problem ranging from the prediction of transcription factor binding sites (TFBS) on DNA to the prediction of DNA-binding proteins and their binding sites. Methods to predict TFBS have been widely reported [27–30] and do not form the subject of this chapter, in which the main focus is on methods that can be employed to predict DNA-binding sites in proteins as well as finding proteins, which are likely to bind DNA. Obviously, alignment search through a data set of known DNA-binding proteins is an efficient way to find DNA-binding sites or proteins [31]. However, alignment-based methods work only if a similar protein–DNA interaction has been previously discovered and are also limited by the fact that some interactions may not be conserved in evolution [32–34]. Thus, we need methods which can detect binding sites and proteins when sequence identity with known DNA-binding proteins is low or absent. Due to their ability to handle large data sets and performance, primary focus in this chapter is on the statistical and machine learning approaches applied to sequence-based predictions or coarse-grained and bulk structural features such as secondary structure and solvent accessibility.

Data Sources

Any bioinformatics prediction method relies on previously annotated data sets. In this regards, large amounts of data have been compiled by researchers, which serves as the secondary but convenient source of information required for building prediction models and benchmarking their performance. The data sets relating to protein–DNA interactions may be grouped into three categories viz. structure databases, thermodynamic or stability databases and sequence-based databases. Brief outline of these databases is provided below.

Protein-DNA Complexes

PDB, NDB and PDBSum: The most reliable source of information on the mode of protein–DNA interactions is arguably the structure of a protein-DNA complex [35, 36]. Often protein and DNA structures in complex differ in their complex and unbound states [37–39]. Protein Data Bank is an ultimate source of all three-dimensional structures of proteins as well as DNA. Similar but more nucleic-acid focused information is available in the Nucleic Acids Database (NDB). Together these databases provide extensive information on the structures of proteins and DNA as a complex or in the unbound form and are well connected to other derived databases. A related database, PDBSum [40] provides additional information on contacting residues, and is well linked to graphical representations of geometrical features e.g. Nucplot [41]. Databases dedicated to protein-nucleic acid complexes have also been developed. BIPA and ProNuc are the most noteworthy among them [42–44].

Biological units and quaternary structure: Most protein–DNA interactions occur by way of multiple proteins acting together or multimeric unit [45] of the same protein recognizing sequence repeats or symmetric parts of DNA. In many cases, detailed cooperative mechanism of interaction is not known. Three-dimensional structures deposited in PDB and NDB, often give structure of the minimal asymmetric unit. Reconstructing a fully functional "biological unit" incorporating the quaternary structure of the complex is not straightforward [46]. PDB makes a preliminary attempt to provide information on biological unit, which is based on certain computational procedures and has certain shortcomings. More elaborate procedures and databases have been developed [47–49]. However, in many cases quaternary structure of protein-protein-DNA complexes remain unresolved and more research in this direction is required.

Thermodynamics and In Vitro Experiments

Quaternary and tertiary structures of protein-DNA complexes provide most crucial information on the DNA-binding modes of proteins and their binding sites. However, three-dimensional structure of complex or unbound DNA-binding protein is often unavailable. Moreover, even the structure of the complex falls short in revealing potential effects of mutation, specificity and stability of protein-DNA complex. Therefore in vitro experiments have to be performed by systematically or randomly mutating individual or several residues in a protein to ascertain their contribution to the stability and hence the biological function of these proteins. A large number of such experiments on various protein families have been reported by researchers. Information on these experiments has been compiled into a convenient database, called ProNIT [50]. At the time of writing this text, ProNIT contains more than 10,000 thermodynamic data entries, coming from 271 proteins, of which nearly 3,000 entries correspond to mutants and others correspond to stability of the Protein-DNA complex in its wild-type. Number of analyses on these data sets have been conducted, which have helped in unraveling the salient features and the big picture of protein–DNA interactions [51, 52].

Functionally Annotated Data Sets

Structural and thermodynamic information on protein–DNA interaction is neither fully available nor sufficient to determine certain informative contexts such as evolutionary footprint. Many times, knowledge of protein-binding sites on DNA (transcription factor binding sites) and evolutionary variation measured in so-called position-dependent weight matrices is handier or is the only source of information on these interactions. Information on binding sites of related families and consensus patterns form the main elements of databases such as TRANSFAC, JASPAR and COTRASIF [53–55]. Although, this chapter is primarily focused on proteins, the knowledge of target binding sites on DNA and their evolutionary patterns are a powerful source to fine-tune DNA-binding proteins, not extensively used yet, but an area in which the subject is likely to grow further.

Control Data Sets

All prediction methods need to be *trained* over a database of known interactions, and require large number of samples of interacting proteins (or binding sites) as well as non-interacting or control data sets from which a discriminating function need to be generated. In the case of binding sites, interface and non-interface residues provide automatic sets of positive and control data. However, problem is more complex in full length proteins. One can never be fully sure for a protein to be non-binding as there may be special conditions in and sequences to which a protein, not yet known to be DNA-binding may actually bind. Moreover, if we treat all proteins not reported to be DNA binding as our control data, we get a very large control data (entire sequence space) compared to a few hundred positive data. To overcome this problem, sampling of control proteins from an structure data sets such as SCOP and Protein Data Bank have been attempted, under the assumption that such data sets approximately sample the control protein sequence space [56].

Computational Techniques

Most bioinformatics prediction methods try to establish a relationship between a set of features which can be directly computed from sequence and binding behavior. Binding behavior (prediction) is usually a single Boolean or real number designating a residue or a protein to be (or its likelihood of being) binding or non-binding. Specific implementations of these methods to DNA-binding problem are described later in this chapter. It suffices here to state that most computational prediction methods attempt to develop a relationship between a set of known features to the binding status of a protein or residue. They differ from each other in the mathematical details of the model as well as they method used in optimizing adjustable parameters in the model. For example features may be simply modeled over a linear regression model or Naïve Bayesian classifier model, assuming mutual independence of features and making prediction by using a weighted summation of feature values. Or, the features may be combined using a polynomial, radial or a logically defined function as is done in a support vector machine. Similarly, features may be combined by a function of unknown, complex and automatically determined shape as is achieved by a neural network. Ability of a model to make accurate prediction however depends much more strongly on the selection of initial features than on the choice of the prediction model as such. For example SVM, neural network and other machine learning methods can potentially model complex decision surfaces well and also take care of non-additive nature of features, whereas linear regression, Bayesian classifier cannot directly account for non-additive nature of features, but are sometimes preferred due to their ability to identify direct contribution of individual features and convenience and efficiency of obtaining an optimal solution. In the following specific implementations to DNA-binding problem are discussed.

Methods for Predicting DNA-Binding Sites

Definition of a Binding Site

Ideally, a binding site must be defined in terms of its involvement in a biologically meaningful interaction. At a single residue level, it means that residues which contribute to the biological function or those which cause change of function when a mutation to replace them is effected, are considered binding. However, such a detailed annotation is hardly possible for all the residues of each complex. Direct structure-based annotation of binding sites is therefore required. From a structural perspective, this involves a number of amino acid residues directly in contact with DNA. Again, from a structural point of view, residues form various kinds of interactions such as hydrogen bond, salt-bridges, van der Waal's interaction and even water-mediated contacts with their target DNA. Identifying each of these interactions and annotating binding residues is not straightforward, especially as crystal structures from which this information is derived are often incomplete and error-prone. More lenient definitions of binding sites are therefore frequently used. These definitions are either based on a difference in the solvent accessibility of complex with DNA compared to protein without DNA. If the difference is more than a cutoff, residues are labeled as binding. An alternative to solvent accessibility change is the nearest contact distance between any of the DNA atoms to any atom of the residue. If the distance is less than a cutoff, residue is annotated as binding. Typical solvent accessibility cutoffs reported in literature are 0.1–1 Å and atomic contact distance cutoffs have varied between 3.5 and 6 Å. More relaxed criterion of distance cutoff are useful in balancing the negative and positive class data for prediction, but may not be as informative as closer contact base-amino acid residues. Solvent accessibility based definitions are intutively similar. A simple comparison can be made by noting that change in solvent accessibility for a water probe (radius 1.4 Å) will happen if any atom from a residue is closer than 2.8 Å from any DNA atom. Thus the two definitions roughly capture the same information. A benchmark between various distance cutoffs for binding site definition shows that even those models which were trained on data generated from other binding site definitions more accurately predicted binding sites defined at 3.5 Å cutoff [57].

Residue Propensities

The first insights into the nature of protein–DNA interactions and determination of initial candidates for further study can be performed by simply looking at the 20 amino acid types and obtaining their statistical preferences to occur in the interface. Such a preference can be described by taking the ratio of the observed relative number of residues of a given type to the relative number of residues of all type in the interface. For example if x% of all Arg are observed to be in the interface compared to y% of all residues being in the interface, the propensity score for Arg is assigned a real number x/y. Propensity score equal to one is an indication of no preference and lower than one is an indication of excluded residues. This score is good for identifying preferred residues in the interface as over-representation in interface is directly related to propensity. However, for residue exclusion the propensity goes as a reciprocal to over-representation (propensity is 0 for no residue of that kind in the interface and 1 means its presence is similar to overall data). Thus the propensity scores used in this way have a skewed distribution between of 20 values assigned to each of the 20 amino-acids. Another issue is to calculate propensity by pooling all the binding/non-binding data of all proteins together and getting a single score or calculating propensity for each protein in the database and getting an average value. In the first case, an unrealistic situation is created because residue populations are specific to individual proteins and in the second option, data within each residue type may be too small to be reliable. All these issues have implications to tests of significance and determining p-values and therefore alternative scales for propensity and methods to compute them have sometimes been used. Despite these efforts, an accurate estimate of single residue preferences in interface and estimating their statistical significance still remains a non-trivial problem and some more work is needed to overcome difficulties outlined above.

Sequence-Based Prediction of DNA-Binding Sites

For a sequence-based prediction model, a binding site will be composed of seemingly disparate regions on the protein sequence, which are geometrically close in

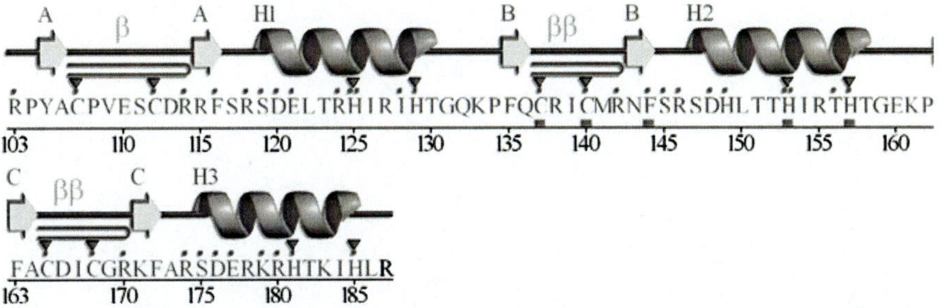

Fig. 1 Distribution of DNA-binding residues over a sequence: multiple binding sites are created on the sequence from a single binding surface on structure. In this example of zinc finger (PDB ID: 1aay), several DNA-binding regions can be seen (residues with a *dot without a triangle*). Although these regions may only interact in presence of each other, challenge of sequence-based predictions is to find all such regions

structure but may be far apart in sequence (see Fig. 1). Thus, multiple "binding sites" made of a single or several contiguous regions need to be predicted. This becomes possible, if each of these binding sites have strong binding signal and their structural contiguity plays little or no role. However, prediction models may implicitly account for structural contiguity. Prediction models also go beyond the direct preferences of single amino acids to interact with DNA, which can be better understood from propensity scores (see previous section). Results of such an analysis expectedly indicate that basic or positive charged residues viz. Arg and Lys are the preferred DNA-binding residues. Analysis of such preferences has been reported [32, 58–60]. Propensity data derived from a recent version of PDB are shown in Fig. 2. It may be noted that sequence-based propensities measure interface enrichment of residue populations compared to all residues and hence may be influenced by a natural tendency of some residues to be on the surface. Thus, Arg and Lys propensities calculated from sequence alone may be slightly exaggerated. However, this is acceptable – even required – because for sequence-based prediction of DNA-binding sites, surface residues may not be a-priori known and these propensities represent a more realistic picture. However, structure-based prediction methods may get biased and hence different set of values must be used in those approaches. Fortunately, our analysis shows that at least in the case of DNA-binding sites, where the most frequently binding residues are Arg and Lys, propensity scores for surface residues are not drastically different from all residues taken together and the two are well-correlated (See Fig. 2). However, subtle differences do exist, which may carry useful information and should be borne in mind, when selecting binding sites from within well known interface residues.

Neither all Arg and Lys residues are located in the interfaces, nor is the interface occupied exclusively of these two residues. It is therefore of utmost importance to know what other factors can be helpful in predicting exact interface residues. If we knew the structure, we may have a fairly good lead from the Arg/Lys enriched regions (or charged patches) on the surface [61]. However, for sequence-based predictions, local sequence environment (or multiple sequence alignments) is the

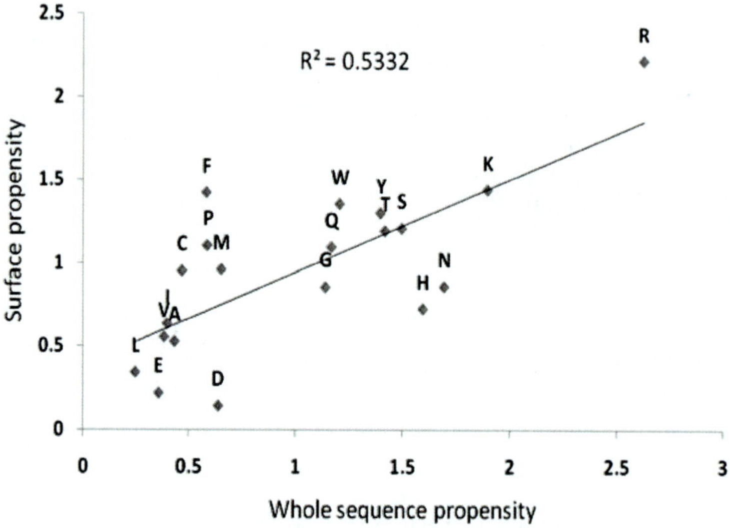

Fig. 2 Whole sequence versus whole surface DNA-binding propensity of 20 amino acid residues

only information available and hence, we need to develop method that can make use of sequence neighbor information within these single sequences or in a multiple alignment of closely related sequences. A combination of these two pieces of information is likely to work even better. This assumption has led to the development of several machine-learning methods to predict DNA-binding sites. In almost all such methods, residue environment is defined by its identity and rows of a position specific substitution matrix derived from multiple alignments. Difference between models lies in the way this information is translated to a prediction i.e. nature of function that links this feature space to target space i.e. binding or no-binding. First method to formulate DNA-binding site problem in this way was reported some time ago [60]. Several other methods have been developed since then, which report a better performance than the original model [60, 62–71]. A typical computational approach to predict DNA-binding sites from sequence is shown in Fig. 3. In general, first a set of features, which are likely to determine a residue's interaction state, is identified. These features may include simply the identity of the amino-acid residue (as in propensity), and its sequence neighbors or may use evolutionary profile of proteins in that residue position (and its neighbors), or a set of averaged biophysical features of local environments, where exact residue identities are replaced by cumulative physical parameters by combining charge, amino acid polarizability, etc. Selected features are then correlated with a set of previously known binding states of residues over a data set. A computational model such as artificial neural network, support vector machine, or Naïve Bayes probabilistic model is than trained over known examples. In principle selected set of features can fit the data in the training examples to an arbitrary degree of accuracy. However, the key to a computational model is its generalization-value i.e. its ability to predict binding sites over a data

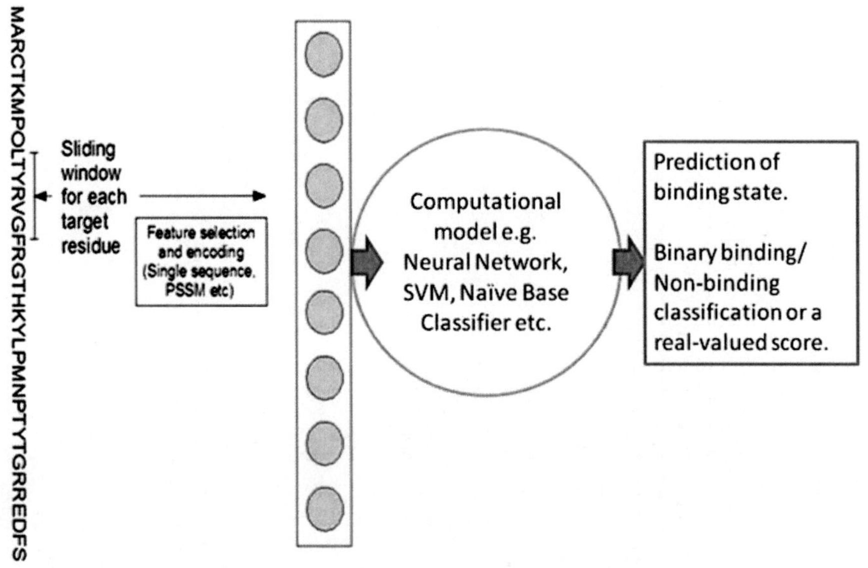

Fig. 3 A typical computational scheme for predicting DNA-binding residues

not used in training. For this purpose models are trained, tested and validated over different data sets. If training data examples contain some of the examples present in the test data sets, the model can give high performance during training, but will underperform over a new data set, which is the actual goal of predictions. Thus careful checks on training/validation procedures are required, when selecting a model for de novo predictions.

Performance is improved due to the use of more complex models as well as availability of more training data sets. Sometimes, reported performance looks exaggerated because issues of redundancy and similarity between training and test data sets are not uniformly treated. Moreover, models can be easily over-fitted, as stated above and hence a simplistic comparison of reported numbers may be quite misleading. From a users' point of view and in the opinion of this author, it is the best to make predictions from multiple publicly available methods and try to reach a consensus. When speed is a greater concern any of these methods can be used as their performance may not be as different as it sounds in the published percentages. A list of DNA-binding site prediction methods, along with their web-based availability is provided in Table 1.

Can We Use Conservation Score to Predict DNA-Binding Sites

Generally speaking, DNA-binding sites can be detected effectively if a homologous protein with known functional information is available. However, to detect a novel DNA-binding site, we need to rely on multiple alignments obtained from

Table 1 Selected machine learning methods for sequence-based prediction of DNA-binding sites in proteins

References	Method	Input	Data redundancy	Reported best performance	Web server URL
[65]	NN	PSSM	25%	76% (AUC of the current version)	http://www.netasa.org/dbs-pssm
[66]	SVM	PSSM, Structure features	20–35%	82% sensitivity 88% specificity	NA
[69, 80]	SVM	PSSM	25%	79% sensitivity 77% specificity	http://lcg.rit.albany.edu/dp-bind/
[64]	SVM	Sequence biophysical features	25%	69% sensitivity 70% specificity	http://bioinformatics.ksu.edu/bindn/
[67]	Naïve bayes	PSSM	30%	44% sensitivity 41% specificity	NA
[81]	NN, SVM	PSSM, predicted structure	20%	67–89% (Q2)	http://cubic.bioc.columbia.edu/services/disis/
[70]	Random forest	PSSM, Secondary structure	90%	~90% (AUC)	http://www.cbi.seu.edu.cn/DBindR/DBindR.htm

Abbreviations: NN: neural network, SVM: support vector machine, PSSM: position-specific substitution matrix, Q2: second quartile of prediction accuracy, AUC: area under the ROC plot, ROC: receiver operative characteristic (plot between false and true positive rate, where positive refers to binding residues and negative to non-binding ones)

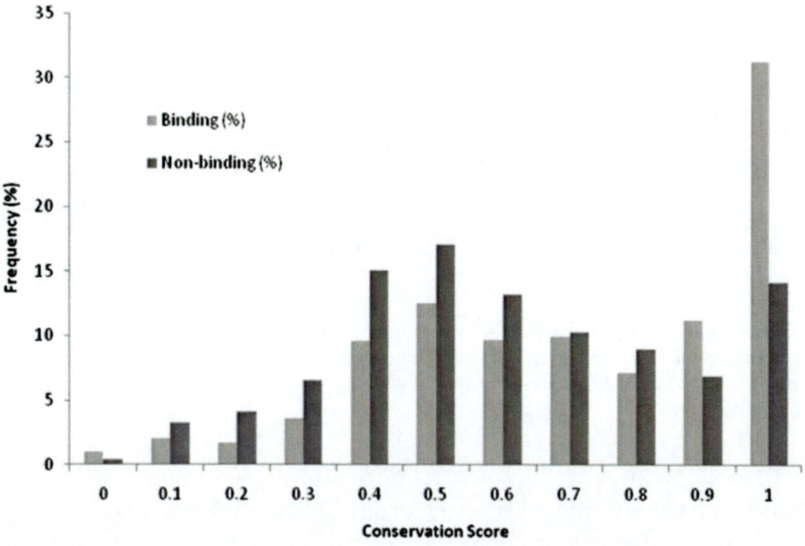

Fig. 4 Relative numbers of conserved residues in interface and non-interface regions of DNA-binding proteins. Data sets and conservation scores are as used in reference [34]

diverse, remotely related proteins, which do not always share a common function and more likely do not contain a known DNA-binding motif. Successful use of PSSM for predicting DNA-binding sites also prompts us to look at the evolutionary aspect of DNA-binding residues. It is obvious that DNA-binding sites, just like any other functionally important residues are more conserved than other residues. Number of studies have confirmed this fact [33, 72]. However, evolvability of proteins also requires that they are not fully conserved. A closer look at histograms of conservation scores in the interface and non-interface (binding or non-binding residues) in Fig. 4, makes clear that interface residues are indeed more conserved than others. Quantitatively, about half of all interface residues are conserved. However, for the remaining half, conservation scores in the interface and non-interface are not significantly different. On the other hand about 15% of non-interface residues are also conserved. In terms of absolute numbers 15% of non-interface residues is much larger number than the absolute number of all residues in the interface. This implies that conservation score alone cannot be used to detect interface residues from sequence with high accuracy and that's why more sophisticated methods, such as machine learning are required and are continuously developed.

Clusters of Conserved Residues (CCRs) and Binding Hot Spots

For many purposes, it may be more useful to predict key functional residues rather than all functional residues. In one sense it means the residues which are most essential for DNA-binding, such that mutations in those residues can disrupt the

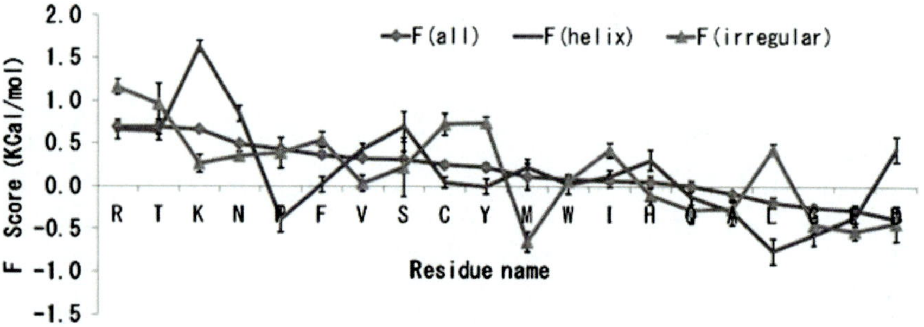

Fig. 5 Average single residue affinity (F-score) of 20 amino acid residues. Although overall stability strongly depends on sequence and structure contexts, average free energy changes in residues occupying interface also shows an interesting trend, which is also biased for helical and irregular secondary structures

formation of a stable complex between protein and DNA. These residues, called hot spots, are well-studied in protein-protein interfaces, in which hot spots are defined as those residues, whose mutations to Ala can cause a free energy change of more than 2.0 kcal/mol. Similar annotations in DNA-binding proteins have also been investigated and it is shown that the hot spots in protein-DNA complexes are made of structurally contiguous conserved residues forming tightly packed clusters or interaction networks [34]. A preliminary look at the 20 amino acid types in proteins reveals that different amino acids contribute differently to free energy on the average [73]. A rough energy scale derived purely from the averages of a thermodynamic data is plotted in Fig. 5, where difference in helical and strand regions is also illustrated. More detailed investigation in terms of conservation scores reveals more complex traits of protein–DNA interactions. It is observed that larger clusters of conserved residues (CCRs) contribute more to stability of protein-DNA complexes. Thus, clustering of conserved residues can be used as a means to predict DNA-binding residues, specially the most significant ones. Figure 6 shows that the CCRs are much more enriched in binding sites than any other residues in proteins, including isolated conserved residues. Organization of these CCRs in proteins depends on their functional and structure classes (see Fig. 7 for example).

Predicting Specificity of Protein–DNA Interaction

While, many methods to predict DNA-binding sites have become available during recent years, not much has been achieved in the direction of predicting specific binding sites. The objective of such predictions would be to pick up partner DNA-binding proteins and their respective targets from a pool of proteins or genes. Independent studies on predicting transcription factor binding sites (TFBS), not covered in this chapter are available. However, simultaneously predicting partners is a difficult task and far from complete. Some of the early efforts have relied on

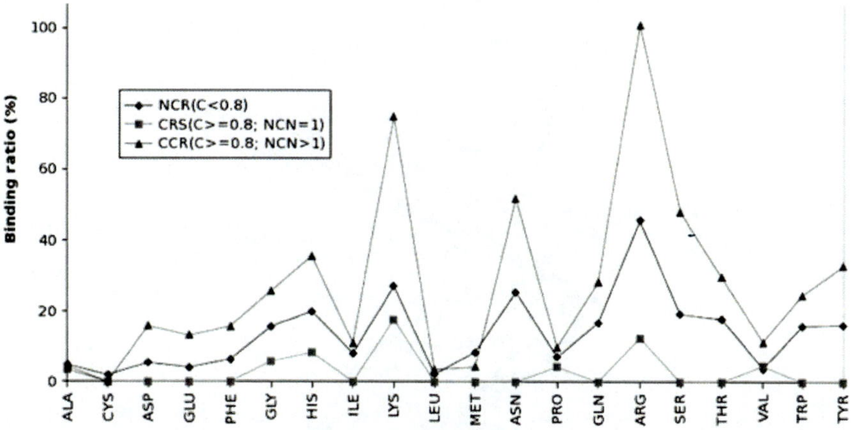

Fig. 6 Relative number of interacting residues is the highest for conserved and clustered residues (CCRs), whereas conserved residue singlets (CRS) are excluded from the interface. Almost 80% of Arg in CCRs are in the interface, which can be used to make first direct prediction of DNA-binding sites. These residues are also typically the hot spot residues (Figure and data taken from [34])

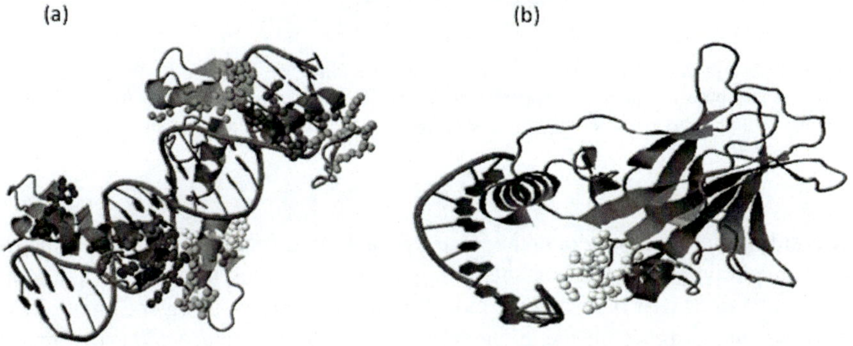

Fig. 7 Typical organization of conserved residues in three-dimensional structure of DNA-binding proteins. Some proteins such as HTH and zinc finger have several small clusters, whereas others have a single cluster crucial for interactions. (**a**) Zinc Finger protein (PDB ID: 1p47) (**b**) p43 core domain in complex with DNA (PDB ID: 3IGK). Images drawn using screenshots from http://ccrxp.netasa.org outputs

knowledge-based potentials resolved between direct and indirect energy models [74–77]. Contribution of indirect readout energy has been successfully developed to predict novel targets for known transcription factors. On the other hand direct interactions have been modeled in terms of base–nucleic acid interactions, as well as DNA–trinucleotide interactions. A novel approach to consider powerful PSSM for predictions was reported recently [78]. In this approach, DNA is broken into overlapping dinucleotide steps and then binding sites corresponding to each of these

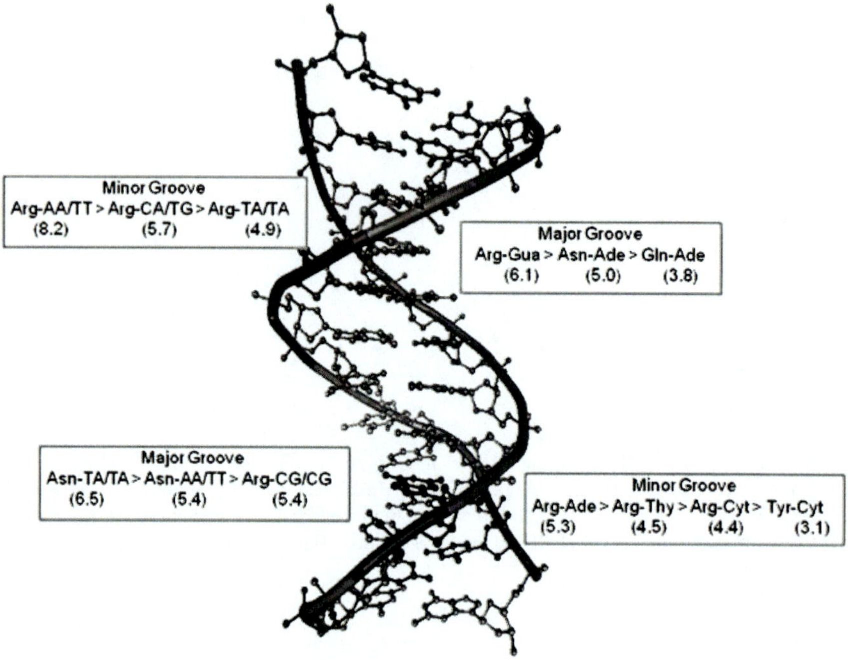

Fig. 8 Base and dinucleotide-specific propensities of amino acid residues in major and minor grooves. From purely sequence-considerations, certain preferences of amino acids for single bases and dinucleotide steps can be inferred. These preferences are different in major and minor grooves. Figure taken from [78]

steps are predicted on protein in the same way as is done for highly successful methods of predicting DNA-binding sites using machine learning methods. A first look at base-amino acid or base-dinucleotide preferences reveals strong specificity conferred by these simple considerations (Fig. 8). Including sequence and evolutionary information is found to be quite successful in predicting interacting pairs at almost the same level of accuracy as non-specific DNA-binding sites (typical area under the ROC plot is ~80%). Although, the performance of these methods are promising, it will take a lot more work to reach a stage where predictions can match the accuracy required to conduct or replace experimental investigations. Preliminary leads are however possible straightaway and shows the usefulness of current state of the art.

Recent Advances and Current Directions

Computational methods have advanced to provide useful inputs and screen results to reduce the experimental search space. However, pace of experiments has also undergone drastic changes during recent years. High throughput sequencing is one technique that has thrown up a huge amount of data giving a huge challenge to the existing computational techniques to scale up [79]. Thus, novel methodologies for analyzing protein–DNA interactions revealed by these experiments are needed. There is a pressing requirement to integrate experimental information with predictions and use multiple sources of information in single computational experiments. Some of these efforts have already proved useful. More work is needed, especially in data management, large-scale modeling and perhaps most importantly catching up with multiple and huge sources of information to unambiguously and accurately elucidate protein–DNA interactions.

Conclusion

DNA-binding sites have been successfully analyzed and machine learning methods have been trained starting with protein-DNA complexes. These methods have proved to be successful and performance has been continuously improving. Computational challenges remain in making the methods more accurate and also in trying to incorporate multiple and seemingly unrelated sources of information. Machine learning approaches are likely to lead efforts in this direction.

References

1. Anderson, W.F., Ohlendorf, D.H., Takeda, Y., Matthews, B.W. Structure of the cro repressor from bacteriophage λ and its interaction with DNA Nature **290**: 754–758 (1982).
2. Benos, P.V., Lapedes, A.S., Stormo, G.D. Is there a code for protein-DNA recognition? Probab(ilistical)ly. BioEssays **24**(5): 466–475 (2002).
3. Benos, P., Bulyk, M.L., Stormo, G.D. Additivity in protein–DNA interactions: how good an approximation is it? Nucleic Acids Res. **30**: 4442–4451 (2002).
4. Berg, O.G., von Hippel, P.H. Selection of DNA binding sites by regulatory proteins-statistical-mechanical theory and application to operators and promoters. J. Mol. Biol. **193**: 723–750 (1987).
5. Garvie, C.W., Wolberger, C. Recognition of specific DNA sequences. Mol. Cell **8**: 937–946 (2001).
6. Seeman, N.C., Rosenberg, J.M., Rich, A. Sequence-specific recognition of double helical nucleic acids by proteins. Proc. Natl. Acad. Sci. **73**: 804–808 (1976).
7. Bewley, C.A., Gronenborn, A.M., Clore, G.M. Minor groove-binding architectural proteins: structure, function and DNA recognition. Structure **27**: 105–131 (1998).
8. Brennan, R., Matthews, B. The helix-turn-helix DNA-binding motif. J. Biol. Chem. **264**: 1903–1906 (1989).
9. Contreras-Moreira, B., Collado-Vides, J. Comparative footprinting of DNA-binding proteins. Bioinformatics **22**(14): e74–e80 (2006).
10. Feng, J.A., Johnson, R.C., Dickerson, R.E. Hin recombinase bound to DNA: the origin of specificity in major and minor groove interactions. Science **263**(5145): 348–355 (1994).
11. Brennan, R.G., Matthews, B.W. Structural basis of DNA-protein recognition. Trends Biochem. Sci. **14**(7): 286–290 (1989).
12. Gilbert, W., Muller-Hill, B. The lac operator is DNA. Proc. Natl. Acad. Sci. **58**: 2415–2421 (1967).
13. Pabo, C.O., Jordan, S.R., Frankel, A.D. Systematic analysis of possible hydrogen bonds between amino acid side chains and B-form DNA. J. biomol. Struct. Dyn. **1**(4): 1039–1049 (1983).

14. Matthews, B.W. Protein–DNA interaction. No code for recognition. Nature **335**: 294–295 (1988).
15. Harrison, S.C. A structural taxonomy of DNA-binding domains. Nature **353**: 715–719 (1991).
16. Pabo, C., Sauer, R. Transcription factors: structural families and principles of DNA recognition. Annu. Rev. Biochem. **61**: 1053–1095 (1992).
17. Luisi, B.F. DNA-transcription – zinc standard for economy. Nature **356**: 379–380 (1992).
18. Berg, J.M., Tymoczko, J.L., Stryer, L. *Biochemistry*, 5th edn. New York, NY: W. H. Freeman and Co. (2002).
19. Larson, C., Verdine, G. The chemistry of protein–DNA interactions. *Bioorganic chemistry: nucleic acids*. Hecht, S.M. (ed.). Oxford: Oxford University Press, pp. 324–346 (1996).
20. Pan, Y., Tsai, C.-J., Ma, B., Nussinov, R. How do transcription factors select specific binding sites in the genome? Nat. Struct. Mol. Biol. **16**: 1118–1120 (2009).
21. Ting, J., Baldwin, A. Regulation of MHC gene expression. Curr. Opin. Immunol. **5**: 8–16 (1993).
22. Struhl, K. Helix-turn-helix, zinc-finger, and leucine-zipper motifs for eukaryotic transcriptional regulatory proteins. Trends Biochem. Sci. **14**: 137–140 (1989).
23. Scheidereit, C., Krauter, P., von der Ahe, D., Janich, S., Rabenau, O., Cato, A., Suske, G., Westphal, H., Beato, M. Mechanism of gene regulation by steroid hormones. J. Steroid Biochem. **24**: 19–24 (1986).
24. Park, R., Haseltine, W., Rosen, C. A nuclear factor is required for transactivation of HTLV-I gene expression. Oncogene **3**: 275–279 (1988).
25. Keller, B., Martini, S., Sedor, J., Kretzler, M. Linking variants from genome-wide association analysis to function via transcriptional network analysis. Semin. Nephrol. **30**(2): 177–184 (2010).
26. Gottesman, S. Bacterial regulation: global regulatory networks. Annu. Rev. Genet. **18**: 415–441 (1984).
27. Bulyk, M.L. Computational prediction of transcription-factor binding site locations. Genome Biol. **5**(1): 201.201–201.211 (2003).
28. Chen, Q., Hertz, G., Stormo, G.D. MATRIX SEARCH 1.0: a computer program that scans DNA sequences for transcriptional elements using a database of weight matrices. Comput. Appl. Biosci. **11**: 563–566 (1995).
29. Workman, C.T., Stormo, G.D. ANN-Spec: a method for discovering transcription factor binding sites with improved specificity. *Pac Symp Biocompu,* vol. 5. Altman, R., Dunker, A.K., Hunter, L., Klein, T.E. (eds.). Palo Alto, CA: Stanford University, pp. 467–478 (2000).
30. Tompa, M., Li, N., Bailey, T.L., Church, G.M., De Moor, B., Eskin, E., Favorov, A.V., Frith, M.C., Fu, Y., Kent, W.J., et al. Assessing computational tools for the discovery of transcription factor binding sites. Nat. Biotechnol. **23**: 137–144 (2005).
31. Frith, M.C., Hansen, U., Spouge, J.L., Weng, Z. Finding functional sequence elements by multiple local alignment. Nuclic Acids Res. **32**(1): 189–200 (2004).
32. Luscombe, N., Thornton, J. Protein–DNA interactions: amino acid conservation and the effects of mutations on binding specificity. J. Mol. Biol. **320**: 991–1009 (2002).
33. Mirny, L.A., Gelfand, M.S. Structural analysis of conserved base pairs in protein-DNA complexes. Nucleic Acids Res. **30**(7): 1704–1711 (2002).
34. Ahmad, S., Keskin, O., Sarai, A., Nussinov, R. Protein–DNA interactions: structural, thermodynamic and clustering patterns of conserved residues in DNA-binding proteins. Nucleic Acids Res. **36**(18): 5922–5932 (2008).
35. Berman, H.M., Westbrook, J., Feng, Z., Gilliland, G., Bhat, T.N., Weissig, H., Shindyalov, I.N., Bourne, P.E. The Protein Data Bank. Nucleic Acids Res. **28**: 235–242 (2000).
36. Berman, H.M., Olson, W.K., Beveridge, D.L., Westbrook, J., Gelbin, A., Demeny, T., Hsieh, S.H., Srinivasan, A.R., Schneider, B. The nucleic acid database: a comprehensive relational database of three-dimensional structures of Nucleic Acids. Biophys. J. **63**: 751–759 (1992).

37. Tama, F., Sanejouand, Y.H. Conformational change of protein arising from normal mode calculations. Proteins Eng. **14**: 1–6 (2001).
38. Dobbins, S.E., Lesk, V.I., Sternberg, M.J.E. Insights into protein flexibility: the relationship between normal modes and conformational change upon protein-protein docking. PNAS **105**(30): 10390–10395 (2008).
39. Boehr, D.D., Nussinov, R., Wright, P.E. The role of dynamic conformational ensembles in biomolecular recognition. Nat. Chem. Boil. **5**: 789–796 (2009).
40. Laskowski, R. PDBsum new things. Nucleic Acids Res. **37**: D355–D359 (2009).
41. Luscombe, N., Laskowski, R., Thornton, J. NUCPLOT: a program to generate schematic diagrams of protein–nucleic acid interactions. Nucleic Acids Res. **25**: 4940–4945 (1997).
42. Lee, S., Blundell, T.L. BIPA: a database for protein–nucleic acid interaction in 3D structures. Bioinformatics **25**(12): 1559–1560 (2009).
43. Bourne, P., Desai, N. PRONUC: a software package for the analysis of protein and nucleic acid sequences. Comput. Methods Programs Biomed. **24**: 27–36 (1987).
44. Prabhakaran, P., Siebers, J.G., Ahmad, S., Gromiha, M.M., Singarayan, M.G., Sarai, A. Classification of protein-DNA complexes based on structural descriptors. Structure **14**: 1355–1367 (2006).
45. Moretti, R., Ansari, A. Expanding the specificity of DNA targeting by harnessing cooperative assembly. Biochimie **90**: 1015–1025 (2008).
46. Poupon, A., Janin, J. Analysis and prediction of protein quaternary structure. Methods Mol. Biol. **609**: 349–364 (2010).
47. Xu, Q., Canutescu, A., Obradovic, Z., Dunbrack Jr R. ProtBuD: a database of biological unit structures of protein families and superfamilies. Bioinformatics **22**: 2876–2882 (2006).
48. Levy, E. PiQSi: protein quaternary structure investigation. Structure **15**(11): 1364–1367 (2007).
49. Yu, X., Wang, C., Li, Y. Classification of protein quaternary structure by functional domain composition. BMC Bioinformatics **7**: 187 (2006).
50. Prabakaran, P., An, J., Gromiha, M., Selvaraj, S., Uedaira, H., Kono, H., Sarai, A. Thermodynamic database for protein–nucleic acid interactions (ProNIT). Bioinformatics **17**: 1027–1034 (2001).
51. Donald, J.E., Chen, W.W., Shakhnovich, E.I. Energetics of protein–DNA interactions. Nucleic Acids Res. **35**(4): 1039–1047 (2007).
52. Potapov, V., Cohen, M., Schreiber, G. Assessing computational methods for predicting protein stability upon mutation: good on average but not in the details. Protein Eng. Des. Sel. **22**(9): 553–560 (2009).
53. Matys, V., Fricke, E., Geffers, R., Gossling, E., Haubrock, M., Hehl, R., Hornischer, K., Aras, D., Kel, A., Kel-Margoulis, O. TRANSFAC: transcriptional regulation, from patterns to profiles. Nucleic Acids Res. **31**: 374–378 (2003).
54. Portales-Casamar, E., Thongjuea, S., Kwon, A., Arenillas, D., Zhao, X., Valen, E., Yusuf, D., Lenhard, B., Wasserman, W., Sandelin, A. JASPAR 2010: the greatly expanded open-access database of transcription factor binding profiles. Nucleic Acids Res. **38**: D105–D110 (2010).
55. Tokovenko, B., Golda, R., Protas, O., Obolenskaya, M., El'skaya, A. COTRASIF: conservation-aided transcription-factor-binding site finder. Nucleic Acids Res. **37**: e49 (2009).
56. Murzin, A., Brenner, S., Hubbard, T., Chothia, C. SCOP: a structural classification of proteins database for the investigation of sequences and structures. J. Mol. Biol. **247**: 536–540 (1995).
57. Andrabi, M., Mizuguchi, K., Sarai, A., Ahmad, S. Benchmarking and analysis of DNA-binding site prediction using machine learning methods. Proceedings of IEEE International Joint Conference Neural Networks, June 1–6, Hong Kong, NN0554, pp. 1746–1750 (2008).
58. Jones, S., van Heyningen, P., Berman, H.M., Thornton, J.M. Protein–DNA interactions: a structural analysis. J. Mol. Biol. **287**: 877–896 (1999).
59. Jones, S., Shanahan, H.P., Berman, H.M., Thornton, J.M. Using electrostatic potentials to predict DNA-binding sites on DNA-binding proteins. Nucleic Acids Res. **31**(24): 7189–7198 (2003).

60. Ahmad, S., Gromiha, M., Sarai, A. Analysis and prediction of DNA-binding proteins and their binding residues based on composition, sequence and structural information. Bioinformatics **20**: 477–486 (2004).
61. Tsuchiya, Y., Kinoshita, K., Nakamura, H. Structure-based prediction of DNA-binding sites on proteins using the empirical preference of electrostatic potential and the shape of molecular surfaces. Proteins **55**: 885–894 (2004).
62. Ahmad, S., Sarai, A. Moment-based prediction of DNA-binding proteins. J. Mol. Biol. **341**: 65–71 (2004).
63. Tjong, H., Zhou, H.X. DISPLAR: an accurate method for predicting DNA-binding sites on protein surfaces. Nucleic Acids Res. **35**(5): 1465–1477 (2007).
64. Wang, L., Brown, S.J. BindN: a web-based tool for efficient prediction of DNA and RNA binding sites in amino acid sequences. Nucleic Acids Res. **34**: W243–W248 (2006).
65. Ahmad, S., Sarai, A. PSSM-based prediction of DNA binding sites in proteins. BMC Bioinformatics **6**: 33 (2005).
66. Bhardwaj, N., Langlois, R.E., Zhao, G., Lu, H. Kernel-based machine learning protocol for predicting DNA-binding proteins. Nucleic Acids Res. **33**(20): 6486–6493 (2005).
67. Yan, C., Terribilini, M., Wu, F., Jernigan, R.L., Dobbs, D., Honavar, V. Predicting DNA-binding sites of proteins from amino acid sequence. BMC Bioinformatics **7**: 262 (2006).
68. Yu, X., Cao, J., Cai, Y., Shi, T., Li, Y. Predicting rRNA-, RNA-, and DNA-binding proteins from primary structure with support vector machines. J. Theor. Biol. **240**: 175–184 (2006).
69. Hwang, S., Gou, Z., Kuznetsov, I.B. DP-Bind: a web server for sequence-based prediction of DNA-binding residues in DNA-binding proteins. Bioinformatics **23**(5): 634–636 (2007).
70. Wu, J., Liu, H., Duan, X., Ding, Y., Wu, H., Bai, Y., Sun, X. Prediction of DNA-binding residues in proteins from amino acid sequences using a random forest model with a hybrid feature. Bioinformatics **25** (1): 30–35 (2009).
71. Zen, A., de Chiara, C., Pastore, A., Micheletti, C. Using dynamics-based comparisons to predict nucleic acid binding sites in proteins: an application to OB-fold domains. Bioinformatics **25**(15): 1876–1883 (2009).
72. Yao-Lin, C., Huai-Kuang, T., Cheng-Yan, K., Yung-Chian, C., Yuh-Jyh, H., Jinn-Moon, Y. Evolutionary conservation of DNA-contact residues in DNA-binding domains. BMC Bioinformatics **9**: S3 (2008).
73. Andrabi, M., Ahmad, S. A single-residue affinity scale for DNA-binding using linear perceptron. Proceedings of International Conference on Pattern Recognition in Bioinformatics, Melbourne (2008).
74. Gao, M., Skolnick, J. DBD-Hunter: a knowledge-based method for the prediction of DNA–protein interactions. Nucleic Acids Res. **36**(12): 3978–3992 (2008).
75. Gromiha, M.M., Siebers, J.G., Selvaraj, S., Kono, H., Sarai, A. Intermolecular and intramolecular readout mechanisms in protein–DNA recognition. J. Mol. Biol. **337**(2): 285–294 (2004).
76. Selvaraj, S., Kono, H., Sarai, A. Specificity of Protein-DNA recognition revealed by structure-based potentials: symmetric/asymmetric and cognate/non-cognate binding. J. Mol. Biol. **322**: 907–915 (2002).
77. Araúzo-Bravo, M.J., Fujii, S., Kono, H., Ahmad, S., Sarai, A. Sequence-dependent conformational energy of DNA derived from molecular dynamics simulations: toward understanding the indirect readout mechanism in protein-DNA recognition. J. Am. Chem. Soc. **127**(46): 16074–16089 (2005).
78. Andrabi, M., Mizuguchi, K., Sarai, A., Ahmad, S. Prediction of mono- and di-nucleotide-specific DNA-binding sites in proteins using neural networks. BMC Struct. Biol. **9**: 30 (2009).
79. Buck, M.J., Lieb, J.D. ChIP-chip: considerations for the design, analysis, and application of genome-wide chromatin immunoprecipitation experiments. Genomics **83**: 349–360 (2004).
80. Kuznetsov, I.B., Gou, Z., Li, R., Hwang, S. Using evolutionary and structural information to predict DNA-binding sites on DNA-binding proteins. Proteins **64**: 19–27 (2006).
81. Ofran, Y., Mysore, V., Rost, B. Prediction of DNA-binding residues from sequence. Bioinformatics **23**(13): 347–353 (2007).

Electrostatic Properties for Protein Functional Site Prediction

Joslynn S. Lee and Mary Jo Ondrechen

Abstract The development of computational tools for the prediction of protein function from the three-dimensional structure is a very important problem in the post-genomic era. To date there are over 9,900 structural genomics protein structures in the Protein Data Bank and most of these are of unknown or uncertain function. Methods for the identification of the residues in a protein structure that participate in the biochemical function provide key information about the function of the protein. We and others have developed computational methods for the prediction of functionally important residues in proteins. The focus of this chapter is on protein function at the atomic level, *i.e.* catalysis and recognition. Methods that utilize computed electrostatic properties, specifically THEMATICS and POOL, are described.

Introduction

The development of computational tools for the prediction of protein function from the three-dimensional structure is a very important problem in the post-genomic era. To date there are over 9,900 structural genomics protein structures in the Protein Data Bank [1–2] and most of these are of unknown or uncertain function. Methods for the identification of the residues in a protein structure that participate in the biochemical function provide key information about the function of the protein. We and others have developed computational methods for the prediction of functionally important residues in proteins. The focus of this chapter is on *protein function at the atomic level*, i.e. *catalysis and recognition*, and on methods that utilize computed electrostatic properties.

Computed electrostatic properties bring special advantages to the quest for functional information about a protein structure. First of all, *they require only the structure of the query protein as input*. Thus they return predictions even for novel

M.J. Ondrechen (✉)
Department of Chemistry and Chemical Biology, Northeastern University, Boston, MA 02115, USA
e-mail: M.Ondrechen@neu.edu

D. Kihara (ed.), *Protein Function Prediction for Omics Era*,
DOI 10.1007/978-94-007-0881-5_10,

folds and engineered structures, as well as for proteins with orphan sequences or with few sequence homologues. Furthermore, these predictions are just as reliable for these difficult cases as they are for the well-characterized proteins in the benchmark sets used for the testing and verification of the methods. Second, *these properties are directly related to the chemistry of individual residues* and thus are well suited to the identification of residues with special catalytic or binding properties. Finally, *electrostatics-based methods are orthogonal to the more common sequence-based methods* that rely on sequence alignments and phylogenetic trees; thus, when information about sequence conservation or evolutionary history is available, combination of the methods can, at least in principle, lead to significant enhancement in the quality of the predictions. Indeed, electrostatics-based methods have proved to be powerful tools for functional site prediction.

In the prediction of functionally important residues, there is always a trade-off between sensitivity (the ability to predict the maximum number of truly important residues) and selectivity (the ability to predict only the truly important residues and not the unimportant residues). The goal is to *maximize sensitivity while minimizing false positives*.

In order to test the performance of predictors of functionally important residues, an annotated dataset is needed as a benchmark. Typically the Catalytic Site Atlas (CSA) [3–4], a referenced compilation of catalytically active residues previously identified in the literature for hundreds of enzymes, is used to obtain the validation set for functional site prediction methods. While no listing of catalytically active residues can possibly be complete, as not all residues have been tested experimentally and reported, the CSA represents the best available compilation of known catalytic residues.

Performance in catalytic residue prediction is defined in terms of true positives (TP), true negatives (TN), false positives (FP), and false negatives (FN). Positives and negatives are defined using the CSA as the reference set. The recall rate for catalytic residue prediction is defined as:

$$\text{Recall} = \text{Sensitivity} = \text{TP}/(\text{TP} + \text{FN}) \tag{1}$$

The false positive rate is defined as:

$$\text{False positive rate} = \text{FP}/(\text{TN} + \text{FP}) \tag{2}$$

Finally the specificity is defined as:

$$\text{Specificity} = \text{TN}/(\text{TN} + \text{FP}) \tag{3}$$

The specificity is related to the false positive rate (FPR) as:

$$\text{Specificity} = 1 - \text{FPR} \tag{4}$$

Previously our group has reported on THEMATICS (for Theoretical Microscopic Titration Curves), an electrostatics-based method for the prediction of functionally important residues in protein 3D structures [5–7]. THEMATICS has been shown to predict functionally important residues with good sensitivity and a low false positive rate [7]. More recently, Partial Order Optimum Likelihood (POOL) [8] utilizes THEMATICS and other input features in a new, monotonicity-constrained maximum likelihood machine learning method, for enhanced performance in prediction of catalytic and binding residues.

Methods

THEMATICS

In the application of THEMATICS, the electrical potential function of the protein structure is first computed using a finite difference Poisson-Boltzmann procedure. Then a hybrid method [9] is used to compute theoretical titration curves for each of the ionizable residues. These titration curves take the form of the proton occupation for each residue as a function of the pH. The shapes of the titration curves are evaluated by an automated procedure, using the curve shape metrics described by Ko et al. [6] to quantify the degree of deviation from the typical Henderson-Hasselbalch (H-H) titration behavior. These curve shape metrics are subjected to statistical analysis in order to identify the residues that deviate most from the ideal H-H curve shape. Note that THEMATICS predictions are based on the *shapes* of the computed titration curves and not on the computed pK_a shifts, although THEMATICS has sometimes been described incorrectly as a pK_a shift method [10]. While pK_a shifts are common in active sites, they also occur too frequently in other parts of protein structures, e.g. salt bridges, to give precise active site predictions.

THEMATICS has been established as a successful, top performing site predictor across a wide range of enzymes from all functional classes [7]. In order to verify its effectiveness in catalytic site prediction, THEMATICS was applied to the entire original, manually curated set of 170 enzymes in the CSA [3–4]. THEMATICS was shown to identify, with high selectivity, all or some of the residues in known interaction sites in 93% of enzymes [7]. When performance in the prediction of annotated catalytic residues was compared with that of other 3D-structure-based methods, THEMATICS showed better sensitivity with much lower false positive rates, as demonstrated by the ROC (Receiver Operator Characteristic – i.e. true positive rate versus false positive rate) curves [7]. A very important characteristic of THEMATICS performance is its selectivity – it predicts precise, highly localized sites [7].

A key feature of THEMATICS is that the query protein does not have to have any similarity, in sequence or in structure, to any other protein. Originally function prediction was based primarily on sequence analysis, although sequence similarity does not always imply functional similarity [11–13]. Other methods use structural relationships in conjunction with sequence analysis [14–32] for improved performance.

There are presently a few approaches in addition to THEMATICS that are based solely on the structure of the query protein and some of these also employ electrostatic properties. Elcock [33] reported that likely functional residues could be identified by their electrostatic folding free energies obtained from solution of the Poisson-Boltzmann equations. Bate and Warwicker [34] later identified a point near the active site using the peak of the electrostatic potential in the solvent space above the protein structure. A graph theoretic approach predicts candidate active site residues based on their closeness of interaction with the other residues in the structure [35]. Another method uses purely geometric features of the protein structure [36]. More recently, ligand binding sites have been predicted through the computational identification of regions where interactions cause a large change in protein conformation distribution [37]. Ligand binding sites can also be detected with the mapping of small solvent-like molecules onto the protein surface, either experimentally [38] or with the corresponding computational docking method [39]. The method of Laurie and Jackson [40] is of this type, but uses only a single van der Waals probe.

THEMATICS, which requires no sequence alignments, has been shown to match performance, or even outperform, the best methods that predict functional sites from sequence alignments *and* the 3D structure. However, it is important to note that the performance of the methods that require a sequence alignment is expected to degrade [24–25] when applied to Structural Genomics (SG) proteins that have fewer, or less diverse, sequence homologues than the well studied proteins in the verification sets. On the other hand, *THEMATICS performance on SG proteins in principle should match its performance on the verification set* because it requires only the 3D structure of the query protein and it treats all input structures in the same fashion; it does not depend on any prior knowledge or relationships to other proteins.

THEMATICS predictions are freely available via the pfweb server: http://pfweb.chem.neu.edu/thematics/submit.html.

Users can either upload a protein structure file in pdb format, or alternatively give the PDB ID for the structure of interest. THEMATICS calculations on the server utilize the optimum statistical and distance cut-offs determined by Wei et al. [7]; these values return the highest Matthews Correlation Coefficient (MCC), as measured using the CSA annotations. The maximum MCC reflects a balance between sensitivity and specificity. Results are returned to the user via e-mail. Results take the form of one or more clusters.

For instance, for the dimer structure with PDB ID 2qe8, an uncharacterized structural genomics protein from *Anabaena variabilis,* THEMATICS returns two clusters for each of the two subunits of the dimer, a seven-member cluster [D123, K246, C249, D250, D293, D306, R342] and a one-member cluster [D202]. Only clusters with two or more residues are considered predictive; thus the seven-member cluster constitutes the functional site prediction and the single-member cluster [D202] is not a part of the predicted active site.

POOL

A new machine learning approach, called Partial Order Optimum Likelihood (POOL) was designed [8] to make significant enhancements in site prediction capability. Originally POOL was applied using THEMATICS input features and later was expanded to include other types of input features. In principle, POOL can use any input feature, provided the probability of the functional importance of residues depends monotonically on that feature.

POOL, a multidimensional, monotonicity-constrained maximum likelihood technique, starts with the hypothesis that the larger the THEMATICS metrics for a given residue, the higher the probability that the residue is important for function. These features consist of two computed properties, called μ_3 and μ_4 [6], of ionizable residues that describe titration curve shape. Extension of the POOL method to include predictions of non-ionizable residues is achieved through the introduction of *environment variables*. While THEMATICS features apply only to the ionizable residues (Arg, Asp, CysH, Glu, His, Lys, Tyr, and the N- and C- termini), the environment variables $\mu_3{}^{env}$ and $\mu_4{}^{env}$ measure the magnitude of the THEMATICS features of the ionizable residues that are spatially close to the residue in question. Note that $\mu_3{}^{env}$ and $\mu_4{}^{env}$ are properties of all residues, not just ionizable residues. Thus the THEMATICS input feature for POOL is the four-dimensional vector (μ_3, μ_4, $\mu_3{}^{env}$, $\mu_4{}^{env}$) for the seven residue types that are ionizable and the two-dimensional vector ($\mu_3{}^{env}$, $\mu_4{}^{env}$) for all of the non-ionizable residue types. This extension to include non-ionizable residues results in even better performance than with the original THEMATICS features alone and constitutes to date the best functional site predictor based on 3D structure only, achieving performance that is as good or nearly as good as methods that use both 3D structure and sequence alignment data [8].

It is interesting to note that the THEMATICS features μ_3 and μ_4 are derived from a function that is related to the binding capacity [41] for protons; μ_3 and μ_4 are also related to the coefficients in the proton binding polynomial [42].

These electrostatics features from THEMATICS are combined with multidimensional isotonic regression to form maximum likelihood estimates of probabilities that specific residues belong to an active site. This allows likelihood ranking of all ionizable residues in a given protein based on THEMATICS features. The corresponding ROC curves and statistical significance tests demonstrate that this method outperforms prior THEMATICS based methods, which in turn have been shown previously [7] to outperform other 3D-structure based methods for identifying active site residues.

POOL generates a value for each residue that is proportional to the probability that the residue is functionally important. One of the advantages of POOL is that it can incorporate any residue-based input feature upon which the probability of functional importance depends monotonically.

One such feature is the cleft size rank, an integer that represents the ordinal size of the surface cleft to which a given residue belongs. Previous studies have shown that active site residues tend to be located in one of the largest clefts in a protein structure [43–45]. Indeed it has been reported that in 83% of single-chain enzymes, the active site is located in the largest cleft [44]. Nearly all active sites are principally located in one of the five largest clefts of a protein structure, with the largest cleft containing the active site for the highest fraction of enzymes and with the fractions decreasing as the size rank progresses to smaller clefts [46]. The cleft size rank is a geometric feature that can be quickly computed for each residue in any protein structure. Although the cleft size rank alone does not perform very well for active residue prediction, its inclusion as input to POOL, as an addition to the THEMATICS input features, does lead to small but statistically significant improvement in site prediction performance [8].

Similarly, POOL easily incorporates sequence conservation scores, for those cases where there are a sufficient number of homologues. When this information is included, the resulting method has been shown to outperform the best methods that use any combination of sequence alignments and 3D structures [8]. It is further demonstrated that when THEMATICS features, cleft size rank, and alignment-based conservation scores are used individually or in combination, THEMATICS features represent the single most important component of such classifiers [8]. The POOL method we have developed is general and is a viable machine learning approach to any problem where a predicted outcome depends monotonically on each of the input variables. Most importantly, POOL is a top-performing site predictor and it *enables THEMATICS to be used to predict all residues, not just the ionizable ones.*

POOL output consists of a list of all residues, rank-ordered according to the probability of functional importance. The top-ranking residues constitute the POOL prediction. The cut-off point in the rank-ordered list may be set according to the intended application. The cut-off value is generally set to select the top 5–8% of all residues, as this returns good sensitivity with excellent specificity. A 5% false positive rate, which corresponds to 95% specificity, returns a recall rate of 70%, which is good enough to characterize a functional site. Full recall (100% sensitivity) is achieved with only a 17% false positive rate. This performance compares quite favorably with other methods, for instance INTREPID achieves 93% sensitivity with a 20% false positive rate [28] on a similar test set. However, false positive rates in the 17–20% range may be too high to be useful, as discussed below. We prefer to work with a little lower sensitivity but much better specificity; this combination is achievable with THEMATICS and POOL.

Discussion

What Is the Basis for the Success of THEMATICS?

As a standalone functional site predictor, THEMATICS has been shown to perform very well [7]. Its performance was measured on the original, manually curated set of 170 proteins in the Catalytic Site Atlas [3–4], where catalytic residues are labeled based on experimental literature citations. The THEMATICS success rate was found to be equal to or better than that of other 3D-structure-based methods, but with better precision and lower false positive rates [7]. This was all achieved with only one type of input, namely the computed titration curve shape metrics.

We attribute the success of the method, in particular its ability to predict highly localized, precise active sites, to its reliance on computed chemical properties. Chemically active residues are predicted with information about their chemistry, specifically their proton binding properties. While there is some error associated with the titration curves computed by electrostatics methods, the statistical [6–7] and machine learning [8] analyses on the curve shape metrics have proved to be highly successful in selecting the outliers, i.e. those residues with titration curve shapes that deviate most from typical Henderson-Hasselbalch behavior.

We have argued [47] that the anomalous titration behavior enables a residue, in a large ensemble of protein molecules, to exist in both protonation states with appreciable population over a wide pH range. This is in contrast to a typical Henderson-Hasselbalch weak acid or base, which is protonated at pH values less than the pK_a and deprotonated at pH values greater than the pK_a, with a very narrow pH range around the pK_a where both protonation states are populated in an ensemble of molecules. For the residues with anomalous titration behavior, this pH range is expanded significantly. This type of non-Henderson-Hasselbalch titration behavior is common for polyprotic acids and a protein is in fact a macromolecular polyprotic system.

Furthermore, for an active site residue, this ability to have both protonation states populated over a wider pH range is an advantage in catalysis [47]. First of all, by definition of a catalyst, a catalytic Brønsted-Lowry acid or base must be able to act as both acid and base because it must regenerate itself for the next turnover cycle. Thus a residue that donates a proton as part of the catalytic mechanism must also accept a proton before the end of each cycle. The anomalous titration behavior also enables catalytic residues to have the correct mix of properties. Consider for example one common first step in an enzyme-catalyzed reaction, the abstraction of a proton from an alpha carbon atom, a reaction that requires a strong base. Suppose that the enzyme in question operates in vivo at pH 7. Suppose that the conjugate acid of the catalytic base has a pK_a of 13 and that it obeys the Henderson-Hasselbalch equation. Such a base may not be strong enough to abstract a proton from a carbon atom, but even if it were, it would not be able to react because at pH 7, it is essentially fully protonated. Only one in one million protein molecules in the ensemble would have this residue deprotonated at pH 7. On the other hand, a base with anomalous titration behavior can be a strong base and at the same time have significant population of the deprotonated state at neutrality. Thus *the anomalous titration behavior helps to facilitate catalysis for active site residues.*

It is our working hypothesis that nature builds enzyme active sites with clusters of neighboring ionizable residues with similar pK_a values, so that there is strong interaction between their protonation events. This strong interaction gives rise to anomalous titration curve shapes and promotes catalysis. The deviations in the titration curve shape are measured by the features computed in a THEMATICS analysis.

The enhanced performance afforded by POOL using THEMATICS input features only is attributed to the ability of *POOL to extract more information from these features than the earlier statistical and machine learning analyses; this leads to better quality predictions of functionally important residues.* First POOL was applied with just THEMATICS features as input, using features similar to those used previously by our Support Vector Machine (SVM) classifier [48] and by our statistical selection [6–7]. Tong et al. showed in 2009 [8] that the POOL analysis outperforms all of these earlier THEMATICS analyses with no cleaning of the training data and no clustering after the classification. This suggests that the underlying monotonicity assumptions of POOL enable better use of the THEMATICS input metrics.

Another obvious reason for the success of POOL is its ability to *predict all residues, not just the ionizable residues.* In the previous statistical versions of THEMATICS, only seven types of residues are predicted: Arg, Asp, CysH, Glu, His, Lys, and Tyr. The N- and C- termini are also included in the original THEMATICS analysis, although these residues are only rarely involved in catalysis. Serine is excluded from the original THEMATICS analysis because its pK_a is generally too high for its deprotonation equilibrium to have significant interactions with those of other residues; attempts to include serine in the original THEMATICS analysis lead to lower quality predictions and thus serine has not been considered an ionizable residue for purposes of THEMATICS analyses. In spite of this, THEMATICS has still performed well compared to other 3D-structure-based methods [7]. This is in part because the seven residue types predicted by THEMATICS are the seven most prevalent catalytic residues. Among the literature-annotated catalytic residues analyzed by Bartlett et al. [3], the most common residue types, in order starting with the most common, are: His, Asp, Arg, Glu, Lys, CysH, and Tyr. Together these seven residue types constitute about 75% of all annotated catalytic residues [3, 7]. However this means that THEMATICS has a maximum residue recall rate, or sensitivity, of 75%, since by its nature it cannot predict the remaining 25% of catalytic residues. POOL is advantageous because it can predict all residue types. For instance, POOL predicts all three residues of the catalytic triad of serine proteases such as subtilisin, including the serine, whereas THEMATICS only predicts the Asp and His residues.

POOL is able to take advantage of a variety of input features, in addition to THEMATICS features. Any property of the residues in a protein structure can be a POOL input variable, provided the probability that a residue is catalytically important is a monotonic function of that variable. The current version of POOL incorporates a geometric feature, the cleft size rank, and the sequence conservation scores.

Table 1 summarizes POOL performance with and without conservation scores. The average specificities achieved at 90, 80, and 70% recall, together with the average recall rates achieved at 95, 90, and 80% specificity are shown. Specificity and recall are reported using all three input features, THEMATICS, geometric, and conservation scores (T, G, and C) and using the 3D-structure-based features (T and G) only, as measured on a 160 protein test set [8]. The figures of merit in Table 1 represent outstanding performance; see for example Table 1 of Sankararaman and

Table 1 POOL performance with and without sequence conservation data

Input features	T, G, and C	T and G only
POOL performance		
Specificity at 90% recall (%)	91	89
Specificity at 80% recall (%)	92	91
Specificity at 70% recall (%)	95	93
Recall at 95% specificity (%)	70	60
Recall at 90% specificity (%)	91	87
Recall at 80% specificity (%)	100	100

Input features: T = THEMATICS, G = geometry (cleft size rank), C = conservation scores. Performance data are for a test set of 160 annotated proteins [8]

Sjölander [28]. Table 1 shows that, even in the absence of sequence conservation information, POOL is able to make good predictions of catalytic residues with input features computed solely from the 3D structure of the query protein.

Applications

Prediction of protein functional residues is a first step toward functional annotation of a protein. One specific application has been to functional assignment within superfamilies, which consist of sets of proteins with similar 3D structure but often with significant functional diversity. Wei et al. have shown [49] that, for the small DJ-1 superfamily, placement of the predicted functional residues onto a 3D structural alignment reveals patterns characteristic of biochemical function; this enables one to sort the superfamily into subclasses according to their function.

Other applications of functional site prediction from electrostatic properties include better understanding ligand binding [50–52] and inhibitor design. These applications all require that the functional residues are predicted with both sensitivity and precision.

Precision

While many functional residue prediction methods boast high recall rates of annotated catalytic residues, often these also correspond to high false positive rates. In some cases, the measures of selectivity, such as precision, specificity, or false positive rates, are not reported at all [30]. Of course, the value of a high-recall prediction is significantly diminished if the corresponding false positive rate is high. While not all electrostatics based methods are capable of good selectivity, those that utilize titration curve shapes are able to return very low false positive rates with good recall.

The CSA-100, a non-redundant subset of 100 enzymes from the CSA, is often used for verification purposes [28]. This set of enzymes consists of a total of 36,230 residues, of which 314 are annotated as functionally important. Thus approximately

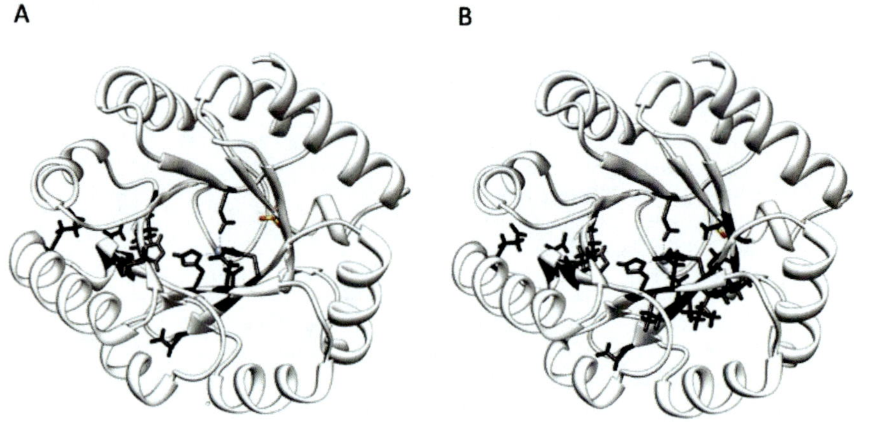

Fig. 1 Predictions for the structural genomics protein Pfal009167. (**a**) 5% POOL cut-off value; (**b**) 8% POOL cut-off value

1% of all residues currently are considered functionally important. While this represents a lower bound, as not all residues have been tested and thus some important residues are not listed in the CSA, it gives a rough idea of the fraction of total residues that should be returned by a site prediction method.

For application purposes, we have found that generally it is less important to predict every single functional residue than it is to predict most of the functional residues with few false positives. We have observed that the fraction of total residues predicted should be in the range of about 5–8% or less. Predictions that return higher fractions of residues are not particularly useful for application purposes, as the predicted region of the protein surface is too large.

This is illustrated in Figs. 1 and 2. Figure 1 depicts typical POOL predictions for the structural genomics protein Pfal00167 (PDB ID 1TQX) [53], a putative D-ribulose 5-phosphate 3-epimerase from *P. falciparum,* a member of the ribulose phosphate binding barrel superfamily [54]. The backbone is shown as a ribbon and the side chains of the predicted residues are shown as dark sticks. The prediction consisting of the top 5% of all residues is shown in Fig. 1a and that of the top 8% of all residues in Fig. 1b. The prediction of Fig. 1a is superimposable on the known active sites of previously characterized D-ribulose 5-phosphate 3-epimerases [54–55] and contains four known catalytic residues H36, D38, H70, and D179. Although this prediction misses one known active site residue, Q177, the similarity of the predicted site to the known binding sites of the well-studied structures with PDB IDs 1RPX [55] and 2FLI [54] is sufficient to confirm the putative functional annotation.

Figure 2 shows the POOL predictions for the same structural genomics protein if higher cut-off values are used. Figure 2a depicts the top 15% of all residues and 2b depicts the top 20% of all residues. These predictions constitute a large fraction of the protein surface area and are less useful.

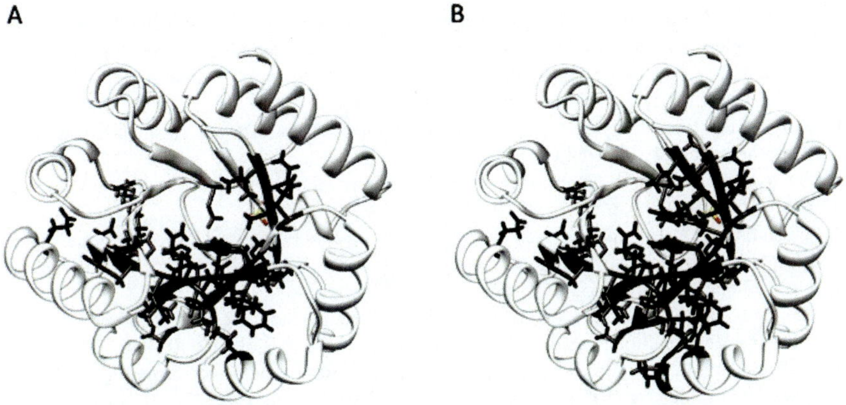

Fig. 2 Predictions for the structural genomics protein Pfal009167. (**a**) 15% POOL cut-off value; (**b**) 20% POOL cut-off value

Future Directions

While POOL in its present form shows excellent performance as a catalytic residue predictor, there are some additional features that could be built in to enhance its performance, including information about evolutionary history obtained from a phylogenetic tree [28, 56].

Acknowledgments This work was supported in part by the National Science Foundation under grant MCB-0843603 awarded to Mary Jo Ondrechen and an IGERT Traineeship and an NSF Graduate Research Fellowship awarded to Joslynn S. Lee.

References

1. Berman, H.M., Westbrook, J., Feng, Z., Gilliland, G., Bhat, T.N., Weissig, H., Shindyalov, I.N., Bourne, P.E. The Protein Data Bank. Nucleic Acids Res. **28**(1): 235–242 (2000).
2. Westbrook, J., Feng, Z., Chen, L., Yang, H., Berman, H.M. The Protein Data Bank and structural genomics. Nucleic Acids Res. **31**: 489–491 (2003).
3. Bartlett, G.J., Porter, C.T., Borkakoti, N., Thornton, J.M. Analysis of catalytic residues in enzyme active sites. J. Mol. Biol. **324**: 105–121 (2002).
4. Porter, C.T., Bartlett, G.J., Thornton, J.M. The Catalytic Site Atlas: a resource of catalytic sites and residues identified in enzymes using structural data.Nucleic Acids Res. **32**(Suppl 1): D129–133 (2004).
5. Ondrechen, M.J., Clifton, J.G., Ringe, D. THEMATICS: a simple computational predictor of enzyme function from structure. Proc. Natl. Acad. Sci. USA **98**: 12473–12478 (2001).
6. Ko, J., Murga, L.F., Andre, P., Yang, H., Ondrechen, M.J., Williams, R.J., Agunwamba, A., Budil, D.E. Statistical Criteria for the identification of protein active sites using theoretical microscopic titration curves. Proteins Struct. Funct. Bioinform. **59**: 183–195 (2005).
7. Wei, Y., Ko, J., Murga, L.F., Ondrechen, M.J. Selective prediction of interaction sites in protein structures with THEMATICS. BMC Bioinformatics **8**: 119 (2007).
8. Tong, W., Wei, Y., Murga, L.F., Ondrechen, M.J., Williams, R.J. Partial order optimum likelihood (POOL): maximum likelihood prediction of protein active site residues using 3D structure and sequence properties. PLoS Comput. Biol. **5**(1): e1000266 (2009).

9. Gilson, M.K. Multiple-site titration and molecular modeling: two rapid methods for computing energies and forces for ionizable groups in proteins. Proteins **15**(3): 266–282 (1993).
10. Gherardini, P.F., Helmer-Citterich, M. Structure-based function prediction: approaches and applications. Brief. Funct. Genomic. Proteomic. (2008).
11. Karp, P.D. What we do not know about sequence analysis and sequence databases. Bioinformatics **14**: 753–754 (1998).
12. Devos, D., Valencia, A. Practical limits of function prediction. Proteins Struct. Funct. Genet. **4**: 98–107 (2000).
13. Wilson, C.A., Kreychman, J., Gerstein, M. Assessing annotation transfer for genomics: quantifying the relations between protein sequence, structure and function through traditional and probabilistic scores. J. Mol. Biol. **297**: 233–249 (2000).
14. Landgraf, R., Xenarios, I., Eisenberg, D. Three-dimensional cluster analysis identifies interfaces and functional residue clusters in proteins. J. Mol. Biol. **307**: 487–502 (2001).
15. de Rinaldis, M., Ausiello, G., Cesareni, G., Helmer-Citterich, M. Three-dimensional profiles: a new tool to identify protein surface similarities. J. Mol. Biol. **284**: 1211–1221 (1998).
16. Aloy, P., E.Querol, Aviles, F.X., Sternberg, M.J.E. Automated structure-based prediction of functional sites in proteins: applications to assessing the validity of inheriting protein function from homology in genome annotation and to protein docking. J. Mol. Biol. **311**: 395–408 (2001).
17. Ota, M., Kinoshita, K., Nishikawa, K. Prediction of catalytic residues in enzymes based on known tertiary structure, stability profile, and sequence conservation. J. Mol. Biol. **327**: 1053–1064 (2003).
18. Gutteridge, A., Bartlett, G., Thornton, J.M. Using a neural network and spatial clustering to predict the location of active sites in enzymes. J. Mol. Biol. **330**: 719–734 (2003).
19. Innis, C.A., Anand, A.P., Sowdhamini, R. Prediction of functional sites in proteins using conserved functional group analysis. J. Mol. Biol. **337**: 1053–1068 (2004).
20. Carter, C.W., LeFebvre, B.C., Cammer, S.A., Tropsha, A., Edgell, M.H. Four-body potentials reveal protein-specific correlations to stability changes caused by hydrophobic core mutations. J. Mol. Biol. **311**(4): 625–638 (2001).
21. Meng, E.C., Polacco, B.J., Babbitt, P.C. Superfamily active site templates. Proteins **55**: 962–976 (2004).
22. Pazos, F., Sternberg, M.J.E. Automated prediction of protein function and detection of functional sites from structure. Proc. Natl. Acad. Sci. USA **101**: 14754–14759 (2004).
23. Cheng, G., Qian, B., Samudrala, R., Baker, D. Improvement in protein functional site prediction by distinguishing structural and functional constraints on protein family. Nucleic Acids Res. **33**(18): 5861–5867 (2005).
24. Petrova, N., Wu, C. Prediction of catalytic residues using Support Vector Machine with selected protein sequence and structural properties. BMC Bioinformatics **7**(1): 312 (2006).
25. Youn, E., Peters, B., Radivojac, P., Mooney, S.D. Evaluation of features for catalytic residue prediction in novel folds. Protein Sci. **16**: 216–226 (2007).
26. Landau, M., Mayrose, I., Rosenberg, Y., Glaser, F., Martz, E., Pupko, T., Ben-Tal, N. ConSurf: the projection of evolutionary conservation scores of residues on protein structures. Nucleic Acids Res. **33**(Web Server issue): W299–302 (2005).
27. Innis, C. siteFiNDER|3D: a web-based tool for predicting the location of functional sites in proteins. Nucleic Acids Res. **35**: W489–W494 (2007).
28. Sankararaman, S., Sjolander, K. INTREPID: INformation-theoretic TREe traversal for protein functional site identification. Bioinformatics **24**: 2445–2452 (2008).
29. Tang, Y.-R., Sheng, Z.-Y., Chen, Y.-Z., Zhang, Z. An improved prediction of catalytic residues in enzyme structures. Protein Eng. Des. Sel. **21**: 295–302 (2008).
30. Bray, T., Chan, P., Bougouffa, S., Greaves, R., Doig, A., Warwicker, J. SitesIdentify: a protein functional site prediction tool. BMC Bioinformatics **10**:379 (2009).
31. Sankararaman, S., Sha, F., Kirsch, J., Jordan, M., K. Sjölander. Active site prediction using evolutionary and structural information. Bioinformatics **26**(5): 617–624 (2010).

32. Wilkins, A., Lua, R., Erdin, S., Ward, R., Lichtarge, O. Sequence and structure continuity of evolutionary importance improves protein functional site discovery and annotation. Protein Sci. **19**: 1296–1311 (2010).
33. Elcock, A.H. Prediction of functionally important residues based solely on the computed energetics of protein structure. J. Mol. Biol. **312**: 885–896 (2001).
34. Bate, P., Warwicker, J. Enzyme/non-enzyme discrimination and prediction of enzyme active site location using charge-based methods. J. Mol. Biol. **340**: 263–276 (2004).
35. Amitai, G., Shemesh, A., Sitbon, E., Shklar, M., Netanely, D., Venger, I., Pietrokovski, S. Network analysis of protein structures identifies functional residues. J. Mol. Biol. **344**: 1135–1146 (2004).
36. Xie, L., Bourne, P.E. A robust and efficient algorithm for the shape description of protein structures and its application in predicting ligand binding sites. BMC Bioinformatics **8**: s4–s9 (2007).
37. Ming, D., Cohn, J.D., Wall, M.E. Fast dynamics perturbation analysis for prediction of protein functional sites. BMC Struct. Biol. **8**(5) (2008).
38. Mattos, C., Ringe, D. Locating and characterizing binding sites on proteins. Nat. Biotechnol. **14**(5): 595–599 (1996).
39. Silberstein, M., Dennis, S., Brown, L., Kortvelyesi, T., Clodfelter, K., Vajda, S. Identification of substrate binding sites in enzymes by computational solvent mapping. J. Mol. Biol. **332**: 1095–1113 (2003).
40. Laurie, A.T.R., Jackson, R.M. Q-SiteFinder: An energy-based method for the prediction of protein-ligand binding sites. Bioinformatics **21**: 1908–1916 (2005).
41. Di Cera, E., Gill, S.J., Wyman, J. Binding capacity: cooperativity and buffering in biopolymers. Proc. Natl. Acad. Sci. USA **85**: 449–452 (1988).
42. Di Cera, E., Chen, Z.-Q. The binding capacity is a probability density function. Biophys. J. **65**: 164–170 (1993).
43. Laskowski, R.A. SURFNET: a program for visualizing molecular surfaces, cavities and intermolecular interactions. J. Mol. Graph. **13**: 323–330 (1995).
44. Laskowski, R.A., Luscombe, N.M., Swindells, M.B., Thornton, J.M. Protein clefts in molecular recognition and function. Protein Sci. **5**: 2438–2452 (1996).
45. Liang, J., Edelsbrunner, H., Woodward, C. Anatomy of protein pockets and cavities: measurement of binding site geometry and implications for ligand design. Protein Sci. **7**: 1884–1897 (1998).
46. Wei, Y. Computed electrostatic properties of protein 3D structure for functional annotation and biomedical application. Boston: Ph.D. Dissertation, Northeastern University, p. 236 (2007).
47. Shehadi, I.A., Yang, H., Ondrechen, M.J. Future directions in protein function prediction. Mol. Biol. Rep. **29**: 329–335 (2002).
48. Tong, W., Williams, R.J., Wei, Y., Murga, L.F., Ko, J., Ondrechen, M.J. Enhanced performance in prediction of protein active sites with THEMATICS and support vector machines. Protein Sci. **17**: 333–341 (2008).
49. Wei, Y., Ringe, D., Wilson, M.A., Ondrechen, M.J. Identification of functional subclasses in the DJ-1 superfamily proteins. PLoS Comput. Biol. **3**(e10): 120–126 (2007).
50. Chan, C.S., Winstone, T.M., Chang, L., Stevens, C.M., Workentine, M.L., Li, H., Wei, Y., Ondrechen, M.J., Paetzel, M., Turner, R.J. Identification of residues in DmsD for twin-arginine leader peptide binding, defined through random and bioinformatics-directed mutagenesis. Biochemistry **47**(9): 2749–2759 (2008).
51. Murga, L.F., Ondrechen, M.J., Ringe, D. Prediction of interaction sites from Apo 3D structures when the holo conformation is different. Proteins **72**(3): 980–992 (2008).
52. Relloso, M., Cheng, T.Y., Im, J.S., Parisini, E., Roura-Mir, C., DeBono, C., Zajonc, D.M., Murga, L.F., Ondrechen, M.J., Wilson, I.A., et al. pH-dependent interdomain tethers of CD1b regulate its antigen capture. Immunity **28**(6): 774–786 (2008).

53. Caruthers, J., Bosch, J., Buckner, F., Voorhis, W.V., Myler, P., Worthey, E., Mehlin, C., Boni, E., DeTitta, G., Luft, J., et al. Structure of a ribulose 5-phosphate 3-epimerase from Plasmodium falciparum. Proteins Struct. Funct. Bioinform. **62**(2): 338–342 (2006).
54. Akana, J., Fedorov, A.A., Fedorov, E., Novak, W.R.P., Babbitt, P.C., Almo, S.C., Gerlt, J.A. d-Ribulose 5-Phosphate 3-Epimerase: functional and structural relationships to members of the ribulose-phosphate binding (β/α)8-barrel superfamily. Biochemistry **45**(8): 2493–2503 (2006).
55. Kopp, J., Kopriva, S., K.-H. Süss, Schulz, G.E. Structure and mechanism of the amphibolic enzyme -ribulose-5-phosphate 3-epimerase from potato chloroplasts. J. Mol. Biol. **287**(4): 761–771 (1999).
56. Lichtarge, O., Bourne, H.R., Cohen, F.E. An evolutionary trace method defines binding surfaces common to protein families. J. Mol. Biol. **257**(2): 342–358 (1996).

Function Prediction of Genes: From Molecular Function to Cellular Function

Kengo Kinoshita and Takeshi Obayashi

Abstract The identification of gene function is a challenging task for molecular biology, since it is quite difficult to assess all of the possible functions experimentally. Therefore, some computational methods to predict or narrow-down the possible functions of genes are needed. There are two different views of functions, namely, molecular functions and cellular functions, and they require completely different approaches. The molecular functions of genes are considered to be tightly coupled with the three-dimensional structures of proteins (i.e. gene products), because molecular functions are archived by a set of chemical reactions, and chemical reactions are realized through molecular interactions among proteins and small molecules. Many methods have been developed based on docking approaches and structural similarity searches; here, we introduce some recent methodologies, including our methods. On the other hand, for cellular functions, the context of the genes or its position in an interaction network should be considered to discern the biological functions, because each biological function is determined by interactions with other gene products. For this purpose, we focus on the co-existence of genes, because co-existence is a necessary condition for the interactions. In this chapter, we will introduce some gene coexpression databases and describe the use of gene coexpression for the identification of cellular functions.

Introduction

According to the rapid progress in genome sequencing techniques, such as the next generation sequencer, more than 1,000 genome sequences have been determined and stored in the public sequence databases. However, many of the genes on the genomes have not been characterized. To fully understand biological systems at the molecular level, the functions of all of the genes must first be clarified. Although

K. Kinoshita (✉)
Graduate School of Information Science, Tohoku University, Sendai, Miyagi, Japan; Institute for Bioinformatics Research and Development, Japan Science and Technology Agency, Chiyoda-ku, Tokyo, Japan
e-mail: kengo@ecei.tohoku.ac.jp

D. Kihara (ed.), *Protein Function Prediction for Omics Era*,
DOI 10.1007/978-94-007-0881-5_11,

the functions of the genes should be verified experimentally in the final stage, it is almost impossible to check all of the possible gene functions. Therefore, some computational methods to refine the possibilities for experimental studies must be developed. In this chapter, we refer to these methods as the function prediction of genes.

In principle, we should be able to predict the functions of genes when we have the genome sequence, because the genome sequence contains the entire information of the living organism. However, this is not currently feasible. Therefore, the usual practice is to search for genes with similar "features", in some sense. The most powerful and popular feature is the genome or amino acid sequence. Sequence similarity implies an evolutional relationship, and an evolutional relationship will lead to a functional relationship, and thus the genes with similar sequences will often encode proteins with similar functions, although distantly related genes and paralogous genes may have different functions [1]. Since the sequence-based approach has been well described in the second section of this book, here we will focus on two different features; namely, the three-dimensional (3D) structure of proteins, and the expression pattern of genes measured by DNA microarray analysis.

Before going into the details, we should first clarify the functions of the genes. It is now widely accepted that there are two extreme views of the function: the molecular function and the cellular function. The molecular function is the function of a single molecule, such as its enzymatic activity, and it is tightly coupled with the 3D structure of the protein. On the other hand, the cellular function is determined by the context of each protein in the interaction network. For example, MEK kinase, existing in the crosstalk point of the EGF and NGF signal transduction pathways (see Fig. 1 in Sasagawa et al. [2]), phosphorylates ERK, and thus its molecular function is phosphorylation, while its cellular function or response to the external signal is determined by the expression patterns of other genes [2]. These two views are completely different, and thus the methods used to predict each of the functions are also different. We will start by discussing the molecular function in the next section, and will then proceed to the cellular function. The molecular function prediction is related to the similarity search of 3D structures of proteins, while the cellular function prediction described in this chapter is based on the similarity of expression patterns.

Molecular Function

It was a very difficult task to determine protein 3D structures during the 1990s, but now it has become one of the standard approaches to analyze protein functions. As a result, more than 60,000 structures have been deposited in the Protein Data Bank (PDB) as of 2010 [3]. According to the rapid increase in protein 3D structures, the similarity of protein structures has become useful to detect the functional relationships of proteins, in a similar manner as sequence similarity. However, in contrast to the sequence similarity, the structural similarities have complicated meanings, because there are several representations in protein structures (Fig. 1) and each representation can have different meanings. In other words, sequence similarity implies

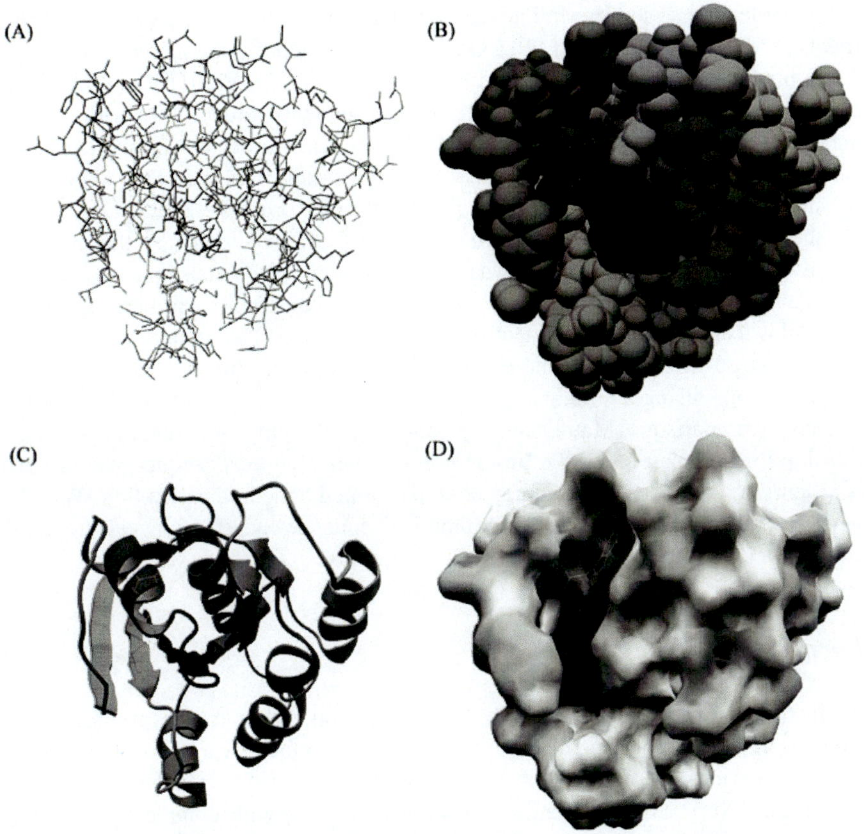

Fig. 1 Several representations of a protein structure (PDB: 5p21, Ras protein, was used). (**a**) Wireframe, (**b**) CPK, (**c**) ribbon, (**d**) molecular surface. Figures were prepared with jV [61]

functional similarity by way of protein evolution, but the relationship between the 3D structure and the function of a protein is sometimes supported by the evolutional relationship, and other times is a result of biochemical principles. Global folds of proteins (Fig. 1c) are considered to be more conserved than amino acid sequences, and thus the fold level similarity often supports the evolutional relationship, except when the fold is a superfold [4], while the similarity of local atomic configurations will exhibit the same or related enzymatic activities, as in the case of the catalytic triad of serine proteases [5]. Therefore, in the case of protein structures, we should first determine the elements for comparison.

Global Fold Similarity

Many methods have been developed to detect global fold similarities. Some of them search for the similar spatial arrangement of Cα or Cβ atoms (Dali [6], SSAP [7], ASH [8], Matras [9]), and some detect similar backbone fragments (SURF [10], CE

[11]), while others identify similar spatial arrangements of secondary structural elements (VAST [12], PROTEP [13], COSEC [14, 15]). Most of the methods introduce restraints of the sequence order, and a few (SURF, VAST, PROTEP and COSEC) can detect the similarity independently of the sequence order [16]. In the 1990s and early 2000s, the similarities without the sequence-order constraints attracted some attention, but recently they have been essentially ignored. This is possibly because the similar global structures without sequence order were quite rare, except for the superfolds due to their internal symmetry [15], and because the interpretation of the kinds of similarities was usually difficult. As a consequence, modern fold comparison methods assumed sequence order constraints and became one of the sophisticated methods to detect remote homologues. This trend is clearly seen in Matras by Kawabata [9], who employed a unique score tuned to detect the distant homologs, using a similar formulation to the Dayhoff substitution matrix for sequence comparisons. Matras can evaluate the probability of an evolutional relationship for each pair of similar proteins, even when significant sequence similarity is absent. In a similar way, Standley et al. developed a method called SeSAW [17], where they introduced a new score, combining sequence conservation and structural similarity, to identify the functional sites of proteins.

Local Atomic Configurations

As in the case of sequence motifs, local structural elements can provide some clues to infer the functions of proteins. The most famous case of local similarity is the catalytic triad of serine protease, where three catalytic residues (Ser, Asp, His) have a very similar arrangement among proteins with completely different folds. Another example can be found among DNA polymerases beta and DNA polymerase I, where three catalytic residues with carboxyl groups (three Asp/Glu) have a similar configuration to catalyze the polymerase reactions. In both cases, the three residues have similar configurations for each catalytic activity. This observation seems to indicate that similarity searches of local atomic configurations will enable us to identify the functions of uncharacterized proteins. To evaluate this possibility, the first systematic analyses of structural comparison were performed by Kobayashi and Go, who found a few new structural elements, four residue fragments, shared by proteins with different folds (ddligase and protein kinase) [18]. They performed a systematic comparison of local structures around the base parts of ATP and GTP, by superimposing the common backbone structures of the adenine and guanine rings. Similar analyses were performed around the phosphate binding sites of mono-nucleotides by Kinoshita et al. [19], and the ddligase and protein kinase also share similar atomic configurations around the phosphate binding sites. Denessiouk et al. further extended the analyses, and found that ddligase and protein kinase share other structural elements in addition to the nucleotide binding site [20]. This example is also impressive, but the main message here is that the structural elements shared with proteins with different functions are found only in limited cases. In other words, a similarity search of local atomic configurations will work well only for limited proteins, and it will often result in the similarity due to a similar fold.

In principle, protein molecular functions are determined by the 3D structures of proteins, especially the local atomic configurations of the catalytic residues, and thus it seemed likely that proteins with similar enzymatic activities would have similar local atomic configurations, but this was not true in natural proteins. This means that the same function can be achieved by several or many different atomic configurations, and when one configuration was invented during the course of evolution, it was retained carefully. This observation indicates that similarity searches of local atomic configuration are not effective, because there are only a few cases with similar atomic configurations and similar functions, beyond the evolutional relationship. However, it should be noted that this observation does not indicate that the proteins with similar atomic configurations cannot have a similar function. A protein can have one or more different functions that have not been experimentally identified. An interesting experiment by Ikura et al. [21] provides a good example that a protein can have a function other than the known function. They used proline isomerase to perform a similarity search of 32 atoms in the active site, and found four proteins with similar atomic configurations and different folds: Chk1, EVEh-1, alpha-amylase, and endopeptidase. The former two proteins are not commercially available, and thus they checked the proline isomerase activity of the latter two proteins. As a result, the two proteins actually had proline isomerase activities, although the total activity measured by kcat/Km was only 0.5% for alpha-amylase and 5% for endopeptidase, as compared with the wild type proline isomerase activities. These activities are quite low as compared with the computationally designed proteins [22, 23], but a discussion about the design, rather than the identification of functions of proteins through the similarity of structures, is beyond the scope of this chapter.

Molecular Surface Similarity

As described above, analyses employing the similarity of atomic configurations have limited effectiveness. To extend the possibility of similarity-based approaches, some researchers have tried to use a different representation of proteins, that is, the molecular surface of proteins. The molecular surface of a protein is the contact surface between the atoms in a protein and a probe sphere with a certain radius (1.4 Å is often used, considering the size of a water molecule), and is also referred to as the Connolly surface [24]. Although the molecular surface itself is a smooth, continuous surface, it is usually represented by a set of triangle meshes for computational approaches, and thus the comparison was performed for a set of vertexes contained in each molecular surface.

The first trial was done by Rosen et al. [25], who used a geometrical hashing (GH) algorithm to find similar shapes of the molecular surfaces of proteins. This approach was extended by Kinoshita et al. in 2002 [26], who incorporated the electrostatic potential on the molecular surface of proteins by using Clique search (CS) algorithms. The GH algorithm is sensitive to small differences in structures, because it makes the superimposition by using three vertexes on a surface, while the CS algorithm performs a superimposition with all of the corresponding vertexes. However, from the viewpoint of calculation time, the GH algorithm is much better than the CS algorithm. More recently, Sael et al. have developed a computationally

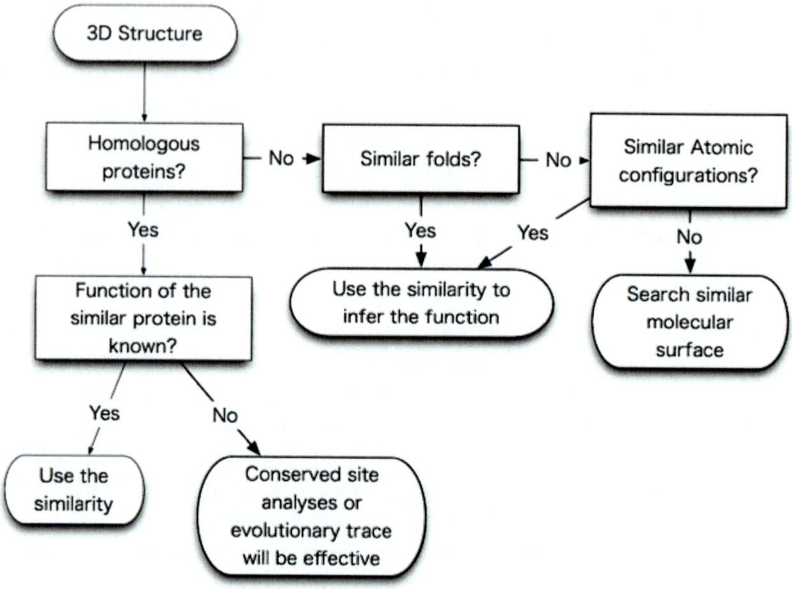

Fig. 2 Flow chart of a structure-based approach

more efficient approach [27], in which a rotational invariant representation of the molecular surface, called 3D Zernike descriptors, was introduced.

Among these methods, Kinoshita et al. have provided a web-based server, called eF-seek (http://ef-site.hgc.jp/eF-seek). The users can upload a PDB formatted file as a query, and they will receive a prediction. This method actually worked well for some cases [28–31], but it will be one of the final options to predict the functions of proteins (Fig. 2).

Beyond the Simple Similarity Search

We have described the similarity-based approaches, where we have shifted from global similarities to local ones, and employed molecular surfaces to address the point that there are small numbers of similar atomic configurations in natural proteins with different folds. Proteins will follow physicochemical rules, even though they are natural products of life, and thus each interaction between a ligand and a protein should be reasonable from a physicochemical viewpoint. In other words, almost all of the elemental interactions will be similar among proteins. Therefore, if we can integrate such elemental interactions, then we will be able to predict the ligand binding sites on proteins.

According to these considerations, Kasahara et al. have developed a method called BUMBLE [32]. They considered "fragments" as the interactive units, and regarded all possible three-successive atoms in ligands as the ligand fragments, and representative three atoms in every amino acid as protein fragments, and then searched for similar fragment-fragment interactions on the query protein. After the superimposition of protein fragments on the query protein, they identified the sites around the query protein where the same types of ligand fragments frequently appeared, which they called hotspots. Finally, their method created putative ligand conformations according to the known fragment interactions and the spatial distribution of the hotspots. They only considered the fragment interactions, and thus their method can detect putative binding sites even when there are no proteins with similar binding sites as a whole. In addition, they create the ligand conformation, and thus their method can predict ligands with novel conformations. This method is also available on the web (http://bumble.hgc.jp).

At first glance, their approach may look like a docking approach with statistical pair potential. However, docking approaches search for possible rotations and translations, and BUMBLE searches for similar fragment interactions. The number of possible freedoms in the former case is far larger than that in the latter case, and thus their method can be more effective. In addition, a pair statistical potential approximates many body interactions with two body interactions, and thus it cannot explicitly consider the motifs, which are important elements of molecular interactions. A comparison of BUMBLE and AutoDock [33] revealed that BUMBLE is more effective to find the correct binding sites, while the latter is better at the prediction of correct ligand conformations. This seems to indicate that the docking approach is suitable for fine-tuning of the complex structure, while the similarity search approach is better for a rough search of the possible binding sites.

Limitations of Structure Based Approaches: Protein Disorder

The fragment based method is very robust to structural changes of proteins, but the so-called "disordered regions" are still a big problem for all structure-based function predictions, because disordered regions are usually invisible in the apo-forms of protein structures, which are prevalent in the current PDB (Fig. 3). The disordered regions are considered to be more abundant in higher organisms, such as eukaryotes [34], and thus they pose a serious obstacle to clarifying the biological systems in eukaryotes. The invisible loops hinder the prediction of complex structures with ligands. However, we can reliably predict the disordered regions, and thus it is possible to speculate about the functional sites of proteins, although we cannot identify the specific roles of the sites.

Many prediction methods have been developed, and various web servers are available [35]. Among them, Ishida and Kinoshita developed a method based on a meta approach [36] (http://prdos.hgc.jp/meta), where they integrated the prediction results from eight independent predictions, including the four best prediction methods in CASP6 [37]. On average, a meta predictor will outperform all of the component predictors, but in each comparison, there are weak and strong cases. For example, this meta predictor is good at short disorder prediction, but is not optimized for long disordered regions. Therefore, for a user who is interested in long

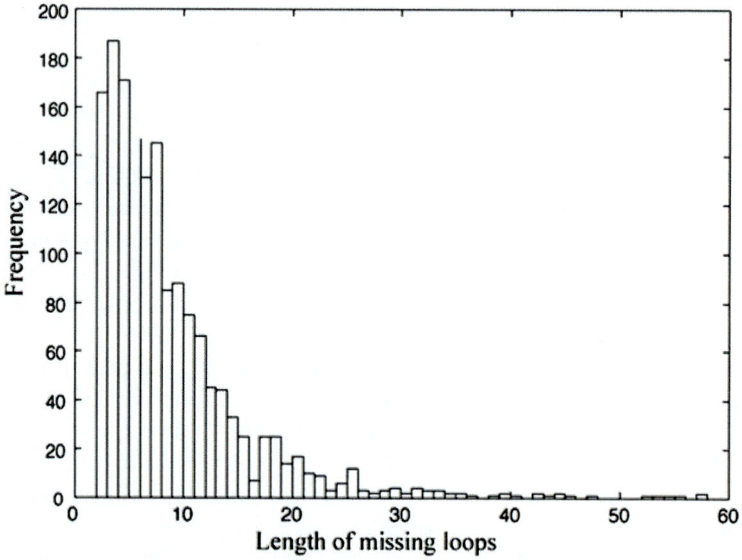

Fig. 3 Length distribution of missing loops in the PDB, as of Dec 25, 2009. The redundancy in the PDB was eliminated by using the PISCES server [62], with a 20% sequence identity and a 2.5 Å resolution threshold. The non-redundant set included 4,565 pdb chains. Missing loops were identified by comparing the amino acid sequence in SEQRES and that in the ATOM record, and 1,590 missing loops other than N- and C-termini among 1,077 chains were found. The mean length and the standard deviation of the length of the missing loops are 8.7 and 9.4, respectively

disordered regions, a method optimized for such regions would be more effective. Noguchi and coworkers have developed a series of disorder prediction methods for short-, long- and mostly disordered proteins [38–40] (http://mbs.cbrc.jp/poodle).

Cellular Function

As described in the previous section, a structural similarity search is a powerful method to detect a functional molecular relationship. However, it is not very useful to infer the relationship of cellular functions, because cellular functions are determined by the contexts of the proteins under consideration. For the cellular function, we should first determine the interaction partners in the protein-protein interaction networks.

Protein–Protein Interactions

Due to the progress in experimental techniques of proteome analyses, a vast amount of information about possible protein interactions is available in public databases, such as IntAct [41], BioGrid [42] and HPRD [43]. These databases contain information about many interactions, and include some interactions obtained by high throughput (HTP) techniques, such as yeast two hybrid experiments. However, HTP data still contain some false positives and false negatives, although they have been improved recently. For example, although proteins with different localizations will not interact with each other in a real biological situation, they may be detected as interacting with each other (false positive cases), while proteins that interact only under limited conditions, such as those with posttranslational modifications like phosphorylation, will not be detected (false negative ones).

The former problem has been tackled by many researchers. One possible approach is to gather different HTP experiments and use a consensus of the data, to enhance the reliability [44]. In a more sophisticated approach, Jansen et al. used mRNA coexpression and co-localization to find reliable interactions [45]. Patil and Nakamura integrated information such as the Gene Ontology annotation, the homologous proteins' interactions and the existence of known interacting Pfam domains [46, 47], and their results are available in a web database at http://hintdb.hgc.jp/htp.

On the other hand, however, the latter problem, the false negative cases, is very difficult to manage, because many possible modification patterns and/or biological environments can affect the conditional interactions of proteins. For this problem, the coexistence of proteins will provide some clues to infer the possible interaction partners for each protein, because coexistence is a necessary condition for interactions. Although we lack comprehensive protein expression level data, we have vast amounts of data regarding mRNA expression measured by DNA microarray techniques. The expression level of a protein can differ from that of the mRNA, but a large amount of DNA microarray data can be useful.

Measuring Gene-Coexpression

Gene coexpression is the similarity of the expression pattern of genes over a number of microarray samples, and it has diverse biological meanings. For example, all of the subunits of a protein complex should be regulated in a coordinated manner to realize the complex structure, and thus they are strictly coexpressed. Actually, new subunits of the chloroplastic NAD(P)H dehydrogenase complex were predicted and verified experimentally [48]. In a similar way, a series of enzymes in a metabolic pathway are often coexpressed [49], and a regulatory relationship between a transcription factor and its target genes can be detected by coexpression [50]. The interaction partners of an Arabidopsis replisome factor, ETG1, were searched by gene coexpression and confirmed by a co-purification experiment [51]. Therefore, gene coexpression can potentially be useful for the identification of functionally related genes.

Gene coexpression has actually been used in the field of plant biology [52], but it has not been widely used in animals, especially in higher organisms such as human and mouse. This is partly because post-transcriptional regulation in higher organisms is more complicated than that in plants, and thus it is difficult to identify coexpression reliably, and also because the tissue organization in animals is far more complex than that in plants. For the former reasons, gene coexpression information reflects transcriptional regulation, and thus the reliability of co-expression will become weaker as the influence of post-transcriptional regulation becomes stronger. One possible approach to overcome this difficulty is to combine PPI information with gene coexpression data, to complement the lack of protein level regulation. The latter reason is more serious, because complex tissue organization requires severe regulatory coordination among tissues, and thus involves more complicated gene regulation. This situation makes it more difficult to extract gene-to-gene functional relationships from a simple similarity index of gene expression patterns between two genes. This possible limitation of gene coexpression was also mentioned by Yanai et al., who studied the expression data of several different mouse tissues [53].

To overcome these difficulties, we have developed a database of gene coexpression, COXPRESdb, for human, mouse and rat [54], and recently extended it with four more species. Pearson's correlation coefficients (PCC) were generally used as the measure of similarity, and higher PCC values are considered to have stronger functional relationships (Fig. 4). However, the distribution of PCC values is quite different for each gene and for each dataset, and thus the PCC value itself is not suitable for a comparison of various genes. Therefore, we used the rank of the Pearson's correlation, rather than the PCC value. The idea of using the rank of a value is very simple, but it was quite effective, as we demonstrated [54]. This is possibly because the number of functionally related genes is not very much different for each gene, while the strength of coexpression is quite divergent, reflecting the fact that gene coexpression has various biological meanings. For example, the gene pairs in a protein complex tend to have higher PCC values than those in metabolic pathways. Furthermore, we also developed a method to extend the single correlation values

From PSMD14		From HOXB2		From HIST1H2BM	
Cor	Gene	Cor	Gene	Cor	Gene
0.54	PSMC2	0.36	HOXB4	0.29	HIST1H2AI
0.51	PSMD1	0.31	HOXB6	0.27	HIST1H1D
0.51	PSMA1	0.28	HOXB7	0.27	HIST1H3D
0.5	PSMA5	0.25	HOXB5	0.27	TMEM132A
0.5	RAN	0.18	SKAP1	0.27	KIAA1652
0.49	PSMB7	0.18	CREB3L4	0.27	CHRND
0.49	PSMA6	0.17	VTCN1	0.26	GRIN3B
0.49	PSMD12	0.17	CRIP1	0.26	TES
0.48	MRPL47	0.17	UCP2	0.26	SSX5
0.48	PSMA3	0.17	PIAS3	0.26	442503

Fig. 4 Example of the relationship between correlation values and rank for three genes. Functionally related genes to each query gene are shown in *shaded boxes*

into a multidimensional correlation, to describe the complex regulation in higher organisms [55]. The multidimensional correlation is based on the subtraction of the principal components of expression patterns. Each subtracted data set virtually corresponds to each biological situation [55].

Systematic analyses of the performance of gene function prediction by coexpression for Arabidopsis, human, mouse and rat were also performed [54]. As a result, the prediction performance was slightly better for Arabidopsis than human and mouse, but all of them were far better than the random prediction. The prediction performance in rat was slightly worse than that for the other three organisms, possibly due to the small number of samples available in the NCBI/GEO. In addition, MAS5 and RMA normalization were better than gcRMA and plier normalization, but the difference in the prediction performance with the different normalizations was smaller than that with the different measures of coexpression, i.e., correlation ranks or values.

Two Approaches in Gene-Coexpression Analyses

There are two approaches for using gene-coexpression data according the different stages of analyses, the "narrow-down" approach, and the "guide-gene approach" [52]. In the early stage of analyses, we usually have little information about the functional relationship, and many genes can become candidates for the analyses, while in the later stage a few genes are specified as target genes, and new relationships with the query genes are investigated. Therefore, in the former stage, the genes under consideration should be "narrowed-down" from many candidate genes, and in the later stage, the new genes are explored according to a "guide" of query genes. Here we will describe the actual steps of these two approaches by using our gene coexpression database, COXPRESdb (http://coxpresdb.jp) [56].

For the narrow-down approach, COXPRESdb provides a tool called "NetworkDrawer" (http://coxpresdb.jp/top_draw.shtml#networkdrawer), where gene networks for a set of genes (user input genes and related genes) are drawn to inspect the internal structure of the coexpression relationships and the known protein-protein interactions. Figure 5 shows an example of an output of "NetworkDrawer", where 10 genes related to a steroid biosynthetic process were selected with the gene annotation of Gene Ontology (GO:0006694) and used as input genes. As seen in the figure, visual inspections could reveal three gene clusters in the networks. The first cluster is composed of the genes needed to synthesize the steroid, and are marked with a small red dot, indicating the existence of a KEGG annotation [57] for the gene. The second cluster is composed of the genes for steroid hormone synthesis. The last cluster corresponds to the PPIs of signaling genes to regulate both clusters. In this way, a large number of genes can be classified based on the gene coexpression information, and each cluster can correspond to some biological functions, although their biological meanings may be diverse.

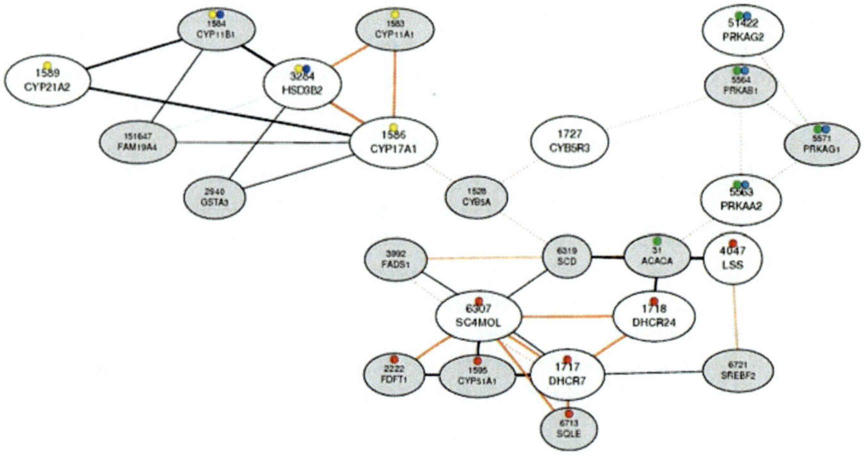

Fig. 5 Gene coexpression network. Each *oval* corresponds to a single gene. *Bright* and *dark ovals* indicate the input genes and the related genes, respectively. The related genes were automatically detected according to the strength of gene coexpression and the existence of PPIs. *Thick, thin*, and *dotted lines* represent strong coexpression, modest coexpression and PPI, respectively. The *small colored circles* in some *ovals* indicate the existence of some annotations in the KEGG database

When the users can narrow down the genes being considered to a few genes, they can use the guide gene approach. If the target gene is a single one, then the user can quickly access the strongly coexpressed genes from the query gene, based on the pre-calculations. Our database provides a gene list for all single genes based on the rank of gene-to-gene correlation of the expression patterns, where the top 300 genes with stronger coexpression for the gene under consideration are available at http://coxpresdb.jp/data/cor/(EntrezGeneID).html. Although several thousand coexpressed genes from a single gene can have statistically significant coexpression [52], 100 genes may be appropriate, from a practical point of view. However, the long gene list will be useful, when the user combines it with other omics data. For this kind of large-scale analysis, all of the coexpression data in COXPRESdb can be downloaded. When the number of genes being considered is two or more, a tool named "CoexSearch" can be used for guide gene analyses, and the user can combine multiple gene lists from each of the genes of interest, based on the correlation ranking.

Although an experimental validation of these approaches for human has not been performed, very successful results are available for Arabidopsis [58, 59]. Experimental validation requires more time than computational evaluations, and these two remarkable results were obtained in the 5 years since we provided the gene coexpression database for plants. Only 2 years have passed since COXPRESdb became available, and thus we believe that successful experimental results will be obtained in the near future, because the computational validation shows comparable performance among the gene coexpression databases for plant and human [54].

Some coexpressed gene databases are also available. They are summarized from the viewpoint of the target species and the available approaches in Table 1. As shown

Table 1 Gene coexpression data, represented in two ways. A gene list representation is used for the guide gene approach, while a gene network representation is used for the narrow-down approach. Ath; Arabidopsis thaliana (thale cress), Osa; Oryza sativa japonica (Japanese rice), Hvu; Hordeum vulgare (barley), Pop; Populus trichocarpa (black cottonwood), Gma; Glycine max (soybean), Mtr; Medicago truncatula (barrel medic), Tae; Triticum aestivum (wheat), Vvi; Vitis vinifera (wine grape), Zma; Zea mays (maize), Hsa; Homo sapiens (human), Mmu; Mus musculus (mouse), Rno; Rattus norvegicus (rat), Gga; Gallus gallus (chicken), Dre; Danio rerio (zebra fish), Dme; Drosophila melanogaster (fly), Cel; Caenorhabditis elegans (nematoda), Sce; Saccharomyces cerevisiae (yeast), Eco; Escherichia coli

Target species (plant)										Target species (animals)									Availability of coexpression data			
Database name	Ath	Osa	Hvu	PoP	Gma	Mtr	Tae	Vvi	Zma	Has	Mmu	Rno	Gga	Dre	Dme	Cel	Sce	Eco	Gene List	Gene Network	Bulk download	Reference
ACT	O																		O			63
ATTED-II	O	O																	O	O	O	64
AraNET	O	O	O	O	O	O	O													O	O	65
BAR	O			O															O			66
CoP	O	O	O	O	O			O	O										O			67
COXPRESdb										O	O	O	O	O	O	O			O	O	O	56
CressExpress	O																		O			68
CSB.DB	O																O	O	O			69
GeneCAT	O	O	O	O															O	O		70
STARNET 2	O	O								O	O	O	O	O	O	O				O		71
Rice Array Database		O																				72
RiceArrayNet		O																		O		73

in Table 1, the number of databases available for plant biology is larger than that for animal researchers, and each of the plant databases has unique features. For example, CressExpress calculates gene coexpression data from the user's specified microarray samples. The gene coexpression strength measured by the PCC value is usually sensitive to the samples, and thus selecting appropriate microarray samples is an important step, but it is not straightforward. Therefore, CressExpress provides an excellent interface for users to select the appropriate microarray samples. Another useful feature is the information about homologous relationships. GeneCAT, ATTED-II and COXPRESdb provide combined coexpressed genes among species, using homologous genes. Such inter-species comparisons will reduce the noise of microarray experiments, especially for the genes with low intensity probes, and also will provide some important clues to infer the evolutionary aspects of the functional modules of genes.

Conclusion

In this chapter, we have described the two different approaches for the function identification of uncharacterized genes on genomes. In the first half of the chapter, we focused on the molecular functions of proteins, which are tightly coupled with their structural information, and also discussed the limitations due to the intrinsically disordered regions found in proteins. In the latter half of this chapter, we switched to the cellular functions, and discussed the construction of interaction networks. For this purpose, we emphasized the strength of gene coexpression, rather than the usual protein interaction networks, due to its potential power. At the same time, PPIs yield valuable information, and thus the combined network approaches, such as the probabilistic functional network by Marcotte [60], will be promising. In addition, user-friendly interfaces for the network will be critically important, because the interpretation of a large-scale network is quite difficult for an information scientist, and thus the perspective of a biologist is indispensable.

References

1. Todd, A.E., Orengo, C.A., Thornton, J.M. Evolution of function in protein superfamilies, from a structural perspective. J. Mol. Biol. **307**, 1113–1143 (2001).
2. Sasagawa, S., Ozaki, Y., Fujita, K., Kuroda, S. Prediction and validation of the distinct dynamics of transient and sustained ERK activation. Nat. Cell. Biol. **7**: 365–373 (2005).
3. Berman, H.M., Westbrook, J., Feng, Z., Gilliland, G., Bhat, T.N., Weissig, H., Shindyalov, I.N., Bourne, P.E. The Protein Data Bank. Nucleic Acids Res. **28**: 235–242 (2000).
4. Orengo, C.A., Jones, D.T., Thornton, J.M. Protein superfamilies and domain superfolds. Nature **372**: 631–634 (1994).
5. Polgar, L. The catalytic triad of serine peptidases. Cell. Mol. Life Sci. **62**: 2161–2172 (2005).
6. Holm, L., Park, J. DaliLite workbench for protein structure comparison. Bioinformatics **16**: 566–567 (2000).
7. Orengo, C.A., Taylor, W.R. SSAP: sequential structure alignment program for protein structure comparison. Methods Enzymol. **266**: 617–635 (1996).

8. Standley, D.M., Toh, H., Nakamura, H. ASH structure alignment package: sensitivity and selectivity in domain classification. BMC Bioinformatics **8**: 116 (2007).
9. Kawabata, T. MATRAS: A program for protein 3D structure comparison. Nucleic Acids Res. **31**: 3367–3369 (2003).
10. Alexandrov, N.N., Takahashi, K., Go, N. Common spatial arrangements of backbone fragments in homologous and non-homologous proteins. J. Mol. Biol. **225**: 5–9 (1992).
11. Shindyalov, I.N., Bourne, P.E. Protein structure alignment by incremental combinatorial extension (CE) of the optimal path. Protein Eng. **11**: 739–747 (1998).
12. Madej, T., Gibrat, J.F., Bryant, S.H. Threading a database of protein cores. Proteins **23**: 356–369 (1995).
13. Grindley, H.M., Artymiuk, P.J., Rice, D.W., Willett, P. Identification of tertiary structure resemblance in proteins using a maximal common subgraph isomorphism algorithm. J. Mol. Biol. **229**: 707–721 (1993).
14. Mizuguchi, K., Go, N. Comparison of spatial arrangements of secondary structural elements in proteins. Protein Eng. **8**: 353–362 (1995).
15. Kinoshita, K., Kidera, A., Go, N. Diversity of functions of proteins with internal symmetry in spatial arrangement of secondary structural elements. Protein Sci. **8**: 1210–1217 (1999).
16. Alexandrov, N.N., Go, N. Biological meaning, statistical significance, and classification of local spatial similarities in nonhomologous proteins. Protein Sci. **3**: 866–875 (1994).
17. Standley, D.M., Yamashita, R., Kinjo, A.R., Toh, H., Nakamura, H. SeSAW: balancing sequence and structural information in protein functional mapping. Bioinformatics **26**: 1258–1259 (2010).
18. Kobayashi, N., Go, N. ATP binding proteins with different folds share a common ATP-binding structural motif. Nat. Struct. Biol. **4**: 6–7 (1997).
19. Kinoshita, K., Sadanami, K., Kidera, A., Go, N. Structural motif of phosphate-binding site common to various protein superfamilies: all-against-all structural comparison of protein-mononucleotide complexes. Protein Eng. **12**: 11–14 (1999).
20. Denessiouk, K.A., Johnson, M.S. When fold is not important: a common structural framework for adenine and AMP binding in 12 unrelated protein families. Proteins **38**: 310–326 (2000).
21. Ikura, T., Kinoshita, K., Ito, N. A cavity with an appropriate size is the basis of the PPIase activity. Protein Eng. Des. Sel. **21**: 83–89 (2008).
22. Jiang, L., Althoff, E.A., Clemente, F.R., Doyle, L., Rothlisberger, D., Zanghellini, A., Gallaher, J.L., Betker, J.L., Tanaka, F., Barbas, 3rd, C.F., Hilvert, D., Houk, K.N., Stoddard, B.L., Baker, D. De novo computational design of retro-aldol enzymes. Science **319**: 1387–1391 (2008).
23. Rothlisberger, D., Khersonsky, O., Wollacott, A.M., Jiang, L., DeChancie, J., Betker, J., Gallaher, J.L., Althoff, E.A., Zanghellini, A., Dym, O., Albeck, S., Houk, K.N., Tawfik, D.S., Baker, D. Kemp elimination catalysts by computational enzyme design. Nature **453**: 190–195 (2008).
24. Connolly, M.L. Solvent-accessible surfaces of proteins and nucleic acids. Science **221**: 709–713 (1983).
25. Rosen, M., Lin, S.L., Wolfson, H., Nussinov, R. Molecular shape comparisons in searches for active sites and functional similarity. Protein Eng. **11**: 263–277 (1998).
26. Kinoshita, K., Furui, J., Nakamura, H. Identification of protein functions from a molecular surface database, eF-site. J. Struct. Funct. Genomics **2**: 9–22 (2002).
27. Sael, L., La, D., Li, B., Rustamov, R., Kihara, D. Rapid comparison of properties on protein surface. Proteins **73**: 1–10 (2008).
28. Handa, N., Terada, T., Kamewari, Y., Hamana, H., Tame, J.R., Park, S.Y., Kinoshita, K., Ota, M., Nakamura, H., Kuramitsu, S., Shirouzu, M., Yokoyama, S. Crystal structure of the conserved protein TT1542 from Thermus thermophilus HB8. Protein Sci. **12**: 1621–1632 (2003).
29. Kinoshita, K., Nakamura, H. Identification of protein biochemical functions by similarity search using the molecular surface database eF-site. Protein Sci. **12**: 1589–1595 (2003).

30. Kinoshita, K., Nakamura, H. Identification of the ligand binding sites on the molecular surface of proteins. Protein Sci. **14**: 711–718 (2005).
31. Standley, D.M., Kinjo, A.R., Kinoshita, K., Nakamura, H. Protein structure databases with new web services for structural biology and biomedical research. Brief Bioinform. **9**: 276–285 (2008).
32. Kasahara, K., Kinoshita, K., Takagi, T. Ligand binding site prediction of proteins based on known fragment-fragment interactions. Bioinformatics **26**: 1493–1499 (2010).
33. Trott, O., Olson, A.J. AutoDock Vina: improving the speed and accuracy of docking with a new scoring function, efficient optimization, and multithreading. J. Comput. Chem. **31**: 455–461 (2010).
34. Ward, J.J., Sodhi, J.S., McGuffin, L.J., Buxton, B.F., Jones, D.T. Prediction and functional analysis of native disorder in proteins from the three kingdoms of life. J. Mol. Biol. **337**: 635–645 (2004).
35. Ferron, F., Longhi, S., Canard, B., Karlin, D. A practical overview of protein disorder prediction methods. Proteins **65**: 1–14 (2006).
36. Ishida, T., Kinoshita, K. Prediction of disordered regions in proteins based on the meta approach. Bioinformatics **24**: 1344–1348 (2008).
37. Jin, Y., Dunbrack, Jr. R.L. Assessment of disorder predictions in CASP6. Proteins **61**(Suppl 7): 167–175 (2005).
38. Hirose, S., Shimizu, K., Kanai, S., Kuroda, Y., Noguchi, T. POODLE-L: a two-level SVM prediction system for reliably predicting long disordered regions. Bioinformatics **23**: 2046–2053 (2007).
39. Shimizu, K., Hirose, S., Noguchi, T. POODLE-S: web application for predicting protein disorder by using physicochemical features and reduced amino acid set of a position-specific scoring matrix. Bioinformatics **23**: 2337–2338 (2007).
40. Shimizu, K., Muraoka, Y., Hirose, S., Tomii, K., Noguchi, T. Predicting mostly disordered proteins by using structure-unknown protein data. BMC Bioinformatics **8**: 78 (2007).
41. Kerrien, S., Alam-Faruque, Y., Aranda, B., Bancarz, I., Bridge, A., Derow, C., Dimmer, E., Feuermann, M., Friedrichsen, A., Huntley, R., Kohler, C., Khadake, J., Leroy, C., Liban, A., Lieftink, C., Montecchi-Palazzi, L., Orchard, S., Risse, J., Robbe, K., Roechert, B., Thorneycroft, D., Zhang, Y., Apweiler, R., Hermjakob, H. IntAct – open source resource for molecular interaction data. Nucleic Acids Res. **35**: D561–565 (2007).
42. Stark, C., Breitkreutz, B.J., Reguly, T., Boucher, L., Breitkreutz, A., Tyers, M. BioGRID: a general repository for interaction datasets. Nucleic Acids Res. **34**: D535–539 (2006).
43. Keshava Prasad, T.S., Goel, R., Kandasamy, K., Keerthikumar, S., Kumar, S., Mathivanan, S., Telikicherla, D., Raju, R., Shafreen, B., Venugopal, A., Balakrishnan, L., Marimuthu, A., Banerjee, S., Somanathan, D.S., Sebastian, A., Rani, S., Ray, S., Harrys, Kishore, C.J., Kanth, S., Ahmed, M., Kashyap, M.K., Mohmood, R., Ramachandra, Y.L., Krishna, V., Rahiman, B.A., Mohan, S., Ranganathan, P., Ramabadran, S., Chaerkady, R., Pandey, A. Human protein reference database – 2009 update. Nucleic Acids Res. **37**: D767–772 (2009).
44. Ito, T., Chiba, T., Ozawa, R., Yoshida, M., Hattori, M., Sakaki, Y. A comprehensive two-hybrid analysis to explore the yeast protein interactome. Proc. Natl. Acad. Sci. USA **98**: 4569–4574 (2001).
45. Jansen, R., Yu, H., Greenbaum, D., Kluger, Y., Krogan, N.J., Chung, S., Emili, A., Snyder, M., Greenblatt, J.F., Gerstein, M. A Bayesian networks approach for predicting protein-protein interactions from genomic data. Science **302**: 449–453 (2003).
46. Finn, R.D., Mistry, J., Tate, J., Coggill, P., Heger, A., Pollington, J.E., Gavin, O.L., Gunasekaran, P., Ceric, G., Forslund, K., Holm, L., Sonnhammer, E.L., Eddy, S.R., Bateman, A. The Pfam protein families database. Nucleic Acids Res. **38**: D211–222 (2010).
47. Patil, A., Nakamura, H. Filtering high-throughput protein-protein interaction data using a combination of genomic features. BMC Bioinformatics **6**: 100 (2005).
48. Takabayashi, A., Ishikawa, N., Obayashi, T., Ishida, S., Obokata, J., Endo, T., Sato, F. Three novel subunits of Arabidopsis chloroplastic NAD(P)H dehydrogenase identified by bioinformatic and reverse genetic approaches. Plant J. **57**: 207–219 (2009).

49. Yonekura-Sakakibara, K., Tohge, T., Niida, R., Saito, K. Identification of a flavonol 7-O-rhamnosyltransferase gene determining flavonoid pattern in Arabidopsis by transcriptome coexpression analysis and reverse genetics. J. Biol. Chem. **282**: 14932–14941 (2007).
50. Hirai, M.Y., Sugiyama, K., Sawada, Y., Tohge, T., Obayashi, T., Suzuki, A., Araki, R., Sakurai, N., Suzuki, H., Aoki, K., Goda, H., Nishizawa, O.I., Shibata, D., Saito, K. Omics-based identification of Arabidopsis Myb transcription factors regulating aliphatic glucosinolate biosynthesis. Proc. Natl. Acad. Sci. USA **104**: 6478–6483 (2007).
51. Takahashi, N., Lammens, T., Boudolf, V., Maes, S., Yoshizumi, T., De Jaeger, G., Witters, E., Inze, D., De Veylder, L. The DNA replication checkpoint aids survival of plants deficient in the novel replisome factor ETG1. EMBO J. **27**: 1840–1851 (2008).
52. Obayashi, T., Kinoshita, K. Coexpression landscape in ATTED-II: usage of gene list and gene network for various types of pathways. J. Plant. Res. **123**: 311–319 (2010).
53. Yanai, I., Korbel, J.O., Boue, S., McWeeney, S.K., Bork, P., Lercher, M.J. Similar gene expression profiles do not imply similar tissue functions. Trends Genet. **22**: 132–138 (2006).
54. Obayashi, T., Kinoshita, K. Rank of correlation coefficient as a comparable measure for biological significance of gene coexpression. DNA Res. **16**: 249–260 (2009).
55. Kinoshita, K., Obayashi, T. Multi-dimensional correlations for gene coexpression and application to the large-scale data of Arabidopsis. Bioinformatics **25**: 2677–2684 (2009).
56. Obayashi, T., Hayashi, S., Shibaoka, M., Saeki, M., Ohta, H., Kinoshita, K. COXPRESdb: a database of coexpressed gene networks in mammals. Nucleic Acids Res. **36**: D77–82 (2008).
57. Kanehisa, M., Araki, M., Goto, S., Hattori, M., Hirakawa, M., Itoh, M., Katayama, T., Kawashima, S., Okuda, S., Tokimatsu, T., Yamanishi, Y. KEGG for linking genomes to life and the environment. Nucleic Acids Res. **36**: D480–484 (2008).
58. Sugano, S.S., Shimada, T., Imai, Y., Okawa, K., Tamai, A., Mori, M., Hara-Nishimura, I. Stomagen positively regulates stomatal density in Arabidopsis. Nature **463**: 241–244 (2010).
59. Bednarek, P., Pislewska-Bednarek, M., Svatos, A., Schneider, B., Doubsky, J., Mansurova, M., Humphry, M., Consonni, C., Panstruga, R., Sanchez-Vallet, A., Molina, A., Schulze-Lefert, P. A glucosinolate metabolism pathway in living plant cells mediates broad-spectrum antifungal defense. Science **323**: 101–106 (2009).
60. Lee, I., Lehner, B., Crombie, C., Wong, W., Fraser, A., Marcotte, E. A single gene network accurately predicts phenotypic effects of gene perturbation in Caenorhabditis elegans. Nat. Genet. **40**: 181–188 (2008).
61. Kinoshita, K., Nakamura, H. eF-site and PDBjViewer: database and viewer for protein functional sites. Bioinformatics **20**: 1329–1330 (2004).
62. Wang, G., Dunbrack, Jr. R.L. PISCES: a protein sequence culling server. Bioinformatics **19**: 1589–1591 (2003).
63. Mutwil, M., Usadel, B., Schutte, M., Loraine, A., Ebenhoh, O., Persson, S. Assembly of an interactive correlation network for the Arabidopsis genome using a novel heuristic clustering algorithm. Plant Physiol. **152**: 29–43 (2009).
64. Manfield, I.W., Jen, C.H., Pinney, J.W., Michalopoulos, I., Bradford, J.R., Gilmartin, P.M., Westhead, D.R. Arabidopsis Co-expression Tool (ACT): web server tools for microarray-based gene expression analysis. Nucleic Acids Res. **34**: W504–W509 (2006).
65. Obayashi, T., Hayashi, S., Saeki, M., Ohta, H., Kinoshita, K. ATTED-II provides coexpressed gene networks for Arabidopsis. Nucleic Acids Res. **37**: D987–D991 (2009).
66. Ogata, Y., Suzuki, H., Sakurai, N., Shibata, D. CoP: a database for characterizing co-expressed gene modules with biological information in plants. Bioinformatics **26**: 1267–1268,
67. Toufighi, K., Brady, S.M., Austin, R., Ly, E., Provart, N.J. The botany array resource: e-Northerns, expression angling, and promoter analyses. Plant J. **43**: 153–163 (2005).
68. Steinhauser, D., Usadel, B., Luedemann, A., Thimm, O., Kopka, J. CSB.DB: a comprehensive systems-biology database. Bioinformatics **20**: 3647–3651 (2004).
69. Mutwil, M., Obro, J., Willats, W.G., Persson, S. GeneCAT – novel webtools that combine BLAST and co-expression analyses. Nucleic Acids Res. **36**: W320–326 (2008).

70. Srinivasasainagendra, V., Page, G.P., Mehta, T., Coulibaly, I., Loraine, A.E. CressExpress: a tool for large-scale mining of expression data from Arabidopsis. Plant Physiol. **147**: 1004–1016 (2008).
71. Jung, K.H., Dardick, C., Bartley, L.E., Cao, P., Phetsom, J., Canlas, P., Seo, Y.S., Shultz, M., Ouyang, S., Yuan, Q., Frank, B.C., Ly, E., Zheng, L., Jia, Y., Hsia, A.P., An, K., Chou, H.H., Rocke, D., Lee, G.C., Schnable, P.S., An, G., Buell, C.R., Ronald, P.C. Refinement of light-responsive transcript lists using rice oligonucleotide arrays: evaluation of gene-redundancy. PLoS One **3**: e3337 (2008).
72. Jupiter, D., Chen, H., VanBuren, V. STARNET 2: a web-based tool for accelerating discovery of gene regulatory networks using microarray co-expression data. BMC Bioinformatics **10**: 332 (2009).
73. Lee, T.H., Kim, Y.K., Pham, T.T., Song, S.I., Kim, J.K., Kang, K.Y., An, G., Jung, K.H., Galbraith, D.W., Kim, M., Yoon, U.H., Nahm, B.H. RiceArrayNet: a database for correlating gene expression from transcriptome profiling, and its application to the analysis of coexpressed genes in rice. Plant Physiol. **151**: 16–33 (2009).

Predicting Gene Function Using Omics Data: From Data Preparation to Data Integration

Weidong Tian, Xinran Dong, Yuanpeng Zhou, and Ren Ren

Abstract In the post-genomic era, the continuing development of high-throughput technologies has led to the explosion of enormous amount of omics data, ranging from genomics, transcriptomics, proteomics, metabolomics, to phenomics. Integration of diverse omics data can help us to understand the complete functions of genes in the cell. However, the complexity, heterogeneity, and large-scale of the omics data have created significant challenges to the gene function prediction field. Currently, the focus of this field is to develop efficient and accurate algorithms to integrate omics data for predicting gene function. In this chapter, we first introduce the various types of omics data, and how they relate to gene functions. Then, we review current algorithms available for integrating omics data for gene function predictions. Next, we use a combined algorithm named Funckenstein as an example to further illustrate the integration process. In the final two sections, we discuss current limitations and potential improvements of this field, and offer perspectives for future directions.

Introduction

Understanding the function of genes, including the molecular function, the biological role it plays in the cell, and the impact of its malfunction on phenotypes and diseases, is a central task in biology. Traditionally, experimentalists study the function of genes by focusing on one or a few at a time. The advent of genomic era has completely revolutionized our approach to study biology. Since the initiation of the Human Genome Project (HGP) in 1990 [1], the breakthrough of modern high-throughput sequencing technologies has allowed for the decoding of the complete genomic DNA sequences of more than a thousand cellular organisms including human genome. Along with the accomplishment of complete genome sequences have emerged a diverse range of high-throughput technologies such as

W. Tian (✉)
School of Life Sciences, Fudan University, Shanghai, China; Institute of Biostatistics, Fudan University, Shanghai, China
e-mail: weidong.tian@fudan.edu.cn

D. Kihara (ed.), *Protein Function Prediction for Omics Era*,
DOI 10.1007/978-94-007-0881-5_12, © Springer Science+Business Media B.V. 2011

oligonucleotide array, cDNA array, high-throughput two hybrid system, mass spectrometry, and so on. Thanks to the continuously reduced cost of the high-throughput technologies, it is now a routine task for many laboratories to study the properties and relationships of thousands of genes in parallel, presenting biologists an unprecedented opportunity to study the function of genes at a system level.

Given the sheer volume of the omics data, how to take advantage of the data to generate biologically meaningful insights about gene functions presents a critical challenge to the field of biology. Computational biology or bioinformatics is thus emerging as a new discipline, aiming to develop computational and statistical algorithms to effectively sort, analyze and interpret the omics data. Gene function prediction is one of the most important goals of computational biology. It can not only provide hypothesis about the function of a particular set of genes of interest that can be verified experimentally, but also uncover important mechanisms of gene function through learning the rules of predicting gene function accurately.

As the genomics era starts with the flood of genomics data, i.e., gene and protein sequences, the computational approaches initially focus on inferring gene functions by sequence comparison [2–6]. The underlying hypothesis of the sequence-based methods is that homologous proteins evolving from the same ancestor are likely to share the same function. The sequence-based methods play important roles in annotation of the newly sequenced genomes, with the majority of genes functionally inferred on the basis of the sequence similarity to previously characterized proteins. However, this approach can provide functional insights to only 50% of the genes in the genome by detecting evolutionary relationship with known proteins [7].

The sequence-based methods on gene function prediction are effective in assigning the molecular function of genes, for instance, the catalytic activity of enzymes. However, it often fails to answer what role a gene plays in a biological process, how it interacts with other genes, and where it functions in the cell, which are fundamental questions in biology. This failure is mostly because those functional aspects of the gene are determined not only by the gene sequence, but also by its relationships with other genes that may not be evolutionarily related with the target. To answer those questions, information beyond sequence alone is required. The rich trove of omics data, ranging from genomic sequence, gene expression, protein–protein interaction, genetic interaction, phenotypic change, to epigenetic information, provide information about the behaviors of a gene from various aspects. Therefore, a current challenge in gene function prediction field is to design computational algorithms to piece together information from various types of omics data, in order to obtain the whole picture of the biological role of genes in the cell.

This chapter is organized as the following sections. In the first section, we focus on omics data preparation by describing the latest high-throughput technologies to generate the data and how each type of omics data is related to gene function. In the second section, we review current algorithms available for integrating omics data to predict gene functions. In the third section, we describe in detail a combined algorithm named Funckenstein to illustrate the process of omics-based gene function prediction [8]. In the final two sections, we discuss current limitations and potential improvements of the field, and offer perspective for future directions.

Omics Data Preparation

Before describing omics data and how they relate to gene functions, let's first clarify the meaning of function. The functions of a gene essentially are observations of its behavior in the cell. For a protein kinase, from a biochemist's point of view, its function can be the phosphorylation of a hydroxyl group of a specific substrate; while in a geneticist's opinion, its function can be the signaling transduction pathway in which the gene is involved, or the disease phenotype appearing when the gene is mutated or knocked out. In order to have a complete picture of the gene function, we need to have an ontology system covering various aspects of gene functions. Gene Ontology (GO) is such an ontology system [9]. It contains three ontologies: molecular function, biological process, and cellular component. Molecular function describes the biochemical activity of a gene product, "protein tyrosine kinase" for example. Biological process refers to the biological role to which a set of genes and gene products contribute, e.g. "DNA damage pathway". Cellular component tells where in the cell a gene operates its function, for instance, "nucleus". The GO terms are organized in a directed acyclic graph, and arranged in a manner from general to specific, making it easy to be parsed by computers. GO has become the most widely used functional annotation scheme, and the current goal of gene function prediction is to predict the GO terms associated with each gene in the genome. GO term annotation of genes in different genomes can be found in the GO database.

Following the central pathway of biological information flow from the genome to cellular phenotype, we classify the omics data into five main categories: genomics, transcriptomics, proteomics, metabolomics, and phenomics (Fig. 1). Genomics represents the whole genome sequence information including gene, regulatory element, and non-coding RNA, etc. Transcriptomics covers the whole RNA transcripts in the cell, while proteomics characterizes all proteins in the cell. Metabolomics consists of proteins, mostly enzymes, and metablotes that are catalyzed or produced by enzymes in the cell. Phenomics is the combined result of genomics, transcriptomics, proteomics, and metabolomics, representing all observable cellular or organism characteristics.

Genomics

The first complete genome of a living organism was sequenced in 1995 [10]. In 2003 the complete sequence of the human genome was finished [11]. Today, there are more than 1,000 completely sequenced genomes in the public domain, and some estimates this number could reach to more than 10,000 by 2012. This owes to the introduction of the next-generation sequencing technologies which employ massively parallel sequencing strategy, capable of sequencing millions of sequence

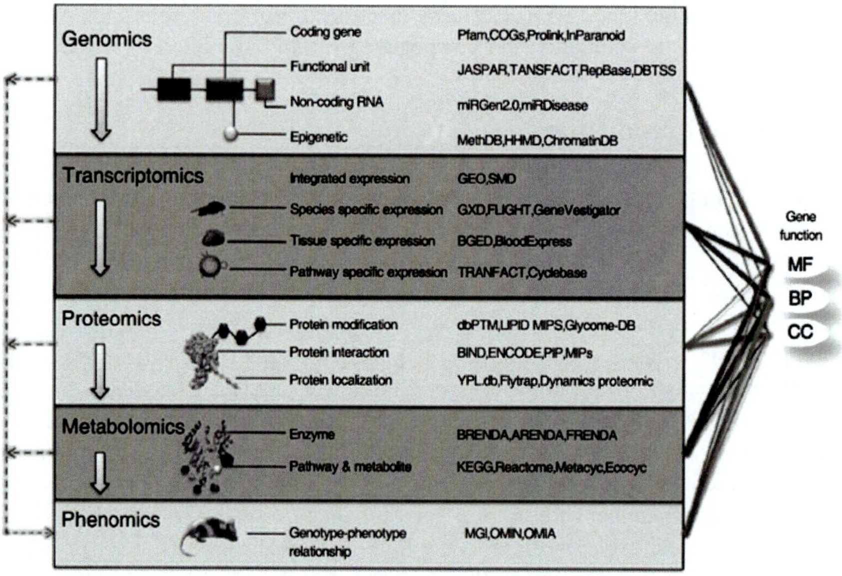

Fig. 1 Omics, database and gene function prediction. Omics data are classified into five main categories: genomics, transcriptomics, proteomics, metabolomics and phenomics. Sub-types of omics data and the representative associated databases under each category are shown. The reference for each database can be found in the main text. The link of each category of omics data to GO function is shown. Here, MF, BP, and CC are abbreviations of molecule function, biological process and cellular component, respectively. The *thickness of the line linking* omics data to gene function represents their empirical relative strength in predicting the corresponding GO functions. For example, among the three GO terms, genomics data are most effective in predicting MF, while phenomics data predict BP better than the other two GO terms. In contrast, proteomics and transcriptomics data predict BP and CC better than MF, while metabolomics data are effective in predicting MF and BP

reads in a single run, such as the 454 [12], Illumina [13] and SOLid system [14]. Using the new technology, the full genome of James Watson, the well-known DNA pioneer, was sequenced and assembled with 7.4 fold coverage in less than 2 months [15]. With such a development pace, the personalized-genomics era will be coming soon.

Model organism databases curate, manage, and store detailed up-to-date information about the gene mapping, annotation, protein domains and structures, expression data, mutant phenotypes, physical and genetic interactions, etc, of the model genomes, such as the Saccharomyces Genome Database (SGD) [16], the Mouse Genome Informatics (MGI) [17], the Arabidopsis Information Resource (TAIR) [18], the Fly Base [19], etc. Such databases are now the researcher's starting point for informed hypothesis generation. There are also databases that store specific genomics data. According to the genome organization, we can classify those databases into coding gene, functional unit such as regulatory sequence, and non-coding RNA databases.

The coding gene and protein sequence databases contain information from a wide range of gene and protein sequence features. They have been the largest sources of training data for gene function prediction. Detection of evolutionary relationships is the first step of functional inference for sequence-based methods. Pfam is a database of evolutionarily related protein sequences [4]. It currently contains more than 10,000 protein families generated from the multiple sequence alignments (MSAs) of evolutionarily related sequences using Hidden Markov Models (HMMs). Those protein families cover more than 70% protein sequences in the protein universe. Evolutionary relationships between sequences can be further distinguished as orthology, paralogy, and inparalogy [20]. Because orthologous sequences are resulted from a speciation event, and likely retain the ancestral function, detection of orthologous relationships can be effective in making gene function prediction [21]. Such databases include Cluster of Othologous Groups (COGs) [22] and InParanoid [23], etc. Phylogenetic profile shows the pattern of the presence and absence of the homolog of a given gene in different genomes. Two genes with similar phylogenetic profile tend to be functionally related, e.g., involved in the same pathway. It provides a non-homology based way to infer functions. ProLinks is a databases of phylogenetic profiles [24]. In addition, there are databases focusing on protein sequence features and patterns related to protein functions, such as Prosite [25] and PRINTS [26]. Direct functional inference can be made when a new sequence matches a known protein feature or pattern.

Functional unit databases include those containing regulatory sequences (e.g., transcription factor (TF) binding sites), repeat elements, and other functional units, such as enhancer, silencer, etc. Those functional units are not genes, but they are located in the vicinity of gene, e.g., in the promoter region, and often evolutionarily conserved. They play important roles in regulating, altering and determining gene functions. Patterns of the functional units can provide important hints about gene function [27]. Identification of genome-wide patterns of TF binding sites can be done by high-throughput technologies including CHIP-chip [28] and CHIP-seq [29]. TF binding sites can also be predicted by in silico methods, mostly based on evolutionary conservation. TRANSFAC [30] and JASPAR [31] are two large TF binding site databases, consisting of both experimentally validated and putative evolutionarily conserved TF binding sites in eukaryotic genomes. In addition, though the functional role of repeat elements remains in speculation, a recent study found that in human genome, functionally similar genes are overrepresented among genes with similar repeat element profiles in the promoter region [32], suggesting that repeat elements information is worth continuing exploration for gene function prediction. RepBase [33] is the database storing repeat element information. The promoter region contains rich information responsible for regulatory role of gene function, and DBTSS (DataBase of Transcription Start Sites (TSS)) [34], which includes precise positional information for TSS and promoter region of the eukaryotic mRNA, can be useful for predicting gene function as well.

More and more evidence have shown that the majority of transcriptome consist of non-coding RNA transcripts [35]. Thus, RNomics, the study of the structure, function, and process of non-coding RNAs, is starting to attract more and more attention. Though the function of most non-coding RNAs remains mysteries, the discovery and extensive studies of microRNA or microRNomics have led to a new paradigm of gene regulation which takes place post-transcriptionally and pre-translationally [36]. MicroRNAs regulate the process of cell development, differentiation, proliferation, mobility, and apoptosis through the regulation of its target genes. Target genes regulated by the same microRNA may be involved in the same biological process. miRGen is a database that provides information about the miRNA target genes and their corresponding TF in human and mouse [37]. miR2Disease provides comprehensive information about human diseases associated with miRNA deregulation from literatures [38].

Besides genomics data, epigenomics that study the epigenetic changes, including DNA methylation and histone modifications, across the entire genome can provide important insights about the function of genes as well [39]. Epigenetics changes can lead to activation or inactivation of genes, and play important roles in cell development, differentiation and tumorigenesis. The DNA Methylation Database (MethDB) [40] and Human Histone Modification Database (HHMD) [41] contain information about DNA methylation and histone modification in human genome, while the ChromatinDB [42] database contains genome-wide ChIP data for histone modifications in yeast genome. With more and more experimental data becoming available, mining epigenomics data will provide a novel approach to predict gene functions.

Transcriptomics

The transcriptome represents the complete set of RNA transcripts in the cell [43]. Both the expression and abundance of RNA transcripts can change in response to cellular development, physiological and environmental condition changes. The microarrays and serial analysis of gene expression (SAGE) represent the most well-used technologies to study transcriptome [44]. Recently, deep sequencing RNA transcripts using the next-generation sequencing technologies has detected RNA transcripts at single base resolution, allowing for the discovery of novel transcripts that cannot be detected with traditional technologies [45]. Transcriptomics data are invaluable to understand gene functions. By focusing on differentially expressed genes under different development stages, one may identify genes responsible for the biological process governing cellular development. In addition, genes with correlated expression patterns under different conditions are likely functionally related [46]. Gene Expression Omnibus (GEO) is the largest public repository of transcriptomics data [47]. It currently contains more than 400 thousands samples submitted from a wide range of platforms on many organisms, and this number is increasing every day. In addition, there are species-specific expression databases, such as GXD for mouse [48], FLIGHT for fly [49], and GeneVestigator [50] for Arabidopsis; tissue specific expression databases, such as BGED for brain [51], and BloodExpress for blood [52]; pathway specific expression databases, such as GermOnline for germ line development [53] and Cyclebase for cell cycle process [54]. The vast amount of transcriptomics data under a wide range of conditions makes mining of transcriptomics data an active field for gene function prediction.

Proteomics

Proteins are the main components of the metabolic pathway, and many proteins interact with each other either in a complex or transiently to function in a biological process. Proteome is the complete set of proteins encoded in the genome. Proteomics is the large-scale study of proteome, focusing on the post-translational modifications of proteins, protein abundance, protein variants, and protein-protein interactions [55]. Depending on the environmental and cellular physiological conditions, proteome may vary significantly from one cell or condition to another. Protein abundance may not be inferable from RNA expression, due to post-transcriptional regulation. Proteins are also subject to post-translational modifications, such as phosphorylation, glycosylation, and acetylation, which are critical for some proteins to be functional.

The most widely used proteomics techniques are two-dimensional gel electrophoresis [56] and mass spectrometry [57]. Both can identify and quantify cellular proteins. New technologies, such as shotgun proteomics, promise to significantly improve the accuracy and coverage of proteome detection [58]. Latest technologies to determine post-translational modifications of proteins include PROTOMAP which combines SDS-PAGE with shotgun proteomics [59]. Databases of post-translational modifications include dbPTM [60], an integrated database containing information about protein phosphorylation, glycosylation and sulfation, etc. Protein subcellular localization is one of the three ontologies of gene functions in GO. There are several species-specific databases of subcellular location, e.g., YPL.db for yeast [61] and Flytrap for Drosophila.

Interactomics is the study of all protein physical interactions in the cell. In a broad sense, the interaction can be extended to refer to the interaction between protein and DNA or RNA, or the genetic interactions between proteins as well. High-throughput interaction technologies include yeast two-hybrid system [62] and tandem Affinity purification followed by mass spectrometry (TAP) [63], etc. Genome-wide protein-protein interactomes have been reconstructed in several model organisms, including yeast [64], worm [65], and human [66]. A number of interaction databases have been established, including BIOGRID [67], MIPS [68], IntAct [69], MINT [70], DIP [71] from published literatures, PIP [72] and OPHID [73] from computational predictions, and the integrated databases, such as BIND [74], HPRD and STRING [75]. Technologies detecting protein-DNA and RNA interactions include Protein-chip [76]. BIND [74] and ENCODE [77] databases contain information about protein-DNA interactions. Genetic interactions can be captured by synthetic genetic array (SGA) [78], diploid-based synthetic lethality analysis with microarrays (dSLAM) [79], synthetic dosage-suppression and lethality and haploinsufficiency [80]. BIOGRID [67] database contains genetic interactions from literatures.

Metabolomics

Metabolomics is the study of small chemical metabolite in the cell. Enzymes are the major components of metabolism that catalyze to convert or give rise to metabolites. In response to change of environmental and cellular condition, the gene expression, translation, and catalytic activity of enzymes can change, which can lead to the change of metabolite profiles. Small metabolites in turn can play important regulatory roles in gene expression, translation, and the biological processes. Therefore, it is necessary to integrate transcriptomics, proteomics and metabolomics data in the same context, in order to obtain a complete picture of gene functions. High-throughput metabolomics technologies include gas chromatographic mass spectrometry (GC/MS) [81], liquid chromatographic mass spectrometry (LC/MS) [82], as well as nuclear magnetic resonance (NMR) [83]. Examples of Enzyme databases include BRENDA (BRaunschweig ENzyme DAtabase) [84] that contains information about classification, nomenclature, reaction, specificity and many features of enzymes. Metabolic pathway databases include KEGG (Kyoto Encyclopedia of Genes and Genomes) [85], MetaCyc [86], and EcoCyc [87].

Phenomics

A phenotype is an observable characteristic of a cell or an organism. It is the consequence of genome, transcriptome, proteome, and metabolome combined. It can be the morphology, development state, biochemical property, physiological condition, or reaction to the external environment. Phenomics, which associates the phenotype with the genotype, investigates genome-wide phenotypic manifestations at cellular and organism level. High-throughput phenotyping (HTP) is critical to phenomics. Current technologies include genome-scale RNAi screens for knock down analysis and phenotype microarray for simple assessment of microbe growth capability. Further advances in experimental technologies and computational algorithms are needed to speed up the phenomics studies. The Online Mendelian Inheritance in Man (OMIM) database has the largest collection of human genotype-disease information [88]. The online Mendelian Inheritance in Animals (OMIA) provides genotype-disease information in animals [89]. PhenomicDB [90] and GeneCards [91] databases provide heterogenous phenotypic information from a number of different model organisms. Phenotype Ontology systems are being developed to store, organize, and manage phenotype in a structured way, similar to that in GO. Mouse Phenotype Ontology (MPO) [92] and PhenoGO [93] provide such framework. Phenotype has been used for gene function prediction. Philip et al. cluster genotype-phenotype data, and assign the overrepresented functions in the cluster to the known gene [94].

In summary, omics data ranging from genomics, transcriptomics, proteomics, metabolomics, to phenomics, are being generated at an unprecedented pace, providing us with tremendous opportunities to tackle the biologically important questions at a whole new level. However, the complexity, heterogeneity, and scale of omics data present significant challenges to the biology community as well. Developing a standard procedure to store, manage, and share omics data is being strongly advocated [95, 96]. The establishment of a common standard will greatly facilitate the process to design better strategies to mine and integrate the omics data.

Computational Algorithms to Integrate Omics Data for Gene Function Prediction

Many computational algorithms have been developed to predict gene functions from omics data. As the omics era starts with completely sequenced genomes, early efforts on algorithm development focused on exploring genomics data for gene function prediction. With diverse sets of omics data introduced by high-throughput technologies continuously emerging, the current focus of the gene function prediction field has switched to omics data integration. Because of the high complexity, heterogeneity, and large-scale of the omics data, it is often difficult to design the integration rules beforehand. Machine learning or statistical algorithms are frequently used to learn from and integrate the complex data to make predictions. Recently, interaction networks or broadly speaking, functional linkage networks, have been used to integrate omics data. In this section, we first briefly summarize sequence-based gene function prediction methods. Then, we introduce several machine learning and statistical algorithms for omics data integration. Finally, we describe in detail the network-based integration, by introducing the construction of functional linkage network and the exploration of network topology for gene function prediction.

Sequence-Based Algorithms for Gene Function Prediction

Most sequence-based gene function prediction methods are based on a simple assumption, i.e., function tends to be conserved among evolutionary related sequences. Thus, detecting evolutionary relationships is a critical step, which is often done by a database search for homologous sequences with powerful tools, such as PSI-BLAST [2]. Function of an unknown gene can be predicted if it is found to share a significant sequence similarity with a known gene. However, this approach is often unreliable, especially for inference of specific functions [3, 6]. For example, systematic analysis of enzyme function inference using homology-based methods reveals that on average, above 60% sequence identity is required for accurate enzyme function inference [6]. With such a restrictive cut-off, however, a significant amount of false negatives would be produced. Modifications of sequence-based methods have been made and achieved significant improvement, including those by distinguishing orthology from paralogy [22, 97], those by inspecting phylogenetic profile information [98], and those by focusing on the functionally important residues in the sequences [5], etc. The sequence-based methods mostly focus on predicting the molecular function aspect of genes. Recently, Hawkins and Kihara investigated the association relationships between different GO terms [99]. They built a Function Association Matrix (FAM) between GO terms from different GO categories. By considering the FAM and PSI-BLAST hit, their PFP algorithm can make predictions of GO terms beyond the molecular function terms. In addition to sequence information, three-dimensional structural information of proteins has also been extensively explored for predicting gene functions [100–102].

Non-network Based Omics Data Integration for Gene Function Prediction

The omics data type can be very different from each other. For example, gene expression is represented by a real value, while a sequence pattern is a binary value, either "present" or "absent" in a gene, and a phenotype can be a categorical value, e.g., "normal", "sick", "very sick". Some machine learning algorithms, such as neural network and Support Vector Machine (SVM), are flexible to the format of the input data. For simplicity, however, the real value and the categorical value can be transformed into binary values. For example, gene expression value can be divided into several bins, with each bin considered as a new feature. After the appropriate coding systems of the omics data are decided, gene function prediction can then be considered as a binary classification problem, for which many machine-learning algorithms are available. Popular machine-learning algorithms include SVM, Bayesian Network (BN), Decision Tree (DT), Neural Network (NN), and so on. Here, we briefly introduce these algorithms, and then focus on examples of using them for omics data integration.

SVMs represent a family of statistical machine-learning methods that aim to optimally separate data into two categories by drawing a hyperplane in an N-dimensional vector space [103]. BN is a representation of a joint conditional probabilistic distribution that encodes the probabilistic relationships among features of interest [104]. DT is essentially a series of questions from which the classification or probability of a gene having a given function can be inferred [105]. NN mimics the human neuron perception system by consisting of a large number of highly interconnected elements to solve a problem [106]. Some of the algorithms can provide the rules of how a prediction is made, making it easier for human to understand, such as BN and DT, while others act like a "black box", such as NN. Yet, all these algorithms have been successfully applied in predicting gene functions.

Pavlidis and coworkers used a kernel-based SVM to combine gene expression profile and phylogenetic profile to infer yeast gene MIPS function categories [107]. Rather than simply concatenating both expression and phylogenetic profiles into a vector space, they used two kernel functions to transform the data into a higher dimension space separately. The new kernels were trained by SVM, with the results simply combined to make a final prediction. Lanckriet and coworkers further improved the kernel-based SVM to combine protein complex, protein domain, protein-protein interaction, genetic interaction, and gene expression information [108]. Instead of simple addition, a weighted linear combination was implemented to combine the results from each kernel. Troyanskaya and coworkers developed a BN-based algorithm named MAGIC to predict functional linked gene pairs from genetic and physical interactions, microarray, and transcription factor binding sites data [109]. Because learning the conditional probability in the BN structure is not an easy task, the authors consulted experimental experts and designed an expert-BN reflecting relationship between different evidence. The results from the BN integration were superior to unsupervised clustering algorithms significantly. Zhang and coworkers used a probabilistic DT to predict co-complex protein pairs from mRNA expression, transcription regulator, subcellular localization, phenotype and some sequence features [110]. Unlike BN, the DT does not rely on any previous assumption about conditional dependence; it automatically weights each data type when building tree. King and coworkers used the DT to make prediction of gene functions from patterns of annotation, and compare the result with that done by BN [111]. The result showed that DT is comparable to BN and in some cases better. NN has been widely used in biological data analysis. Jensen and coworkers developed a NN to predict protein function from various types of predicted potein features, including post-translational modification, sub-location and sorting [112]. Mateos and coworkers used a NN to predict gene function from gene expression data [113]. In addition, they pointed out that the poor performance of machine learning can be attributed to incomplete protein function annotations.

The algorithms introduced above employ a single model to integrate omics data. Multiple models can also be applied. Then, a new model is used to combine the prediction results. Hibbs and coworkers employed three different algorithms, bioPIXIE, MEFIT and SPELL, to predict genes involved in the process of mitochondrion organization and biogenesis [114]. bioPIXIE is a BN model aiming to integrate diverse sets of omics data. MEFIT focuses on integration of only microarray data. SPELL focuses on identifying coexpressed genes associated with the target biological process. The results of the three algorithms were combined with different weights determined based on their association with functional relationships. The combined algorithm achieved better performance than any single classifier did. Tian and coworkers developed a combined algorithm named Funckenstein which has two component classifiers [8]. The two classifies use different sets of omics data to predict gene function independently. A regression model is used to combine the results from these two classifiers. We will describe this algorithm in detail in the third section.

Network-Based Omics Data Integration for Gene Function Prediction

The wide use of high-throughput interaction technologies has allowed for the reconstruction of genome-scale protein physical interaction network in several organisms [64–66]. Extensive studies have been conducted on the interaction network, including using it to integrate omics data and for gene function prediction. In protein interaction network, the nodes are genes, while the edges are protein physical interactions (PPI). The edge can be any sorts of functional relationships as well, including genetic interaction, correlated gene expression, homologous relationship, etc. Thus, the network can be conveniently used as a framework to integrate various sources of omics data. The integrated network is often called functional linkage network (FLN) to indicate the functional links between genes. In addition, the network structure can be explored to obtain more information for gene function prediction. Here, we first introduce the reconstruction of FLN for omics data integration and gene function prediction. Then, we review current algorithms available to explore network structure, in particular the network module, for gene function prediction.

The concept of FLN was first introduced by Marcotte et al. in 1999 [115]. In their work, the functional links between proteins were constructed by combining protein-protein links from various sources: experimentally derived PPI, correlated gene expressions, related domain fusion, correlated phylogenetic profiles, and related metabolic function. Different evidences were simply combined without weight. High confidence protein links were defined as those with more than two evidences. Marcotte group further extended the idea of functional linkage by introducing a probabilistic FLN in yeast genome [116]. They computed a likelihood score of whether a pair of genes has a functional linkage defined by a common KEGG pathway given the evidence. The final FLN was a result of the integration of eight types of omics data, including physical interactions, genetic interactions, mRNA coexpression, functional linkages from literature mining, and computational linkages from gene-fusion and phylogenetic profiles. The resulted functional linkages showed a comparable accuracy in predicting KEGG pathway relationships to that by protein-protein interactions determined by small-scale experiments. Linghu and coworkers employed machine-learning algorithms to automatically integrate five types of omics data: PPI, genetic interaction, expression data, sequence similarity, phylogenetic profile and domain fusion to generate a FLN in yeast genomes [117]. The functional linkage was defined as the presence in the same KEGG pathway. Then, they designed a decision rule to infer protein pathway function from the FLN. Karaoz and coworkers constructed a FLN by using protein-protein interactions as the edges, with the weight determined by the correleated expression value of the interacting genes. A GAIN (Gene Annotation using Integrated Network) algorithm was then used to predict protein functions, by systematically propagating the labels of genes with known GO terms to unlabelled genes across the FLN [118]. Tian and coworkers applied a probabilistic decision tree (PDT) to construct FLN from various sources of experimentally determined protein physical and genetics interactions, and use this FLN to predict candidate gene with specific function annotations [8]. Reconstruction of FLN can also be found in other recent works [119, 120].

Besides integrating multiple sources of omics data into a single FLN, multiple FLNs can also be constructed. The final results can be either from the integration of the result from individual FLN, or from a new FLN integrated from multiple FLNs. For example, GeneMANIA [121], an algorithm developed by Mostafavi and coworkers, first builds multiple FLNs from various sources of omics data. Then, it employs a fast heuristic algorithm derived from linear regression to integrate multiple FLNs into a composite FLN. Finally, it applies a Gaussian field label propagation algorithms to predict gene function from the composite FLN. This algorithm was ranked one of the best methods in predicting gene function in the first critical assessment of mouse gene function based on the evaluation measurement of area under the ROC [122].

Given an interaction network or a FLN, network information can be explored to assist in the prediction of gene functions. The approaches exploring network information can be generally classified into two categories: the direct approach and the module-assisted approach. The direct approach utilizes the local or global network information to predict function. The module-assisted approach is inspired by the observation that interacting or functional linked genes tend to be localized in a dense region in the network, i.e., module [123]. It involves two steps: the first step is to identify the module, and the second step is to predict the function of unknown genes based on the distribution of known genes present in the same module. Here, we introduce the algorithms for both approaches.

The simplest method of the direct approach is the neighbor counting method. For example, Schwikowski et al. counted the neighbor proteins of an unknown protein, and simply assigned the three most frequent functions of the known neighbor proteins to the unknown protein [124]. Hishigaki et al. implemented a χ^2 test for the enrichment of known functions among the neighbor interacting proteins, and assigned the statistically significant functions to the known [125]. Further optimization was done by considering not only the direct interacting proteins, but also the near-neighbor proteins and their distances in the network graph [126]. These methods consider the local information and employ simple statistical test to make predictions. More sophisticated models that consider the global network information have also been developed, including the graph theory based methods. Graph theorey-based methods take the global and full topology of the network into account and employ either a cut- or flow-based algorithm to assign function, which can be generalized as a minimum multi-way cut problem. Vazquez et al. applied this theory to the yeast protein physical interaction network to predict functional class of unknown proteins, by minimizing the number of protein interactions among different functional categories with simulated annealing [127]. In contrast to Vazquez's approach that considers multiple functions at once, Karaoz et al. handled one function at a time, and employ a propagation algorithm to allow the flow of functional information in the network, and assign a score to candidate genes of having the function. Other attractive methods include the Markov Random Field (MRF) theory-based method, which assumes the function of a protein is dependent only on its neighbors and independent of all other proteins. Deng at el was the first group to formalize the idea of MRF in predicting protein function from protein interaction network [128]. Their approach was further generalized by allowing for the use of multiple networks, such as protein physical interaction, genetic interaction and coexpression network [129]. The MRF model is based on a sophisticated statistical theory, and mathematically sound. However, a number of machine learning algorithms have been reported to outperform the MRF model with the same benchmark data used by Deng et al. [8, 130].

The module-assisted approaches involve the identification of modules or dense local structure in the network, which was originally proposed by HartWell [123]. A number of algorithms have been developed for module identification, which can generally be classified as clustering based and non-clustering based. Clustering based methods include algorithms based on the pairwise distance of protein pairs defined as the shortest path length in the network [131], or more sophisticated ways, e.g., using the graph theory. Spirin and Mirny developed two algorithms, SPC (superparamagnetic) and a Monte Carlo-base method, to maximize the density of the obtained clusters [132]. Bader and Hogue developed a molecular complex detection algorithm (MCODE) to isolate the dense regions into modules [133]. The MCODE consists of three steps: vertex weighting based on the core clustering coefficient, prediction of complex memberships, and an optional post-processing filtering or addition of proteins based on connectivity data. Sharan et al. developed a NetworkBlast algorithm to assign a likelihood ration score for each candidate set of proteins in the network [134]. This method uses a greedy network search algorithm and can identify conserved region over several networks. The non-clustering based methods involve the use of prior information about protein-protein interaction or complex information. This information is used to seed a module, which is then expanded based on network connectivity. The Complexpander software developed by Asthana et al., first produces a rank of core proteins from complex data; then, it assigns a probability to the involvement of each protein in the core, and then computes a weighted score for each pair of proteins in the end [135]. Information other than protein physical interactions can also be utilized to identify network modules. For example, Segal et al. proposed a probabilistic model to identify modules not only enriched for interactions, but also enriched for high sequence similarity [136]. Hanisch et al. used the expression information as a filtering process [137], while Tanay et al. integrated the PPI data with gene expression, phenotypic sensitivity and TF binding site, to identify modules [138]. Once the modules are identified, usually statistical tests of the enrichment of known functions are conducted to infer function of the unknown proteins.

Funckenstein, a Combined Algorithm for Omics-Based Gene Function Prediction

Having described various types of omics data and a number of algorithms available for predicting gene function by integrating omics data, here we use a combined algorithm named Funckenstein [8] as an example to further illustrate the process of integrating omics data for gene function prediction. Most algorithms described in the previous section can generally fall into two categories: the "guilt-by-profiling" approach and the "guilt-by-association" approach. The "guilt-by-profiling" approach focuses on mining the gene characteristics, e.g., a conserved sequence motif. The "guilt-by-association" approach explores the relationships between genes for functional association, e.g., orthologous relationship, correlated expression profile, etc. Either approach has its own merit. Funckenstein is an algorithm that combines both approaches to achieve a synergistic performance better than either approach alone does. It has been applied for predicting gene functions (GO) in both yeast and mouse genomes [8, 139].

There are three steps in Funckenstein (see Fig. 2 for the flow chart). The first step is to classify omics data. Following the definition of guilt-by-profiling and guilt-by-association approaches, the collected diverse sources of omics data are classified into two categories: one describing gene characteristics, and another

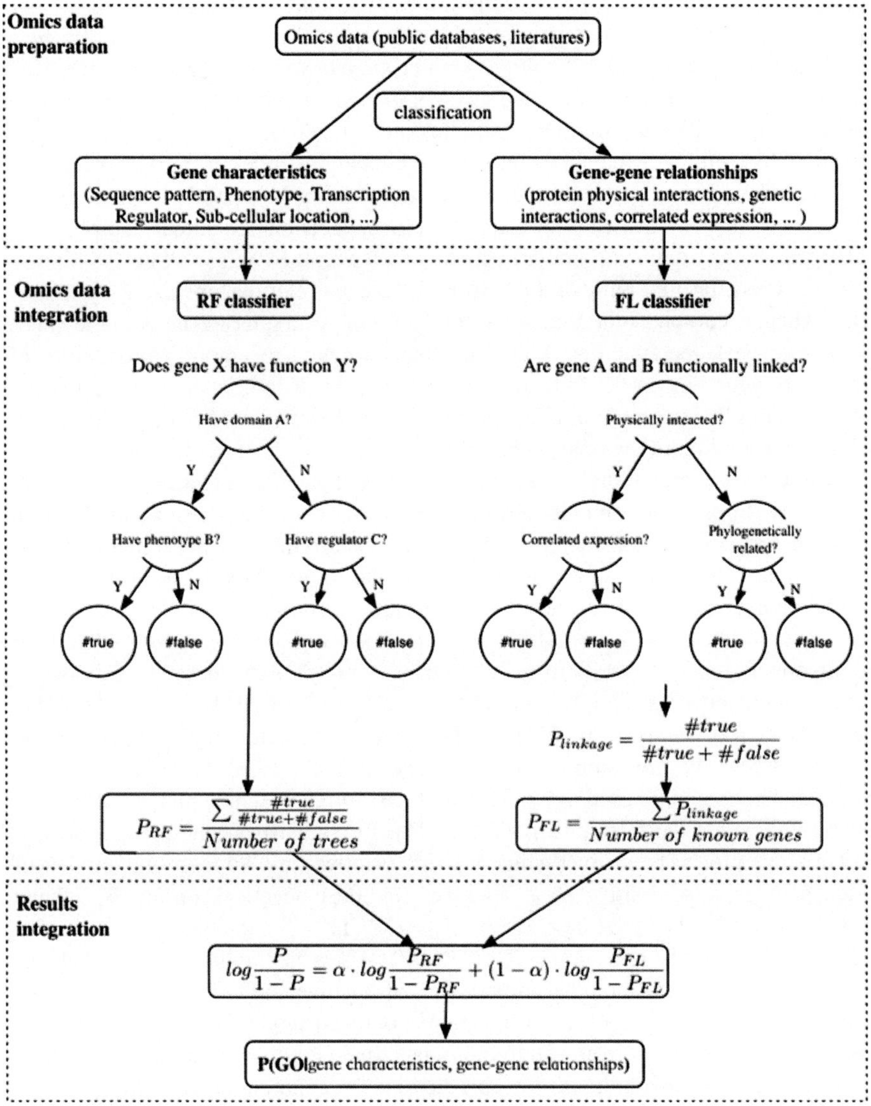

Fig. 2 Flowchart of the Funckenstein algorithm. There are three steps in Funckenstein: omics data preparation, omics data integration, and results integration. RF and FL refer to the random forest and functional linkage classifiers, respectively. The decision tree under the RF is an example of many decision trees in the forest, while that under the FL is an example of 12 decision trees specific for different GO categories

describing gene-gene relationships. Take yeast gene function prediction for an example, the gene characteristics include protein sequence patterns, gene phenotypes, the common transcriptional regulators, protein sub-cellular localization, and protein complex memberships. Some of those characteristics were collected from databases, such as UniProt database, while others were obtained from the supplementary materials of the published literatures. The gene-gene relationships include various types of protein-protein interactions (both physical and genetics) determined by different experimental technologies, which were downloaded from the BIOGRID database directly. In the second step, two component classifiers of Funckenstein (the random forest (RF) and the functional linkage (FL) classifier) are trained to make predictions from the gene characteristics and gene-gene relationships, respectively. The RF classifier employs a random forest algorithm [140] to build hundreds of decision trees from the gene characteristics. Each decision tree outputs a probability of a gene having a given function, which is then averaged across all decision trees. The FL classifier first builds a FLN from gene-gene relationships using a decision tree. Then, it computes the functional linkage score of a query gene with the genes known to have the function, which are then averaged to output a probability of the query gene having the function. In the final step of Funckenstein, a regression model is implemented to combine the probability scores from both the RF and FL classifiers and output the final probability.

There are several things about Funckenstein that need attention. First of all, Funckenstein predicts each GO term independently, i.e., the parent-child GO term relationships are not considered. Secondly, Funckenstein does not allow GO term annotation to be used as a feature in the training to avoid the issue of circularity. Third, rather than building one FLN, Funckenstein builds 12 FLNs by considering the type of ontology, i.e., Molecular Function, Biological Process and Cellular Component, and the specificity of GO terms which is defined by the number of genes annotated with the GO term and ranges from 3 to 10, 11 to 30, 31 to 100, and 101 to 300, respectively. Fourth, when measuring the prediction performance, the area under the precision-recall curve instead of the ROC curve is used. ROC curve has been widely used as a measure of performance, which plots the true positive rate against the false positive rate [141]. In comparison, the precision-recall curve is the plot of precision against the true positive rate. Suppose the number of true positives, false positives, true negatives, and false negatives are TP, FP, TN and FN, respectively, then the true positive rate = TP/(TP+FN), the false positive rate = FP/(FP+TN), and the precision = TP/(TP+FP). When the number of real negatives, (FP+TN), is far more than the number of real positives, (TP+FN), the false positive rate can be very small, even though FP is much larger than TP. In that case, the predictions may not be useful to biologists. In fact, to most biologists, they may be concerned more with the positive predictions the computational biologists made than the negatives. In contrast, the precision-recall curve is independent of the number of real negatives, and is more intuitive to biologists. Accordingly, Funckenstein is optimized based on the area under the precision-recall curve.

Funckenstein has been benchmarked with the same dataset used by a previous integrated algorithm for yeast gene function prediction. That algorithm, developed by Deng et al., uses a Markov Random Field (MRF) to integrate protein-protein interaction, coexpression, and genetic interaction networks, and estimates the prior probability of a gene having a given function by a Naïve Bayes method from protein complex memberships [128]. Funckenstein outperformed this algorithm by a significant margin in predicting yeast gene MIPS functions [8]. In the first critical assessment of the mouse gene function prediction which was participated by nine leading groups in the omics-based gene function prediction field, on average, for most GO categories evaluated, Funckenstein outperformed all other groups in terms of the precision at 20% recall [122]. In sum, Funckenstein achieves state-of-the-art performance in integrating omics data for gene function prediction.

Here we'd like to describe several interesting points during the development of Funckenstein. First of all, to achieve best synergistic effects in performance, it is better to use as different omics data as possible to train the guilt-by-profiling and the guilt-by-association methods separately. For example, a sequence pattern can be considered as a gene characteristic, but it can also be used to link two genes that have the same pattern. In yeast gene function prediction, we tested to code gene characteristics as additional gene relationships to train the FL classifier. Although we could improve the performance of the FL classifier greatly with the new additions, the combined results were worth than before. This suggests that the same omics data should not be utilized more than once. Second, more interactions data can substantially improve the performance of the FL classifier and consequently that of Funckenstein. In the benchmark with Deng et al.'s dataset, there were only a few thousands interactions available; while in the BIOGRID database, there are nearly a hundred thousands interactions curated from various high-throughput studies. The relative contribution of the FL classifier to Funckenstein's performance is significantly increased in the latter benchmark. This suggests by adding more gene-relationships from new omics data, we could further improve Funckenstein's performance. Third, building a FLN helps the FL classifier play a bigger role in predicting specific gene functions. When a GO term is associated with only a few known genes, it is difficult to train from the "positive" samples. In contrast, the 'transfer rules' are learned from the many GO terms within the specific GO category in the FLN. This stresses the importance of reconstructing a FLN in predicting gene functions.

Current Limitations and Potential Improvements

Omics Data Are Not Thoroughly Used

Figure 3 shows the frequency of different types of omics data used in the published "gene function prediction" algorithms since 2001. It is apparent that protein sequence, gene expression, and protein-protein interaction are the dominant omics data for gene function prediction, with the rest of omics data seldom or not used. For the three most used types of omic data, protein-protein interaction and gene expression data are becoming the focus in current algorithm development, which is

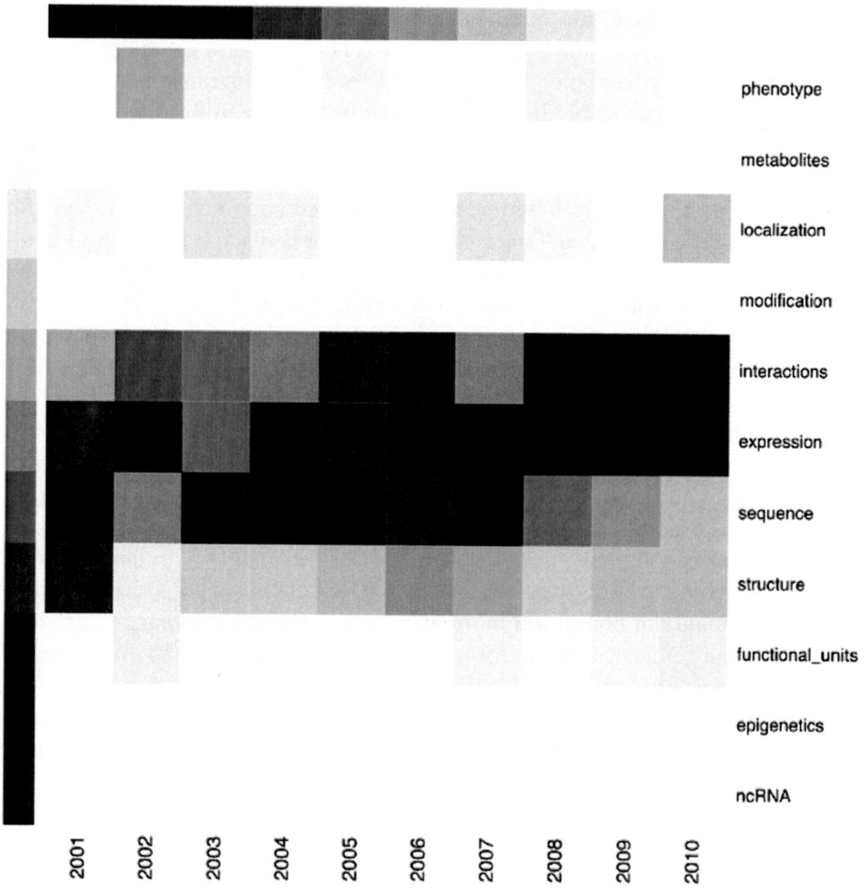

Fig. 3 Heat-map of the number of published gene function prediction algorithms using different types of omics data from 2001 to 2010. A Pubmed search with the "eutils.pl" script obtained from NCBI using different synonyms of "gene function prediction" from 2001 to 2010 results in over 800 literatures. Synonyms corresponding to different types of omics data are then used to count the number of publications using the corresponding omics data each year. The number is plotted in the heatmap. The *blackness* of each square in the heatmap represents the relative frequency of the corresponding publications each year

consistent with the trend that microarray and two-hybrid high-throughput technologies are becoming widely used. The lack of use of other omics data by current algorithms can be attributed to the fact that some omics data are not abundant enough. For example, the phenomics data are still lacking because developing an efficient high-throughput screen for phenotypic change is not an easy task. However, even the genomics data are not fully used. For example, although some algorithms integrate the TF binding site information, only the presence or absence information of the TF binding sites is used. In fact, the combination of TF binding sites, its relative position and the number of occurrence of TF sites in the promoter region all contribute to the target gene functions. In addition, the 5′ UTR and 3′ UTR of the target gene may also contain important functional unit information necessary for the function of the genes. Therefore, a more thorough use of omics data should be done in order to make further improvements. On the other hand, the metabolite, non-coding RNA, and epigenetics information are completely ignored by the current algorithms, which also points out where a potential improvement of the current algorithms can be made.

Omics Data Sharing Is Urgent and Needs to Be Standardized

Another reason why the omics data are not thoroughly used by current algorithms is because of the problem of omics data sharing. Although we have listed a large number of databases storing specific omics data in the first section, these databases may not be updated frequently enough to include the most recent high-throughput studies. In those cases, computational biologists often have to collect a large fraction of omics data from the supplementary of the published literatures by themselves, which is very time-consuming and laborious. In some cases, it may deter computational biologist from using the data. For example, the gene-naming system is often inconsistent from one high-throughput study to another, making automatic cross comparison almost impossible. With more and more omics data accumulated, this issue has become so serious that a number of algorithms for gene name translation have been published lately [142, 143]. In addition, the omics data are often lack of appropriate annotation, making it difficult for computational biologists to use or to interpret the results. With large amount of omics data being generated every day, standardization of omics data for sharing has never been so urgent. The advocate for a guideline like Minimum Information Requested In the Annotation of biochemical Models (MIRIAM) for omics data sharing is becoming louder than ever [95, 144]. The establishment and enactment of such a common standard for omics data sharing will greatly facilitate the improvement of current algorithms.

Omics data sharing is also an issue among computational biologists. A common benchmark omics dataset is important for computational biologists to test their algorithms and compare with others, so that they can make proper improvement. However, most times the benchmark omics dataset used by one algorithm is not accessible to others. CASP (Critical Assessment of Techniques for Protein Structure Prediction) has been successfully conducted for evaluating protein structure prediction methods [145]. A similar project can be extremely useful to the gene function prediction community. The first critical assessment of mouse gene function prediction project (MouseFunc) has been conducted [122], and more such project should follow. However, unlike protein structure prediction which can be compared with an experimentally determined structure, function is difficult to measure in an objective and timely manner, making effective benchmark for function prediction comparison not an easy task.

Is a Complex Model Better than a Simple Model?

The network-based data integration for gene function prediction has attracted the attention of many computational biologists. Various sophisticated algorithms have been developed to explore global network information, including those based on graph theory, and those based on identification of network modules. However, Murali et al. found that a simple local guilt-by-association method outperforms a graph-theory based global method to predict gene function from protein interactions [146]. In addition, Song and Singh recently tested the efficacy of various clustering algorithms in clustering protein interaction networks and predicting protein function [147]. They also compared the clustering algorithms with a simple guilt-by-association algorithm based on neighbor counting. Surprisingly, the simple guilt-by-association algorithm outperformed the sophisticated clustering algorithms in predicting gene functions. This thus raises an interesting question: Is a complex model better than a simple model?

The sophisticated algorithms are often backed by strong mathematics and statistics theories, while a simple model is usually based on empirical observations. However, the sophisticated algorithms often have to make an assumption that the current knowledge about the protein interaction network is complete, which is usually not the case. Take protein interaction network for an example, the interaction network is reconstructed by collecting interactions from various experiments and literatures; i.e., it is an ensemble of protein interactions all kinds of cellular conditions. However, in reality, it is unlikely that all protein interactions in the network are present in the cell at the same time. For example, protein A interacts with both B and C according to current knowledge. But it is possible that B and C may be expressed at different developmental stages. In such case, the presumed information flow from B→A→C or from C→A→B based on network structure is not be true. Accordingly, the label propagations based on network structure would lead to the wrong answer. Therefore, it is not that a simple model is better than a complex model; instead, it is whether a complex model is applicable to the omics data.

Model Driven or Biology Driven?

We have described many machine-learning and statistical algorithms for omics based gene function prediction. A beginner may be confused of which algorithm to choose. Should he choose SVM, BN, DT, ..., or RF? In fact, before any model is applied, the raw omics data has to be pre-processed or selected. Different groups may use different tricks to treat the raw omics data, which would lead to different outcomes. Take Funckenstein for an example, it classify the omics data into gene characteristics and gene-gene relationships categories before the application of the

RF and FL classifiers. This classification is critical to Funckenstein's success, as can be shown in yeast gene function in which the performance is worse without such classification. But how to process the raw omics data? The rational behind Funckenstein's classification is that the biological function of a gene is not only determined by its sequence, but also by what other genes it "interacts" with. As we can see from Funckenstein, perhaps a thorough understanding of the biology behind omics data and make appropriate treatment of omics data may be more effective than trying out a different model.

Prospective of Future Directions

Non-coding RNA Function Prediction

With more omics data emerging, the future of gene function prediction field will be continually focused on integrating newly added the data. However, coding gene sequence only accounts for a tiny fraction in the genome, while current results have shown that more than 70% of the genome are transcribed, with most of the transcripts being non-coding RNA [35]. The important biological role of non-coding RNA in the cell needs to be investigated. Many algorithms have been dedicated to coding gene function prediction. With the development of non-coding RNA experimental technologies, the next wave of gene function prediction will be the omics driven non-coding RNA function prediction.

Gene Function in a Dynamic Context

Gene Ontology provides an excellent system to describe the functions of a gene at three aspects, i.e., molecular function, biological process, and cellular component. However, these definitions do not take the dynamic cellular environment into account. Take catching a terrorist as an example, it is important to know what and where he is going to take actions. But it will be even more useful if we know when he is going to take actions. Similarly, besides knowing that two proteins interact with each other, it would be more interesting for biologists to know at what developmental stage, or by what environmental stimuli, they will interact with each other? Therefore, put gene functions in a dynamic context should be one of the most important and challenging directions in the future.

Acknowledgements This work was supported by the National Natural Science Foundation of China (Grant No. 30971643), and Shanghai Pujiang Program (Grant No. 09PJ1401000) to WT.

References

1. Adams, M., Kelley, J., Gocayne, J., Dubnick, M., Polymeropoulos, M., Xiao, H., Merril, C., Wu, A., Olde, B., Moreno, R. Complementary DNA sequencing: expressed sequence tags and human genome project. Science **252**(5013): 1651 (1991).

2. Altschul, S., Madden, T., Schaffer, A., Zhang, J., Zhang, Z., Miller, W., Lipman, D. Gapped BLAST and PSI-BLAST: a new generation of protein database search programs. Nucleic Acids Res. **25**(17): 3389 (1997).
3. Rost, B. Enzyme function less conserved than anticipated. J. Mol. Biol. **318**(2): 595–608 (2002).
4. Sonnhammer, E., Eddy, S., Durbin, R. Pfam: a comprehensive database of protein domain families based on seed alignments. Proteins Struct. Funct. Genet. **28**(3): 405–420 (1997).
5. Tian, W., Arakaki, A.K., Skolnick, J. EFICAz: a comprehensive approach for accurate genome-scale enzyme function inference. Nucleic Acids Res. **32**(21): 6226–6239 (2004).
6. Tian, W., Skolnick, J. How well is enzyme function conserved as a function of pairwise sequence identity? J. Mol. Biol. **333**(4): 863–882 (2003).
7. Hawkins, T., Kihara, D. Function prediction of uncharacterized proteins. J. Bioinform. Comput. Biol. **5**(1): 1–30 (2007).
8. Tian, W., Zhang, L., Ta an M, Gibbons, F., King, O., Park, J., Wunderlich, Z., Cherry, J., Roth, F. Combining guilt-by-association and guilt-by-profiling to predict Saccharomyces cerevisiae gene function. Genome Biol. **9**(Suppl 1): S7 (2008).
9. Ashburner, M., Ball, C., Blake, J., Botstein, D., Butler, H., Cherry, J., Davis, A., Dolinski, K., Dwight, S., Eppig, J. Gene ontology: tool for the unification of biology. Nat. Genet. **25**(1): 25–29 (2000).
10. Fleischmann, R., Adams, M., White, O., Clayton, R., Kirkness, E., Kerlavage, A., Bult, C., Tomb, J., Dougherty, B., Merrick, J. Whole-genome random sequencing and assembly of Haemophilus influenzae Rd. Science **269**(5223): 496 (1995).
11. ConsortiumInternational, H. G. S. Finishing the euchromatic sequence of the human genome. Nature **431**(7011): 931–945 (2004).
12. Rothberg, J., Leamon, J. The development and impact of 454 sequencing. Nat. Biotechnol. **26**(10): 1117–1124 (2008).
13. Oliphant, A., Barker, D., Stuelpnagel, J., Chee, M. BeadArray technology: enabling an accurate, cost-effective approach to high-throughput genotyping. Biotechniques (JUN): 56–61 (2002).
14. Hultman, T., Stahl, S., Homes, E., Uhlen, M. Direct solid phase sequencing of genomic and plasmid DNA using magnetic beads as solid support. Nucleic Acids Res. **17**(13): 4937 (1989).
15. Levy, S., Sutton, G., Ng, P.C., Feuk, L., Halpern, A.L., Walenz, B.P., Axelrod, N., Huang, J., Kirkness, E.F., Denisov, G., et al. The diploid genome sequence of an individual human. PLoS Biol. **5**(10): e254 (2007).
16. Cherry, J., Adler, C., Ball, C., Chervitz, S., Dwight, S., Hester, E., Jia, Y., Juvik, G., Roe, T., Schroeder, M. SGD: saccharomyces genome database. Nucleic Acids Res. **26**(1): 73 (1998).
17. Blake, J., Richardson, J., Bult, C., Kadin, J., Eppig, J. MGD: the mouse genome database. Nucleic Acids Res. **31**(1): 193 (2003).
18. Rhee, S., Beavis, W., Berardini, T., Chen, G., Dixon, D., Doyle, A., Garcia-Hernandez, M., Huala, E., Lander, G., Montoya, M. The Arabidopsis Information Resource (TAIR): a model organism database providing a centralized, curated gateway to Arabidopsis biology, research materials and community. Nucleic Acids Res. **31**(1): 224 (2003).
19. Drysdale, R., Crosby, M. FlyBase: genes and gene models. Nucleic Acids Res. **33**(Database Issue): D390 (2005).
20. Sonnhammer, E.L., Koonin, E.V. Orthology, paralogy and proposed classification for paralog subtypes. Trends Genet. **18**(12): 619–620 (2002).
21. Clark, A.G., Glanowski, S., Nielsen, R., Thomas, P.D., Kejariwal, A., Todd, M.A., Tanenbaum, D.M., Civello, D., Lu, F., Murphy, B., et al. Inferring nonneutral evolution from human-chimp-mouse orthologous gene trios. Science **302**(5652): 1960–1963 (2003).
22. Tatusov, R., Galperin, M., Natale, D., Koonin, E. The COG database: a tool for genome-scale analysis of protein functions and evolution. Nucleic Acids Res. **28**(1): 33 (2000).

23. O'Brien K, Remm, M., Sonnhammer, E. Inparanoid: a comprehensive database of eukaryotic orthologs. Nucleic Acids Res. **33**(Database Issue): D476 (2005).
24. Bowers, P., Pellegrini, M., Thompson, M., Fierro, J., Yeates, T., Eisenberg, D. Prolinks: a database of protein functional linkages derived from coevolution. Genome Biol. **5**(5): R35 (2004).
25. Hulo, N., Bairoch, A., Bulliard, V., Cerutti, L., De Castro, E., Langendijk-Genevaux, P., Pagni, M., Sigrist, C. The PROSITE database. Nucleic Acids Res. **34**(Database Issue): D227 (2006).
26. Attwood, T., Beck, M. PRINTS-a protein motif fingerprint database. Protein Eng. Des. Sel. **7**(7): 841 (1994).
27. Berman, B., Nibu, Y., Pfeiffer, B., Tomancak, P., Celniker, S., Levine, M., Rubin, G., Eisen, M. Exploiting transcription factor binding site clustering to identify cis-regulatory modules involved in pattern formation in the Drosophila genome. Proc. Natl. Acad. Sci. USA **99**(2): 757 (2002).
28. Buck, M., Lieb, J. ChIP-chip: considerations for the design, analysis, and application of genome-wide chromatin immunoprecipitation experiments. Genomics **83**(3): 349–360 (2004).
29. Schmid, C., Bucher, P. ChIP-Seq data reveal nucleosome architecture of human promoters. Cell **131**(5): 831–832 (2007).
30. Wingender, E., Dietze, P., Karas, H., Knüppel, R. TRANSFAC: a database on transcription factors and their DNA binding sites. Nucleic Acids Res. **24**(1): 238 (1996).
31. Sandelin, A., Alkema, W., Engstrom, P., Wasserman, W., Lenhard, B. JASPAR: an open-access database for eukaryotic transcription factor binding profiles. Nucleic Acids Res. **32**(Database Issue): D91 (2004).
32. Huda, A., MariÒo-RamÌrez, L., Landsman, D., Jordan, I. Repetitive DNA elements, nucleosome binding and human gene expression. Gene **436**(1–2): 12–22 (2009).
33. Jurka, J. RepBase update: a database and an electronic journal of repetitive elements. Trends Genet. **16**(9): 418–420 (2000).
34. Suzuki, Y., Yamashita, R., Nakai, K., Sugano, S. DBTSS: database of human Transcriptional Start Sites and full-length cDNAs. Nucleic Acids Res. **30**(1): 328 (2002).
35. Guttman, M., Amit, I., Garber, M., French, C., Lin, M., Feldser, D., Huarte, M., Zuk, O., Carey, B., Cassady, J. Chromatin signature reveals over a thousand highly conserved large non-coding RNAs in mammals. Nature **458**(7235): 223 (2009).
36. Bartel, D. MicroRNAs genomics, biogenesis, mechanism, and function. Cell **116**(2): 281–297 (2004).
37. Megraw, M., Sethupathy, P., Corda, B., Hatzigeorgiou, A.G. miRGen: a database for the study of animal microRNA genomic organization and function. Nucleic Acids Res. **35**(Suppl 1): D149–D155 (2006).
38. Jiang, Q., Wang, Y., Hao, Y., Juan, L., Teng, M., Zhang, X., Li, M., Wang, G., Liu, Y. miR2Disease: a manually curated database for microRNA deregulation in human disease. Nucleic Acids Res. **37**(Database issue): D98 (2009).
39. Bernstein, B., Meissner, A., Lander, E. The mammalian epigenome. Cell **128**(4): 669–681 (2007).
40. Grunau, C., Renault, E., Rosenthal, A., Roizes, G. MethDB – a public database for DNA methylation data. Nucleic Acids Res. **29**(1): 270 (2001).
41. Zhang, Y., Lv, J., Liu, H., Zhu, J., Su, J., Wu, Q., Qi, Y., Wang, F., Li, X. HHMD: the human histone modification database. Nucleic Acids Res. **38**(Suppl 1): D149–D154 (2009).
42. O'Connor T, Wyrick, J. ChromatinDB: a database of genome-wide histone modification patterns for Saccharomyces cerevisiae. Bioinformatics **23**(14): 1828 (2007).
43. Caron, H., Schaik, B., Mee, M., Baas, F., Riggins, G., Sluis, P., Hermus, M., Asperen, R., Boon, K., Voute, P. The human transcriptome map: clustering of highly expressed genes in chromosomal domains. Science **291**(5507): 1289 (2001).

44. Velculescu, V., Zhang, L., Vogelstein, B., Kinzler, K. Serial analysis of gene expression. Science **270**(5235): 484 (1995).
45. Jarvie, T. Next generation sequencing technologies. Drug Discov. Today Technol. **2**(3): 255–260 (2005).
46. Le Roch, K., Zhou, Y., Blair, P., Grainger, M., Moch, J., Haynes, J., De la Vega, P., Holder, A., Batalov, S., Carucci, D. Discovery of gene function by expression profiling of the malaria parasite life cycle. Science **301**(5639): 1503 (2003).
47. Edgar, R., Domrachev, M., Lash, A. Gene expression omnibus: NCBI gene expression and hybridization array data repository. Nucleic Acids Res. **30**(1): 207 (2002).
48. Ringwald, M., Mangan, M., Eppig, J., Kadin, J., Richardson, J. GXD: a gene expression database for the laboratory mouse. The Gene Expression Database Group. Nucleic Acids Res. **27**(1): 106 (1999).
49. Sims, D., Bursteinas, B., Gao, Q., Zvelebil, M., Baum, B. FLIGHT: database and tools for the integration and cross-correlation of large-scale RNAi phenotypic datasets. Nucleic Acids Res. **34**(Database Issue): D479 (2006).
50. Zimmermann, P., Hirsch-Hoffmann, M., Hennig, L., Gruissem, W. GENEVESTIGATOR. Arabidopsis microarray database and analysis toolbox. Plant Physiol. **136**(1): 2621 (2004).
51. Kato, K., Matoba, R., Saito, S., Matsubara, K. BGED-Brain Gene Expression Database. http://genome.mc.pref.osaka.jp/BGED/index.html
52. Miranda-Saavedra, D., De, S., Trotter, M., Teichmann, S., Gottgens, B. BloodExpress: a database of gene expression in mouse haematopoiesis. Nucleic Acids Res. **37**(Database issue): D873 (2009).
53. Primig, M., Wiederkehr, C., Basavaraj, R., Sarrauste de Menthiere, C., Hermida, L., Koch, R., Schlecht, U., Dickinson, H.G., Fellous, M., Grootegoed, J.A., et al. GermOnline, a new cross-species community annotation database on germ-line development and gametogenesis. Nat. Genet. **35**(4): 291–292 (2003).
54. Gauthier, N., Larsen, M., Wernersson, R., de Lichtenberg, U., Jensen, L., Brunak, S., Jensen, T. Cyclebase org a comprehensive multi-organism online database of cell-cycle experiments. Nucleic Acids Res. **36**(Database issue): D854 (2008).
55. Gorg, A., Weiss, W., Dunn, M. Current two-dimensional electrophoresis technology for proteomics. Proteomics **4**(12): 3665–3685 (2004).
56. Raymond, S., Aurell, B. Two-dimensional gel electrophoresis. Science **138**(3537): 152 (1962).
57. Perkins, D., Pappin, D., Creasy, D., Cottrell, J. Probability-based protein identification by searching sequence databases using mass spectrometry data. Electrophoresis **20**(18): 3551–3567 (1999).
58. Wu, C., MacCoss, M. Shotgun proteomics: tools for the analysis of complex biological systems. Curr. Opin. Mol. Ther. **4**(3): 242–250 (2002).
59. Yona, G., Linial, N., Linial, M. ProtoMap: automatic classification of protein sequences and hierarchy of protein families. Nucleic Acids Res. **28**(1): 49 (2000).
60. Lee, T., Huang, H., Hung, J., Huang, H., Yang, Y., Wang, T. dbPTM: an information repository of protein post-translational modification. Nucleic Acids Res. **34**(Database Issue): D622 (2006).
61. Habeler, G., Natter, K., Thallinger, G., Crawford, M., Kohlwein, S., Trajanoski, Z. YPL. db: the Yeast Protein Localization database. Nucleic Acids Res. **30**(1): 80 (2002).
62. Ito, T., Chiba, T., Ozawa, R., Yoshida, M., Hattori, M., Sakaki, Y. A comprehensive two-hybrid analysis to explore the yeast protein interactome. Proc. Natl. Acad. Sci. **98**(8): 4569 (2001).
63. Puig, O., Caspary, F., Rigaut, G., Rutz, B., Bouveret, E., Bragado-Nilsson, E., Wilm, M., Séraphin, B. The tandem affinity purification (TAP) method: a general procedure of protein complex purification. Methods **24**(3): 218–229 (2001).

64. Yu, H., Braun, P., Yildirim, M., Lemmens, I., Venkatesan, K., Sahalie, J., Hirozane-Kishikawa, T., Gebreab, F., Li, N., Simonis, N. High-quality binary protein interaction map of the yeast interactome network. Science **322**(5898): 104 (2008).
65. Li, S., Armstrong, C., Bertin, N., Ge, H., Milstein, S., Boxem, M., Vidalain, P., Han, J., Chesneau, A., Hao, T. A map of the interactome network of the metazoan C. elegans. Science **303**(5657): 540 (2004).
66. Rual, J., Venkatesan, K., Hao, T., Hirozane-Kishikawa, T., Dricot, A., Li, N., Berriz, G., Gibbons, F., Dreze, M., Ayivi-Guedehoussou, N. Towards a proteome-scale map of the human protein®Cprotein interaction network. Nature **437**(7062): 1173–1178 (2005).
67. Stark, C., Breitkreutz, B., Reguly, T., Boucher, L., Breitkreutz, A., Tyers, M. BioGRID: a general repository for interaction datasets. Nucleic Acids Res. **34**(Database Issue): D535 (2006).
68. Mewes, H., Heumann, K., Kaps, A., Mayer, K., Pfeiffer, F., Stocker, S., Frishman, D. MIPS: a database for genomes and protein sequences. Nucleic Acids Res. **27**(1): 44 (1999).
69. Hermjakob, H., Montecchi-Palazzi, L., Lewington, C., Mudali, S., Kerrien, S., Orchard, S., Vingron, M., Roechert, B., Roepstorff, P., Valencia, A. IntAct: an open source molecular interaction database. Nucleic Acids Res. **32**(Database Issue): D452 (2004).
70. Zanzoni, A., Montecchi-Palazzi, L., Quondam, M., Ausiello, G., Helmer-Citterich, M., Cesareni, G. MINT: a Molecular INTeraction database. FEBS Lett. **513**(1): 135–140 (2002).
71. Xenarios, I., Salwinski, L., Duan, X., Higney, P., Kim, S., Eisenberg, D. DIP, the database of interacting proteins: a research tool for studying cellular networks of protein interactions. Nucleic Acids Res. **30**(1): 303 (2002).
72. Yang, L., Jin, G., Zhao, X., Zheng, Y., Xu, Z., Wu, W. PIP: a database of potential intron polymorphism markers. Bioinformatics **23**(16): 2174 (2007).
73. Brown, K., Jurisica, I. Online predicted human interaction database. Bioinformatics **21**(9): 2076 (2005).
74. Bader, G., Donaldson, I., Wolting, C., Ouellette, B., Pawson, T., Hogue, C. BIND – the biomolecular interaction network database. Nucleic Acids Res. **29**(1): 242 (2001).
75. Mering, C., Huynen, M., Jaeggi, D., Schmidt, S., Bork, P., Snel, B. STRING: a database of predicted functional associations between proteins. Nucleic Acids Res. **31**(1): 258 (2003).
76. Zhu, H., Snyder, M. Protein chip technology. Curr. Opin. Chem. Biol. **7**(1): 55–63 (2003).
77. Thomas, D., Rosenbloom, K., Clawson, H., Hinrichs, A., Trumbower, H., Raney, B., Karolchik, D., Barber, G., Harte, R., Hillman-Jackson, J. The ENCODE Project at UC Santa Cruz. Nucleic Acids Res. **35**(Database issue): D663 (2007).
78. Tong, A., Evangelista, M., Parsons, A., Xu, H., Bader, G., Page, N., Robinson, M., Raghibizadeh, S., Hogue, C., Bussey, H. Systematic genetic analysis with ordered arrays of yeast deletion mutants. Science's STKE **294**(5550): 2364 (2001).
79. Pan, X., Yuan, D., Ooi, S., Wang, X., Sookhai-Mahadeo, S., Meluh, P., Boeke, J. dSLAM analysis of genome-wide genetic interactions in Saccharomyces cerevisiae. Methods **41**(2): 206–221 (2007).
80. Boone, C., Bussey, H., Andrews, B. Exploring genetic interactions and networks with yeast. Nat. Rev. Genet. **8**(6): 437–449 (2007).
81. Dauner, M., Sauer, U. GC-MS analysis of amino acids rapidly provides rich information for isotopomer balancing. Biotechnol. Prog. **16**(4): 642–649 (2000).
82. Jemal, M. High-throughput quantitative bioanalysis by LC/MS/MS. Biomed. Chromatogr. **14**(6): 422–429 (2000).
83. Laskowski, R., Rullmann, J., MacArthur, M., Kaptein, R., Thornton, J. AQUA and PROCHECK-NMR: programs for checking the quality of protein structures solved by NMR. J. Biomol. NMR **8**(4): 477–486 (1996).
84. Schomburg, I., Chang, A., Schomburg, D. BRENDA, enzyme data and metabolic information. Nucleic Acids Res. **30**(1): 47 (2002).

85. Kanehisa, M., Goto, S., Hattori, M., Aoki-Kinoshita, K.F, Itoh, M., Kawashima, S., Katayama, T., Araki, M., Hirakawa, M. From genomics to chemical genomics: new developments in KEGG. Nucleic Acids Res. **34**(Database issue): D354–357 (2006).
86. Krieger, C., Zhang, P., Mueller, L., Wang, A., Paley, S., Arnaud, M., Pick, J., Rhee, S., Karp, P. MetaCyc: a multiorganism database of metabolic pathways and enzymes. Nucleic Acids Res. **32**(Database Issue): D438 (2004).
87. Karp, P., Riley, M., Paley, S., Pellegrini-Toole, A., Krummenacker, M. EcoCyc: enyclopedia of Escherichia coli genes and metabolism. Nucleic Acids Res. **25**(1): 43 (1997).
88. Hamosh, A., Scott, A., Amberger, J., Bocchini, C., McKusick, V. Online Mendelian Inheritance in Man (OMIM), a knowledgebase of human genes and genetic disorders. Nucleic Acids Res. **33**(Database Issue): D514 (2005).
89. Nicholas, F. Online Mendelian Inheritance in Animals (OMIA): a comparative knowledgebase of genetic disorders and other familial traits in non-laboratory animals. Nucleic Acids Res. **31**(1): 275 (2003).
90. Kahraman, A., Avramov, A., Nashev, L., Popov, D., Ternes, R., Pohlenz, H., Weiss, B. PhenomicDB: a multi-species genotype/phenotype database for comparative phenomics. Bioinformatics **21**(3): 418 (2005).
91. Rebhan, M., Chalifa-Caspi, V., Prilusky, J., Lancet, D. GeneCards: a novel functional genomics compendium with automated data mining and query reformulation support. Bioinformatics **14**(8): 656 (1998).
92. Gkoutos, G., Green, E., Am Mallon, J., Davidson, D. *Building mouse phenotype ontologies*. Singapore: World Scientific, p. 178 (2004).
93. Lussier, Y., Borlawsky, T., Rappaport, D., Liu, Y., Friedman, C. PhenoGO: assigning phenotypic context to Gene Ontology annotations with natural language processing. Pac. Symp. Biocomput. **2006**: 64–75 (2006).
94. Philip, G., Bertram, W., Hans-Dieter, P., Ulf, L. Mining phenotypes for gene function prediction. BMC Bioinformatics **9**: 136.
95. Field, D., Sansone, S., Collis, A., Booth, T., Dukes, P., Gregurick, S., Kennedy, K., Kolar, P., Kolker, E., Maxon, M. ′Omics data sharing. Science **326**(5950): 234 (2009).
96. Laibe, C., Le Novère, N. MIRIAM Resources: tools to generate and resolve robust cross-references in Systems Biology. BMC Syst. Biol. **1**(1): 58 (2007).
97. Goodstadt, L., Ponting, C.P. Phylogenetic reconstruction of orthology, paralogy, and conserved synteny for dog and human. PLoS Comput. Biol. **2**(9): e133 (2006).
98. Date, S.V., Marcotte, E.M. Discovery of uncharacterized cellular systems by genome-wide analysis of functional linkages. Nat. Biotechnol. **21**(9): 1055–1062 (2003).
99. Hawkins, T., Kihara, D. PFP: automatic annotation of protein function by relative GO association in multiple functional contexts. ISMB, June 25–29, Detroit, Michigan. pp. 117: 1471–2105 (2005).
100. Watson, J., Sanderson, S., Ezersky, A., Savchenko, A., Edwards, A., Orengo, C., Joachimiak, A., Laskowski, R., Thornton, J. Towards fully automated structure-based function prediction in structural genomics: a case study. J. Mol. Biol. **367**(5): 1511–1522 (2007).
101. Sadowski, M., Jones, D. The sequence-structure relationship and protein function prediction. Curr. Opin. Struct. Biol. **19**: 357–362 (2009).
102. Vaidehi, N., Floriano, W., Trabanino, R., Hall, S., Freddolino, P., Choi, E., Zamanakos, G., Goddard, W. Prediction of structure and function of G protein-coupled receptors. Proc. Natl. Acad. Sci. **99**(20): 12622 (2002).
103. Hearst, M., Dumais, S., Osman, E., Platt, J., Scholkopf, B. Support vector machines. IEEE Intell. Syst. **13**(4): 18–28 (1998).
104. Jensen, F. *An introduction to Bayesian networks*. London: UCL press (1996).
105. Quinlan, J. Induction of decision trees. Mach. Learn. **1**(1): 81–106 (1986).
106. Funahashi, K. On the approximate realization of continuous mappings by neural networks. Neural Netw. **2**(3): 183–192 (1989).

107. Pavlidis, P., Weston, J., Cai, J., Grundy, W. Gene functional classification from heterogeneous data. New York, NY: ACM, pp. 249–255 (2001).
108. Lanckriet, G., De Bie, T., Cristianini, N., Jordan, M., Noble, W. A statistical framework for genomic data fusion. Bioinformatics **20**(16): 2626–2635 (2004).
109. Troyanskaya, O., Dolinski, K., Owen, A., Altman, R., Botstein, D. A Bayesian framework for combining heterogeneous data sources for gene function prediction (in Saccharomyces cerevisiae). Proc. Natl. Acad. Sci. **100**(14): 8348 (2003).
110. Zhang, L., Wong, S., King, O., Roth, F. Predicting co-complexed protein pairs using genomic and proteomic data integration. BMC Bioinformatics **5**(1): 38 (2004).
111. King, O., Foulger, R., Dwight, S., White, J., Roth, F. Predicting gene function from patterns of annotation. Genome Res. **13**(5): 896 (2003).
112. Jensen, L., Gupta, R., Blom, N., Devos, D., Tamames, J., Kesmir, C., Nielsen, H., Staerfeldt, H., Rapacki, K., Workman, C. Prediction of human protein function from post-translational modifications and localization features. J. Mol. Biol. **319**(5): 1257–1265 (2002).
113. Mateos, A., Dopazo, J., Jansen, R., Tu, Y., Gerstein, M., Stolovitzky, G. Systematic learning of gene functional classes from DNA array expression data by using multilayer perceptrons. Genome Res. **12**(11): 1703 (2002).
114. Hibbs, M.A., Myers, C.L., Huttenhower, C., Hess, D.C., Li, K., Caudy, A.A., et al. Directing experimental biology: a case study in mitochondrial biogenesis. PLoS Comput. Bio. **5**(3): e1000322 (2009).
115. Marcotte, E., Pellegrini2 M, Thompson, M., Yeates, T., Eisenberg, D. A combined algorithm for genome-wide prediction of protein function. Proc. Natl. Acad. Sci. USA **93**: 4787–4792 (1996).
116. Lee, I., Date, S., Adai, A., Marcotte, E. A probabilistic functional network of yeast genes. Science **306**(5701): 1555 (2004).
117. Linghu, B., Snitkin, E., Holloway, D., Gustafson, A., Xia, Y., DeLisi, C. High-precision high-coverage functional inference from integrated data sources. BMC Bioinformatics **9**(1): 119 (2008).
118. Karaoz, U., Murali, T., Letovsky, S., Zheng, Y., Ding, C., Cantor, C., Kasif, S. Whole-genome annotation by using evidence integration in functional-linkage networks. Proc. Natl. Acad. Sci. **101**(9): 2888 (2004).
119. Guan, Y., Myers, C., Lu, R., Lemischka, I., Bult, C., Troyanskaya, O. A genomewide functional network for the laboratory mouse. PLoS Comput. Biol. **4**(9) (2008).
120. Kim, W., Krumpelman, C., Marcotte, E. Inferring mouse gene functions from genomic-scale data using a combined functional network/classification strategy. Genome Biol. **9**(Suppl 1): S5 (2008).
121. Mostafavi, S., Ray, D., Warde-Farley, D., Grouios, C., Morris, Q. GeneMANIA: a real-time multiple association network integration algorithm for predicting gene function. Genome Biol. **9**(Suppl 1): S4 (2008).
122. Pena-Castillo, L., Tasan, M., Myers, C.L., Lee, H., Joshi, T., Zhang, C., Guan, Y., Leone, M., Pagnani, A., Kim, W.K., et al. A critical assessment of Mus musculus gene function prediction using integrated genomic evidence. Genome Biol. **9**(Suppl 1): S2 (2008).
123. Hartwell, L., Hopfield, J., Leibler, S., Murray, A. From molecular to modular cell biology. Nature **402**(6761): 47 (1999).
124. Schwikowski, B., Uetz, P., Fields, S. A network of protein-protein interactions in yeast. Nat. Biotechnol. **18**(12): 1257–1261 (2000).
125. Hishigaki, H., Nakai, K., Ono, T., Tanigami, A., Takagi, T. Assessment of prediction accuracy of protein function from protein-protein interaction data. Yeast **18**(6): 523–531 (2001).
126. Chua, H., Sung, W., Wong, L. Exploiting indirect neighbours and topological weight to predict protein function from protein-protein interactions. Bioinformatics **22**(13): 1623 (2006).
127. Vazquez, A., Flammini, A., Maritan, A., Vespignani, A. Global protein function prediction from protein-protein interaction networks. Nature Biotechnology **21**: 697–700 (2003).

128. Deng, M., Zhang, K., Mehta, S., Chen, T., Sun, F. Prediction of protein function using protein-protein interaction data. J. Comput. Biol. **10**(6): 947–960 (2003).
129. Deng, M., Chen, T., Sun, F. An integrated probabilistic model for functional prediction of proteins. J. Comput. Biol. **11**(2–3): 463–475 (2004).
130. Lanckriet, G.R., Deng, M., Cristianini, N., Jordan, M.I., Noble, W.S. Kernel-based data fusion and its application to protein function prediction in yeast. Pac. Symp. Biocomput. **2004**: 300–311 (2004).
131. Arnau, V., Mars, S., Marín, I. Iterative cluster analysis of protein interaction data. Bioinformatics **21**(3): 364 (2005).
132. Spirin, V., Mirny, L. Protein complexes and functional modules in molecular networks. Proc. Natl. Acad. Sci. **100**(21): 12123 (2003).
133. Bader, G., Hogue, C. An automated method for finding molecular complexes in large protein interaction networks. BMC Bioinformatics **4**(1): 2 (2003).
134. Sharan, R., Ideker, T., Kelley, B., Shamir, R., Karp, R. Identification of protein complexes by comparative analysis of yeast and bacterial protein interaction data. J. Comput. Biol. **12**(6): 835–846 (2005).
135. Asthana, S., King, O., Gibbons, F., Roth, F. Predicting protein complex membership using probabilistic network reliability. Genome Res. **14**(6): 1170 (2004).
136. Segal, E., Wang, H., Koller, D. Discovering molecular pathways from protein interaction and gene expression data. Bioinformatics **19**(1): 264–272 (2003).
137. Hanisch, D., Zien, A., Zimmer, R., Lengauer, T. Co-clustering of biological networks and gene expression data. Bioinformatics **18**: 145–154 (2002).
138. Tanay, A., Sharan, R., Kupiec, M., Shamir, R. Revealing modularity and organization in the yeast molecular network by integrated analysis of highly heterogeneous genomewide data. Proc. Natl. Acad. Sci. **101**(9): 2981 (2004).
139. Tasan, M., Tian, W., Hill, D.P., Gibbons, F.D., Blake, J.A., Roth, F.P. An en masse phenotype and function prediction system for Mus musculus. Genome Biol. **9**(Suppl 1): S8 (2008).
140. Breiman, L. Random forests. Mach. Learn. **45**(1): 5–32 (2001).
141. Hanley, J.A., McNeil, B.J. The meaning and use of the area under a receiver operating characteristic (ROC) curve. Radiology **143**(1): 29–36 (1982).
142. Berriz, G., Roth, F. The Synergizer service for translating gene, protein and other biological identifiers. Bioinformatics **24**(19): 2272 (2008).
143. van Iersel, M., Pico, A., Kelder, T., Gao, J., Ho, I., Hanspers, K., Conklin, B., Evelo, C. The BridgeDb framework: standardized access to gene, protein and metabolite identifier mapping services. BMC Bioinformatics **11**(1): 5 (2010).
144. Le Novore, N., Finney, A., Hucka, M., Bhalla, U., Campagne, F., Collado-Vides, J., Crampin, E., Halstead, M., Klipp, E., Mendes, P. Minimum information requested in the annotation of biochemical models (MIRIAM). Nat. Biotechnol. **23**(12): 1509–1515 (2005).
145. Moult, J., Fidelis, K., Rost, B., Hubbard, T., Tramontano, A. Critical assessment of methods of protein structure prediction (CASP) – round 6. Proteins **61**(Suppl 7): 3–7 (2005).
146. Murali, T.M., Wu, C.J., Kasif, S. The art of gene function prediction. Nat. Biotechnol. **24**(12): 1474–1475; author reply 1475–1476 (2006).
147. Song, J., Singh, M. How and when should interactome-derived clusters be used to predict functional modules and protein function? Bioinformatics **25**(23): 3143–3150 (2009).

Protein Function Prediction Using Protein–Protein Interaction Networks

Hon Nian Chua, Guimei Liu, and Limsoon Wong

Abstract Proteins perform biological functions by participating in a large number of interactions, ranging from transient interactions in signaling pathways to permanent interactions within stable complexes. Studies have shown that the immediate interaction neighborhood of a protein can be used to infer its functions. While using only such direct interactions limits prediction coverage, extending the interaction neighborhood to include indirect interaction partners reduces precision significantly, making functional inference unviable. In a series of studies, we find that the extent of partner-sharing between two non-interacting proteins makes a good estimator for their co-participation in similar function. This allows us to include indirect interactions in network-based functional inference with little compromise in precision. We also extend this idea to the related problems of protein complex prediction and interaction data cleansing.

Introduction

Proteins are important building blocks that contribute to key processes within cells. The elucidation of mechanisms underlying protein functionality is an active and important pursuit in biology, and remains a challenging task. Unlike protein sequences or protein-protein interactions, there is currently no systematic experimental technique that can characterize the functions of proteins in a high-throughput fashion. With various sources of biological data being made available at an unprecedented rate, efforts intensify for computational methods that can tap into this growing pool of information for reliable functional characterization of proteins. In this chapter, we summarize our efforts towards this area of research. We will describe our work on the use of protein–protein interactions for computational protein function prediction, protein complex discovery, and improving the reliability of protein–protein interactions.

H.N. Chua (✉)
Institute for Infocomm Research, Singapore 138632
e-mail: hnchua@i2r.a-star.edu.sg

D. Kihara (ed.), *Protein Function Prediction for Omics Era*,
DOI 10.1007/978-94-007-0881-5_13,

Protein–Protein Interactions

Protein–protein interactions generally refer to associations between protein molecules, which include direct physical binding and genetic interactions, amongst other definitions.

Physical Interactions

Physical binding between proteins can be detected in a high-throughput manner using a variety of assays such as co-immunoprecipitation, tandem affinity purification [1, 2], and two-hybrid systems [3–5]. In yeast two-hybrid assays, the GAL4 transcriptional activator is split into two fragments, one containing the binding domain and the other containing the activating domain. To detect an interaction (or lack thereof) between two proteins, one protein is fused to the fragment containing the binding domain (also referred to as the bait) while the other protein is fused to the other fragment (the prey). An interaction between the bait and prey proteins indirectly connects the two fragments of the transcription factor, bringing the activating domain close to the transcription start site, and results in the expression of the downstream reporter gene. In co-immunoprecipitation experiments, proteins that are suspected to interact directly or indirectly with a protein of interest are isolated together with the protein using an antibody, and subsequently identified using western blot. Tandem affinity purification involves creating fusion proteins with one end that can be bound to beads coated with a specific antibody. The modified proteins, along with the unknown proteins that they bind, are isolated over two rounds of purification and identified. The use of fusion proteins makes this technique suitable for systematic genome-wide studies [2, 6]. Datasets of large numbers of physical protein–protein interactions have been experimentally derived using two hybrid systems for a number of species, particularly for the model organisms *Saccharomyces cerevisiae* (budding yeast), *Drosophila melanogaster* (fruit fly) and *Caenorhabditis elegans* (nematode).

Genetic Interactions

Genetic interactions, on the other hand, capture functional dependency between genes from observations of phenotypes exhibited upon two or more gene deletions. The departure of observed phenotypes (usually cell viability) of double-deletion mutants from that expected of the two independent genes (based on the phenotypes of each single-deletion mutant) is used to identify such interactions. While there have been attempts to reconcile such observations with biological models such as parallel or serial pathways, these are insufficient to explain the complex relationships between genes that are reflected in these experiments. Nonetheless, genetic interactions provide great insight into the functional organization of gene products. Positive genetic interactions are often associated with proteins within complexes, while negative genetic interactions often capture redundancy between pathways [7]. Several large-scale genetic interaction experiments have been conducted for yeast [8–10] using the Synthetic Genetic Array technology [8], which allows systematic, unbiased screening for genetic relationships of a large number of array genes against a query gene in a high throughput fashion. Systematic screening for genetic interactions between essential genes is also possible using hypomorphic alleles [10]. The BioGRID database [11] is one of the largest collections of published protein–protein interactions, both physical and genetic, making it a valuable resource for researchers who are interested in studying protein–protein interactions.

Function Prediction Using Protein–Protein Interactions

A protein's functional behavior is intuitively related to its physical interactions with other proteins. Genetic interactions, on the other hand, capture functional dependencies between genes (and the proteins they encode for), such as serial genes in a biosynthesis pathway, or genes in parallel transport pathways. Hence protein–protein interactions potentially enrich for information about functional relationships between proteins that may not be obvious or detectable from other genomic data such as primary or higher level sequence structure.

Many computational approaches have been developed to utilize protein interactions for the functional characterization of proteins. One of the earliest approaches is the neighbor counting method proposed by [12]. The simple method, which assigns a protein with the function that is annotated most frequently to its interaction partners, was applied to a large-scale physical interaction dataset generated from yeast two-hybrid experiments, and performs reasonably well. The approach, however, did not take into account the background frequency of different function annotations. The mere observation of a very common functional annotation assigned to the majority of a group of proteins does not necessary suggests enrichment unless its prior probability is taken into account. Hishigaki and colleagues addressed this limitation by using the Chi-square statistic to estimate the enrichment of functional annotations in each protein's interaction neighborhood [13].

An obvious limitation in both the Neighbor Counting and Chi-square approaches is the inability to infer functional annotations to a protein that do not interact with annotated proteins. These approaches will also be biased in making inference when the majority of the proteins in the interaction neighborhood of a protein are not annotated. To overcome these limitations, some methods cleverly made assumptions along the lines that the "correct" set of functional annotations to unannotated proteins in an interacting network is the one in which functional association between adjacent proteins is best upheld. While it is unfeasible to find such a best solution in the vast space of possible configurations, many stochastic inference techniques can be used to find a reasonably good solution. Such "global" inference methods also have the advantage of being more resilient against errors in functional annotations and in the interaction network.

One such "global" inference approaches is the Markov Random Field method described in [14], which proposes that the probability of a set of inferred annotations to proteins in an interaction network is inversely related to the amount of annotation inconsistencies between interacting proteins. This probability is formally defined for each functional annotation to be a function of its prior probabilities, the number of functionally associated interactions, and the number of functionally unassociated interactions. A Gibbs sampler is then used to find a near optimal set of annotation assignments that maximizes the probability. A similar approach is used in [15]. Vazquez et al. also proposed another optimization method based on Simulated Annealing [16].

Indirect Association of Protein Function

Functional Association Between Indirect Neighbors

In 2006, we proposed the hypothesis of indirect association of protein function [17]. The motivation behind the hypothesis is the observation that many proteins do not share similar function with any of their interaction partners. In the study, we investigated the functional relationships between interacting proteins in the *Saccharomyces cerevisiae* (bakers' yeast) genome using physical and genetic interactions deposited in the BioGRID [11], as well as FunCat functional annotations from MIPS [18]. We observed that there are proteins that do not share any functional annotation with their immediate interaction partners (i.e., level-1 neighbours) and yet share some function similarity with the interaction partners of their immediate partners (i.e., level-2 neighbours). Two examples of such proteins are shown in Fig. 1. Among 4162 annotated yeast proteins in the dataset studied, only 48.0% share some function with its level-1 neighbours. 22.7% of the annotated proteins shared functional annotations with their level-2 neighbours but not their level-1 neighbours. Less than 2% of the annotated proteins share functions with level-1 neighbours without sharing functions with their level-2 neighbours. This suggested that many existing approaches to functional inference based on protein–protein interaction, whether in a local or global fashion, may be somewhat limited by making only assumptions of functional linkage between directly interacting proteins. Local inference methods will not be able annotate a protein with a function that is not observed in its direct neighbors. Global inference methods may erroneously propagate function in an indiscriminative way.

The observation left us pondering if it is possible to make predictions for more proteins by explicitly taking into account the functional annotations of the level-2 neighbors of proteins. Hishigaki et al. [13] studied the use of larger interaction neighborhoods (which they termed n-neighbouring proteins, analogous to our definition of n-level neighbors) by using their Chi-square based method on the functional classification used in the Yeast Proteome Database (YPD), and concluded that the value of n for the best prediction performance is small (1 for cellular role and subcellular localization, and 2 for biochemical function). Such observation is

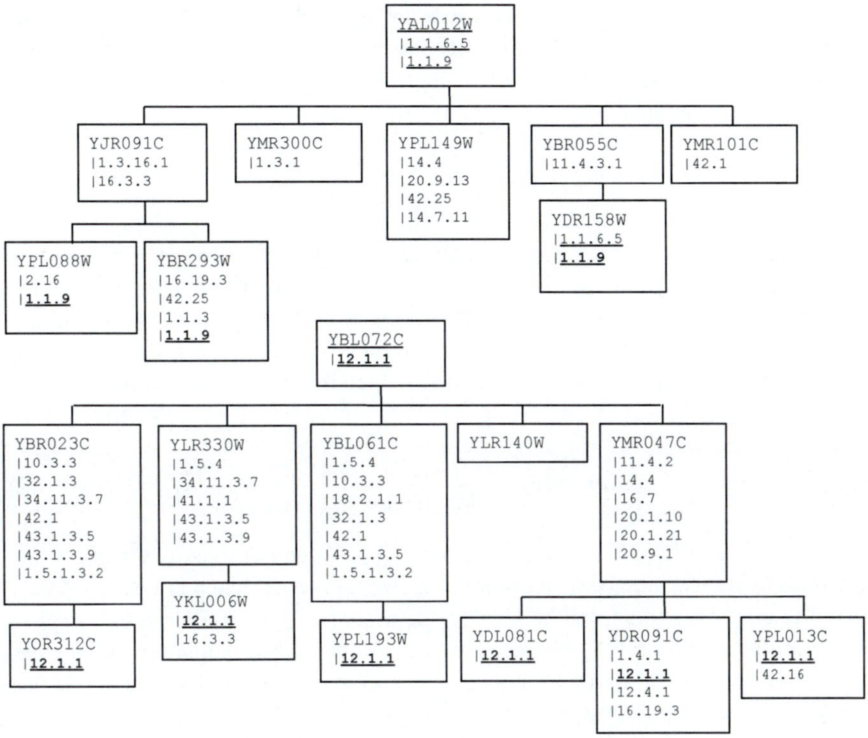

Fig. 1 Two examples of proteins that do not share functional annotations with their direct interaction neighbor, but share functional annotations with their indirect (level-2) neighbors (indirect neighbors that share no annotation are not shown). Figure from [17]

expected as we expect functional relationship to diminish with the interaction distance. The number of neighboring proteins also often increases quickly with the size of the neighborhood, and the predictive powers of the closer (and more functionally related) neighbors tend to be diminished as a result. Moreover, errors in the lower level interaction neighborhood will spill over and propagate to the higher levels, resulting in more errors introduced in each level. Hence the number of functionally irrelevant interactions is expected to be higher when more levels of interactions are used.

Estimating Function Similarity Between Interacting Proteins

To be able make use of the indirect neighbors for increasing prediction coverage without severely affecting precision, some form of filtering has to be employed to avoid including functionally unrelated neighbors in the prediction process. At that time, there have already been some studies that observe functional similarity between proteins with overlapping interaction neighborhood [19, 20]. These independent observations motivated us to study the possibility of using the observation of common interaction partners as a way to identify functionally related protein pairs from the large number of indirectly interacting proteins. We initially adopted the Czekanowski-Dice distance (CD distance) used in [20] for this purpose. The CD-Distance is defined as:

$$D(u,v) = \frac{|N_u \Delta N_v|}{|N_u \cup N_v| + |N_u \cap N_v|} \tag{1}$$

where N_p refers to the set that contains p and proteins that interact with it, and $X \Delta Y$ refers to the symmetric difference between two sets X and Y. $D(u, v) < 1$ if proteins u and v interacts with each other, or with at least one common protein. If $N_u = N_v, D(u, v)$ will be 0. On the other extreme, if $N_u \cap N_v = \emptyset, D(u, v)$ will be 1. This distance function can be trivially converted into its corresponding similarity function:

$$S_{CD}(u,v) = \frac{|N_u \cap N_v|}{|N_u \cup N_v| + |N_u \cap N_v|} \tag{2}$$

The similarity function captures the overlap between two sets reasonably when the sets N_u and N_v are not very different in size. However, when one set is greater than the other, $S_{CD}(u, v)$ will be small even when $N_u \cap N_v$ is a large or complete subset of the smaller set. Since the sets represent interaction neighbors in this case, this means that the similarity score between a protein with low degree and one that is well connected will always be low. As protein interactions are subjected to systematic biases due to experimental design and incomplete coverage, this similarity function is likely to underestimate functional relationships in such cases. Hence we proposed a variant of the similarity function, which we refer to as the Functional Similarity weight (FS-weight) to place greater weight on the overlap between the two sets:

$$S_{FS}(u,v) = \frac{2|N_u \cap N_v|}{|N_u - N_v| + 2|N_u \cap N_v| + \lambda_{u,v}} \times \frac{2|N_u \cap N_v|}{|N_v - N_u| + 2|N_u \cap N_v| + \lambda_{v,u}} \tag{3}$$

$\lambda_{u,v} = \max\left(0, n_{avg} - (|N_u - N_v| + |N_u \cap N_v|)\right)$ where n_{avg} is the average number of interactions that a protein participates in.

Functional Association and Experimental Assays

As described earlier, protein-protein interactions can be observed in a variety of experimental assays. While the different assays are capable of identifying interactions between proteins (and genes), they often rely on very diverse mechanisms. Consequently, each assay comes with its limitations. In yeast two-hybrid systems, false positives (interactions observed that are non-existent) can arise due to a wide number of factors such as background transcriptional activity of baits, mutation of the host yeast strain, bait proteins that binds directly to the DNA upstream of the reporter genes, and "sticky" bait or prey proteins that easily binds a large number of proteins in a non-specific manner [21]. In tandem affinity purification experiments, false negatives (interactions that exist but not observed) may arise due to the TAP tag interfering with interaction, and not all proteins within the complex may bind tightly enough to be detected [22]. While there is no simple way to take into account such differences in the nature and limitations of different experimental assays, we can moderate the impact of such differences to the function prediction process by estimating the confidence we have in each type of experiment with regard to their ability to associate proteins with similar functions. For each type of experiment, this can be a simple estimate of the prior probability that protein interactions observed by such experiments involve protein pairs that share some function:

$$r_i = \frac{\sum\limits_{(u,v)\in E_i} \delta(u,v)}{|E_i|} \tag{4}$$

E_i is the set of interactions observed in experiment i; $\delta(u,v)$ is 1 when protein u and v share some function, 0 otherwise.

For interactions that are observed in multiple experiments, we would expect the confidence to be much higher since it is reproducible and less likely to be a false positive due to random experimental errors. Taking into account the confidence of individual experimental types, as well as reproducibility over multiple experiments of the same or different nature, we can naively combine the prior probabilities for each experimental type to compute the probability that an observed interaction is associated with sharing of function:

$$r_{u,v} = 1 - \prod_{i\in E_{u,v}} (1 - r_i)^{n_{i,u,v}} \tag{5}$$

r_i is the estimated reliability of experimental type i; $E_{u,v}$ is the set of experiments in which interaction between u and v is observed;
$n_{i,u,v}$ is the number of times interaction (u,v) is observed from experimental type i.

With a quantifiable estimate of the confidence of different experimental sources of interaction data, we can incorporate this information into the FS-weight formulation:

$$S_{FS}(u,v) = \frac{2\sum\limits_{w\in(N_u\cap N_v)} r_{u,w}r_{v,w}}{\left(\sum\limits_{w\in(N_u-N_v)} r_{u,w} + \sum\limits_{w\in(N_u\cap N_v)} r_{u,w}(1-r_{v,w})\right) + 2\sum\limits_{w\in(N_u\cap N_v)} r_{u,w}r_{v,w} + \lambda_{u,v}} \times \frac{2\sum\limits_{w\in(N_u\cap N_v)} r_{u,w}r_{v,w}}{\left(\sum\limits_{w\in(N_v-N_u)} r_{v,w} + \sum\limits_{w\in(N_u\cap N_v)} r_{v,w}(1-r_{u,w})\right) + 2\sum\limits_{w\in(N_u\cap N_v)} r_{u,w}r_{v,w} + \lambda_{v,u}} \tag{6}$$

We find the FS-weight measure to correlate positively with function similarity between interacting proteins (pearson's correlation coefficient = 0.53). The measure also exhibit a positive, abeit weaker correlation with function similarity between level-2 interaction neigbors (correlation coefficient = 0.38).

Function Prediction Using Indirect Association

With an appropriate function to estimate the strength of functional relationships between directly and indirectly interacting proteins, it is now more plausible to include the level-2 neighborhood for functional prediction. We proposed the FS-weighted averaging function that uses the weighted frequency of a function x in both the direct (N_u) and indirect (N_v) neighbors of a protein u to compute a normalized score to estimate the likelihood of protein u to participate in function x:

$$f_x(u) = \frac{1}{Z}\left[\lambda r_{\text{int}}\pi_x + \sum_{v \in N_u}\left(S_{FS}(u,v)\delta(v,x) + \sum_{w \in N_v} S_{FS}(u,w)\delta(w,x)\right)\right]$$

Z is the sum of all weights:

$$Z = 1 + \sum_{v \in N_u}\left(S_{FS}\,(u,v) + \sum_{w \in N_v} \max(S_{FS}(u,v)S_{FS}(v,w), S_{FS}(u,w))\right) \quad (7)$$

Evaluation on Function Prediction

The FunCat annotation scheme is a tree-like structure with each child term being a more specific form of its parent. Some fuctional aspects of proteins tend to be better studied than others, and hence some annotation branches tend to be deeper and annotated to a larger number of proteins. To minimize biases when evaluating prediction performance, we want to avoid evaluating redundant annotations (e.g. a functional term and its parent function, as well as more distant ancestor terms). A simple way to achieve this would be to decide on an arbitary level of annotation (e.g. all nodes with a depth of 5), but due to large variations in the depth of different branches, we may end up evaluating very general functions of some branches and very specific functions of others. To overcome this problem, we adopt the informative functional classes approach proposed in [23]. A functional term is designated as *informative* if it is annotated to n or more proteins (we use $n = 30$), and does not have a child term that is annotated to n or more proteins. In this way, an informative term will be the only informative term among all its ancestors or descendants. By using only informative terms, we can ensure that there is no redundancy between the functions that are used for evaluation. Moreover, since these informative terms are annotated to a sufficiently large number of proteins, we will avoid evaluating functional terms that are too rare to be inferred practically. Using a 10-fold cross validation procedure, we benchmarked our proposed method against several published approaches at the time of the study on the prediction of informative FunCat terms using protein-protein interactions from BioGRID and showed that it performed significantly better (Fig. 2). We also benchmarked our method against other approaches using a dataset compiled in an earlier study comprising YPD functional categories

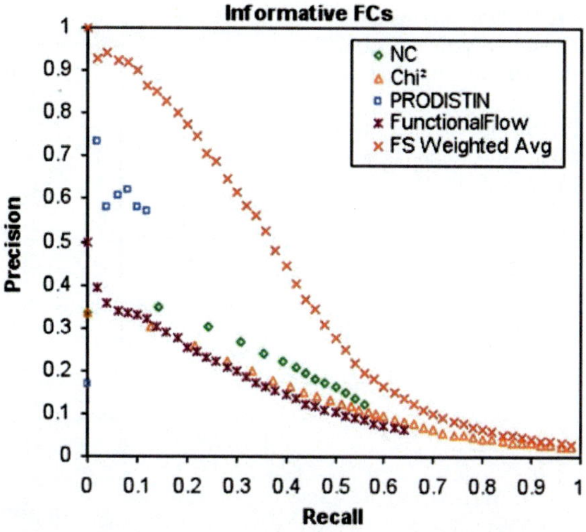

Fig. 2 Precision–recall curves for prediction of FunCat functions for proteins from *S. cerevesiae* from BioGRID interactions using various approaches. Figure from [17]

and protein–protein interactions from MIPS [14], and showed that the predictions made using our method achieved a better precision at nearly all levels of recall for the three YPD categories (Fig. 3).

Prediction of Gene Ontology Functional Annotations on Multiple Species

While we had some success showing that indirect association of FunCat functional annotations are abundant between non-interacting proteins, the annotation scheme that was, and still is most widely adopted is the Gene Ontology (GO). Similar to FunCat, this comprehensive functional annotation scheme organizes functional annotations into a hierarchical structure that explicitly describes parent-child relationships between annotations, where the children of an annotation are more specific annotations that fall under it. The hierarchy structure adopted by GO, however, is a Directed Acyclic Graph (DAG), instead of the tree structure used by Funcat. The main implication of this is that a GO term can have more than one parent term. The GO annotation scheme constitute a DAG structure for each of the 3 namespaces *molecular_function*, *biological_process*, and *cellular_component*, that provide different aspects of biological characterization of a gene and its protein product. To study if the use of indirect functional association is general enough to be beneficial for functional prediction based on the GO scheme, and for species other than *S. cerevisiae*, we performed a follow-up computational study in 2007 on 7 species [24]. The objective of the study was to answer 3 key questions about using protein-protein interactions and indirect functional association for protein function prediction: (1) Does the use of protein-protein interactions provide any additional coverage over the

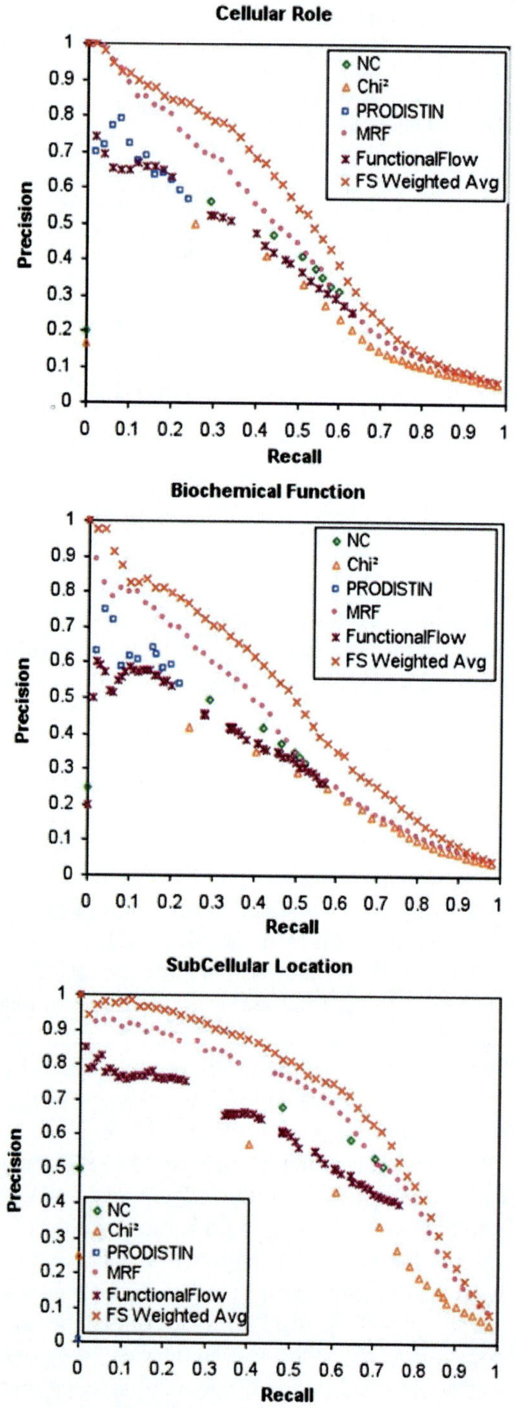

Fig. 3 Precision–recall curves for prediction of YPD functions for proteins from *S. cerevesiae* from MIPS protein–protein interactions using various approaches. Figure from [17]

conventionally accepted use of sequence homology for protein function prediction; (2) Does the use of indirect functional association provide any additional enhancement in coverage over direct guilt by association; and (3) Are the conclusions made for indirect functional association on FunCat terms applicable to function prediction using GO terms over different species with differences in quantity and even quality of data?

Data Availability

At the time of study, protein-protein interaction data was available for 7 species in the BioGRID database: *S. cerevisiae*, *D. melanogaster*, *A. thaliana*, *H. Sapiens*, *M. Musculus*, *R. norvegicus* and *C. elegans*. Gene Ontology annotations were also available for these species. The amount of interaction data available to perform the study is summarized in Table 1. As we can only evaluate prediction performance on annotated proteins, we present the number of interactions that involve annotated proteins as a proxy for data availability.

Table 1 Annotation and interaction data statistics for different species at time of study. Table from [24]

Genome	Interactions involving annotated proteins	Annotated proteins	Avg. no. of annotated neighbours per protein
S. cerevisiae	50,434	4005	21.6654
D. melanogaster	24,991	2763	4.2823
A. thaliana	909	382	1.8386
H. Sapiens	5784	5784	1.6761
M. Musculus	1892	1892	1.3595
R. norvegicus	590	590	0.9803
C. elegans	4349	382	0.7382

Protein–Protein Interactions vs. Sequence Homology

To answer our first question on the usefulness of protein–protein interaction data as an additional source of data to complement conventional sequence homology for protein function inference, we examine the number of known functional annotations that can already be inferred using the top hits of a BLAST search against all sequences from the Gene Ontology Database. The analysis is only done for *S.cerevisiae* and *D. melanogaster* as the amount of protein–protein interaction data is too little for meaningful analysis on the other species. The fraction of known annotations that can be annotated in this way for each species is computed using E-value cut-offs between 1 and 1e–10, and summarized as white bars in Fig. 4. As one would expect, coverage decrease with more stringent E-value cut-offs, possibly in exchange for better precision (not shown). For each E-value cut-offs, we next compile the number of additional functional annotations that can be transfer in a guilt-by-association fashion based on protein–protein interactions as a fraction of the total number of known annotations (light blue bars in Fig. 4). We find

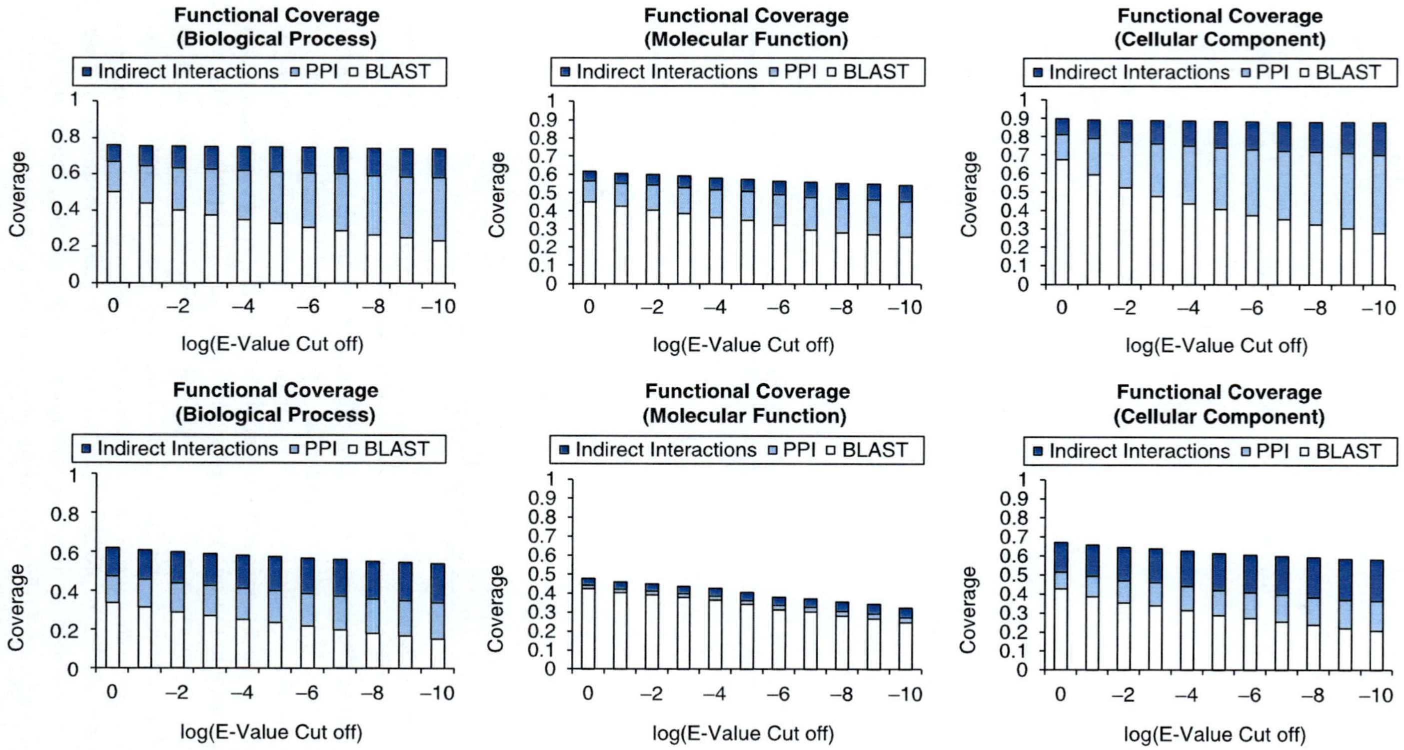

Fig. 4 Fraction of known functional annotations that can be suggested through BLAST homology search; and the additional annotations that can be suggested through: (1) direct-protein interactions (PPI); and (2) indirect-protein interactions. A range of BLAST E-value cut-off between 1 and 1e–10 is used. BLAST is performed on sequences from the gene ontology database. Proteins with very close homologs (E-value <= 1e–25) are excluded from analysis. The *top row* shows the results from *S. Cerevisiae* and the *bottom row* shows the results from *D. melanogaster.* The three columns depict results on the biological process (*left*), molecular function (*centre*) and cellular component (*right*) categories of the Gene Ontology. Figure from [24]

that protein–protein interactions provided some additional coverage (around 20% for *S.cerevisiae* and 10% for *D. melanogastor*) even at relaxed BLAST E-value cutoffs of >=0.01 for inferring *biological_process* and *cellular_component* annotations. Finally, we also compute any further coverage that may be gleaned if we also allow functional inference using indirect functional associations between level-2 interaction neighbors. We found that there is substantial additional coverage that may be gained in this way (dark blue bars in Fig. 4) for both species. This analysis addressed the first two questions we seek to answer, that is: (1) There are a fair number of GO annotations that cannot be inferred through simple sequence homology, but can potentially be predicted from protein-protein interactions; and (2) Extending functional predictions to level-2 neighbors helps to further increase coverage by including functional annotations that cannot be associated to a protein via sequence homology or direct protein–protein interactions.

Function Prediction Performance

Finally, we investigate if the function prediction method that we proposed earlier can be used to make better predictions for GO terms for the seven species by using functional association with indirect interaction neighbours. Again, we used the informative functional classes concept to identify informative GO terms to be used for evaluation for each species. Comparing FS-weighted averaging with the Neighbor-Counting and Chi-Square approaches, we found that FS-weighted averaging achieved superior precision–recall performance in all seven species (Fig. 5).

Indirect Functional Association and Complex Discovery

Protein Complex Discovery

Proteins often perform function by aggregating into complexes to perform sophisticated biological tasks. Many well-conserved protein complexes perform key biological functions such as transcription, splicing, mRNA export and protein synthesis. Through complex formation, the primary molecular functions of individual proteins (such as the ability to bind DNA or RNA, shuttle between membranes, transport certain materials and interact with particular proteins) are recruited in a coordinated fashion to perform highly specialized functions. RNA polymerases, ribosomes and spliceosomes are some examples of widely studied protein complexes with well-understood functionalities. Therefore to better understand the higher-level biological processes in which proteins participate, it is necessary to look beyond individual protein features such as sequences and structures and observe how proteins form larger functional units. While experimental assays such as tandem affinity purification and co-immunoprecipitation can be used to identify protein complexes, these are usually suitable for capturing stable complexes. Many weak or transient complexes are likely to be missed.

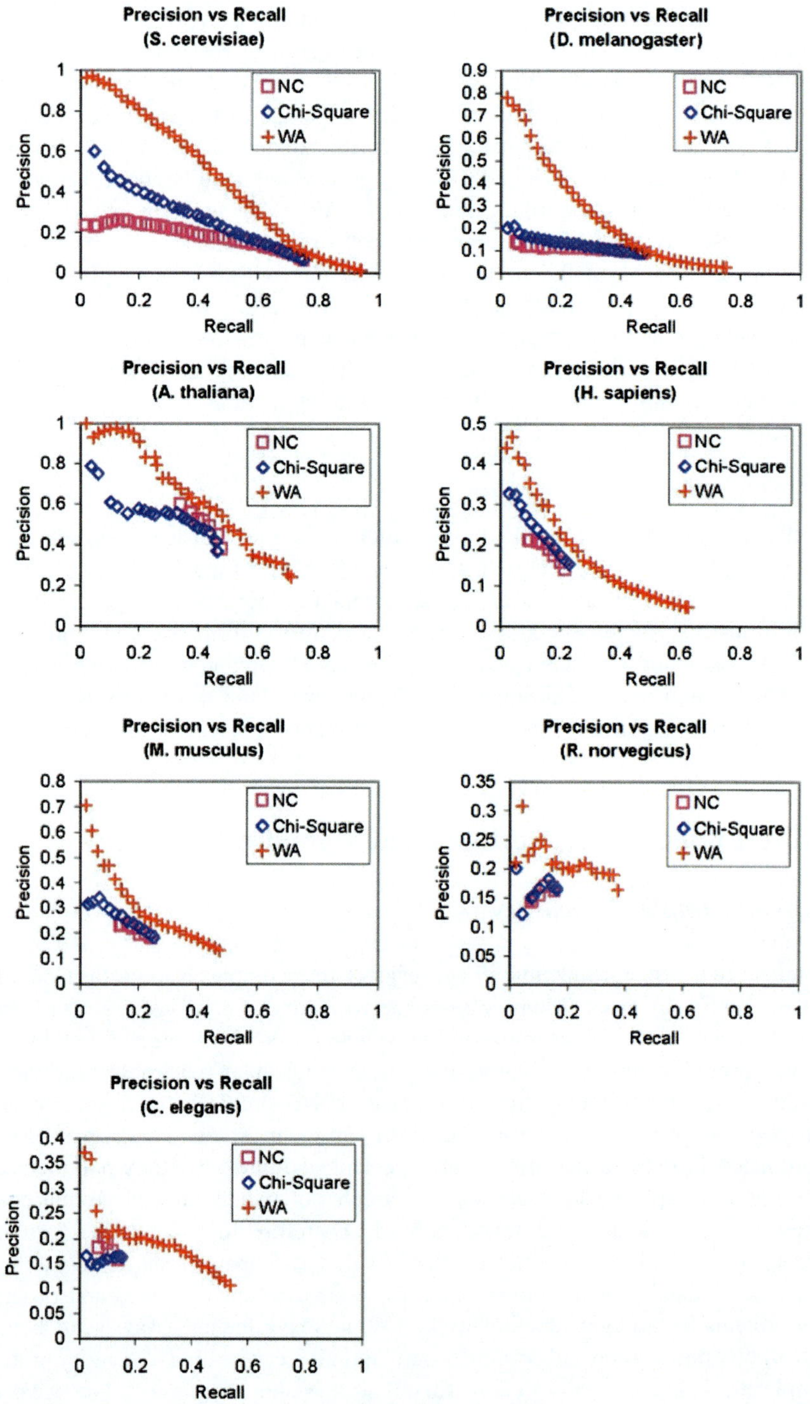

Fig. 5 Precision vs. recall graphs of the predictions of informative GO terms from the gene ontology biological process category using (1) *Neighbour Counting(NC)*; (2) *Chi-Square*; and (3) *FS-Weighted Averaging(WA)*, for seven genomes. Figure from [24]

The importance of identifying protein complexes motivated many bioinformatics approaches to identify protein complexes computationally from protein–protein interactions. Several insightful studies contributed significantly in motivating research in this area. Spirin and Mirny [25] investigated highly connected proteins in a physical protein–protein interaction network, and found functionally related proteins to be highly connected with each other, but sparsely connected with the rest of the network. Some of these densely connected proteins coincide with known stable protein complexes, while many others are found to be related to dynamic functional units involved in activities such as signaling cascades and cell cycle regulation. Bu and colleagues studied topological structures (quasi-cliques and quasi-bicliques) in protein–protein interactions and found that many of these structures involved functionally related proteins [26]. Bader and Hogue [27] proposed a computational method of protein complex discovery from protein–protein interaction networks by "growing" complexes from "seed proteins" with dense local network. The algorithm, MCODE, was subsequently implemented as a plug-in for the popular bioinformatics visualization software Cytoscape [28]. The recurring theme among these studies is that function modularity in biological systems may be encoded in protein–protein interactions, and identifying such functional modules allows us to better understand how proteins function together.

Protein Complexes with Limited Interactions

From our earlier studies, we found that many indirectly interacting proteins share functional annotations from different schemes including YPD, FunCat and GO. These indirectly interacting proteins that perform similar biological functions could in reality be forming protein complexes, with their common interacting proteins acting as adaptors that bring them into close proximity. This is especially likely for larger complexes since proteins have limited binding pockets and usually have reasonably high binding specificity. Since these proteins do not interact, there may not be sufficient overlap between their local interaction neighborhoods for conventional clustering approaches based on network density to associate them. As the FS-weight measure has been demonstrated to provide some estimation to functional similarities between two indirectly interacting proteins, we are interested to see whether including indirect interactions with high FS-weight scores into the protein interaction network can help improve discovery of complexes that involve less physical inter-connections. On the other hand, since the FS-weight can also provide some estimation of functional similarity between proteins that interact, we may be able to remove possibly spurious interactions that are likely to be functionally unrelated from the interaction network. We explore these ideas in a subsequent work [29, 30] that study how complex prediction performance is affected by (1) applying existing clustering methods on modified physical protein–protein interactions; and (2) proposing a clustering algorithm that implicitly take FS-weight into account.

Approaches for Protein Complex Prediction

At the time of the study there are two general approaches to protein complex prediction from protein-protein interactions. The first approach, which we refer to as *clique finding*, imposes a stringent requirement on what constitutes a protein complex. A *clique* is a fully connected subgraph in which each node is connected to all other nodes in the subgraph. Spirin and Mirny [25] explored two methods of finding densely connected subgraphs in a protein interaction network, one of which is to enumerate all cliques in the network. The strict constraint imposed by clique finding keeps false positives low and makes the approach robust to noise in the interaction network. However, sensitivity is likely to be severely limited. Bu and colleagues used a more relaxed constraint for complex prediction by looking for *quasi-cliques*, which are dense subgraphs that are almost complete [26]. The other general approach to complex prediction, which we refer to as clustering, involves the use of heuristic algorithms to find groups of densely connected proteins, usually based on network properties such as network density. Brohee and colleagues [31] evaluated some of these clustering methods, namely the Restricted Neighborhood Cost-Based Clustering (RNSC) [32], MCODE, Markov Clustering (MCL) [33], and Super Paramagnetic Clustering (SPC) [34] for protein complex prediction from protein–protein interaction networks. Using 6 protein–protein interaction networks from [2, 5, 35–38] and cataloged complexes from MIPS [39], the authors optimized the parameters for each clustering algorithm and benchmarked them over several performance metrics.

Modifying the Interaction Network with FS-Weight

Given a input interaction network, FS-weight is applied to assign a score to all interactions as well as level-2 indirect interactions. By applying a threshold $FS\text{-}Weight_{\min}$, we include indirect interactions that surpass this threshold into the original interaction network. On the other hand, direct interactions in the original interaction network that does not meet this threshold are removed from the interaction network. Since the FS-Weight measure exhibit positive correlation with functional similarity, we expect connected proteins in the modified network to be more functionally related than that of the original network. In the study we performed experiments using the 6 protein–protein interaction networks studied in [31], which comprises 2 datasets derived from large-scale yeast two-hybrid studies, and 4 datasets from affinity purification and mass spectrometry. We refer to this combined network as the "combined" dataset. We also used a larger dataset comprising all physical protein-protein interactions from BioGRID which is a superset of the 6 networks.

As a preliminary study of the feasibility of this approach, we compute the fraction of all interactions that involve a pair of proteins that belong to some common complex for the two interaction networks, as well as the modified versions of these networks. We find that if we introduce level-2 indirect interactions indiscriminately, the fraction of interactions that involve co-complex proteins decreases drastically (Fig. 6, L1 & L2). However, if we only include level-2 interactions with high FS-weight scores, we are able to maintain these fractions at similar levels (L1 & Filtered L2) as that for the original interaction networks (L1). Finally, when we also remove direct interactions with low FS-weight after including level-2 interactions with high

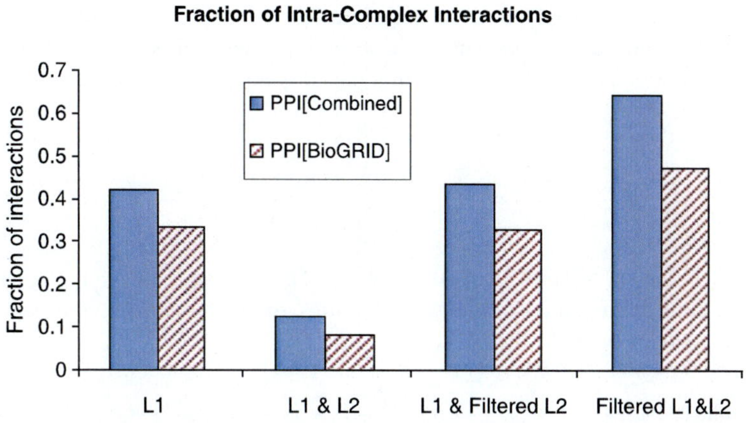

Fig. 6 Fraction of intra-complex interactions with nodes sharing some complex membership for different PPI networks. Figure from [30]

FS-weight, the fractions of the interactions that involve proteins from common complex increased significantly (Filtered L1 & L2). These observations are encouraging and suggest that we could possibly make the network more amenable to complex discovery in this manner.

A New Complex Prediction Approach

Since the FS-weight can provide a decent estimate of the functional relatedness of an interaction, we may be able to exploit this information in the complex prediction process. Taking this idea into consideration, we proposed a novel complex prediction approach and benchmark it alongside with the 4 existing clustering algorithms evaluated in [31]. Our approach, PCP (Protein Complex Prediction), is a heuristic algorithm that involves a three-step iterative process:

Maximal Clique Finding

The first step involves finding all maximal cliques of at least size 2 from the network. This can be done efficiently on a sparse graph using the algorithm described in [40]. For nodes that belong to multiple cliques, we assign them to only one clique using a heuristic method to maximize the average FS-Weight scores of the edges in each non-overlapping clique. Since this is also the performance bottleneck of the algorithm, we also proposed an alternative heuristic method for finding non-overlapping cliques as a replacement for this step which did not have any significant impact on prediction performance.

Computing InterClusterDensity

The clique finding step will return very dense subgraphs that are completely connected. A clique is unlikely to represent a complete real complex, but rather a densely-connected subset of it. To associate less densely connected parts of the complex, we can merge cliques that are well-connected. To provide a quantitative measure of interconnectedness between a pair of subgraphs (S_a, S_b), we define the *InterClusterDensity* (ICD) as follows:

$$ICD(S_a, S_b) = \frac{\sum S_{FS}(i,j)|i \in V_a, j \in V_b, (i,j) \in E}{|V_a| \cdot |V_b|} \tag{8}$$

where V_x is the set of vertices of subgraph S_x. This is simply the weighted sum of all edges between members of the two subgraphs, divided by the maximum number of possible edges between them.

Subgraphs Merging

Using the ICD measure, we can now imagine each clique as a node in a new graph, and insert an edge between two nodes that has a ICD score greater than an arbitary threshold ICD_{min}. We can now perform the maximal clique finding step again on the new graph. The nodes in the cliques found will no longer be proteins, but rather groups of proteins. By reiterating this process, smaller groups of proteins will gradually be merged into large groups in a hierarchical manner. To allow the better connected nodes to be merged first, we start from a high ICD_{min} threshold, and gradually reduce the threshold whenever no further merging can be made.

Performance Evaluation

Known protein complexes from MIPS is used as the gold standard against which performance is evaluated. In order to see if novel predictions are indeed made, we also used MIPS complexes released 2 years apart, in 2004 and 2006. Unlike function prediction, the practical usefulness of complex prediction lies in the ability to predict a set instead of a pair. Therefore to make quantitative evaluation meaningful, we must first define what constitute a correct prediction, that is, the critria for a predicted cluster to be considered as matching a known complex. We adopt the overlap measure from [27]:

$$Overlap(S, C) = \frac{|V_s \cap V_c|}{|V_s| \cdot |V_c|} \tag{9}$$

In [27], and overlap score of 0.2 or more is considered a match. We used a slightly higher threshold of 0.25 in our study. Since there may be more than one cluster matching a complex and vice versa, we used a slightly modified version of the conventional precision and recall measure. We defined precision here as the number of predicted clusters that matched a complex:

$$\text{Precision} = \frac{\text{matched}_{\text{clusters}}}{\text{predicted}_{\text{clusters}}} \tag{10}$$

Similarly, we defined recall as the number of known complex that matched a cluster:

$$\text{Recall} = \frac{\text{matched}_{\text{complexes}}}{\text{known}_{\text{complexes}}} \tag{11}$$

Complex Prediction Performance

We performed protein complex prediction using RNSC, MCL, MCODE and PCP on the original interaction networks as well as the modified networks. For the RNSC, MCL and MCODE algorithms we used the optimal parameters that are derived by the authors in [31]. We determined optimal parameters for PCP empircally. Compared to predictions made on the orignal network (Fig. 7 top row), we found that the precision–recall performance for MCL, MCODE and PCP improved significantly after the networks are augmented and filtered using FS-weight (Fig. 7 middle row) for both the combined and BioGRID datasets. The performance of RNSC, however, did not changed significantly. PCP performed the best among the clustering algorithms studied for both interaction datasets. We also evaluated the predictions made for the modified network against the newer 2006 MIPS complex dataset (Fig. 7 bottom row), and found that precision–recall performance has generally improved for all the algorithms, which suggested that some of the predictions made which are "novel" based on the 2004 complex dataset were indeed identified to be real complexes a couple of years later.

Improving the Reliability of Interactions

Efforts in computational protein function prediction and protein complex discovery are plagued by the common challenges of false positives, and perhaps more seriously, false negatives in protein–protein interactions. Much work has been done to assess the error rates of interaction data [41–44], and estimates based on overlaps in datasets indicated yeast two-hybrid datasets to contain false positives as high as 50%. More recent work [45] suggested that such estimation are likely to be flawed, and a more recent estimate [46] placed the false discovery rate of yeast two-hybrid interactions at around 10% and false negative rate at around 50% for *S.cerevisiae*. Nonetheless, false positives and false negatives is an important concern, and much effort has been made to improve the quality of interaction data by computationally assessing the confidence of individual interactions. Some of these methods involve using independent, biologically relevant data such as gene expression and sequence homology [43, 47], while others solely used topological properties inherent in the network [48–51].

For methods that derive confidence for each interaction using a topological measure, the weighted interactions can be seen as a being more representative of the underlying "real" network. Hence intuitively it would make sense to use this weighted network to re-compute the confidence for each interaction. We showed in

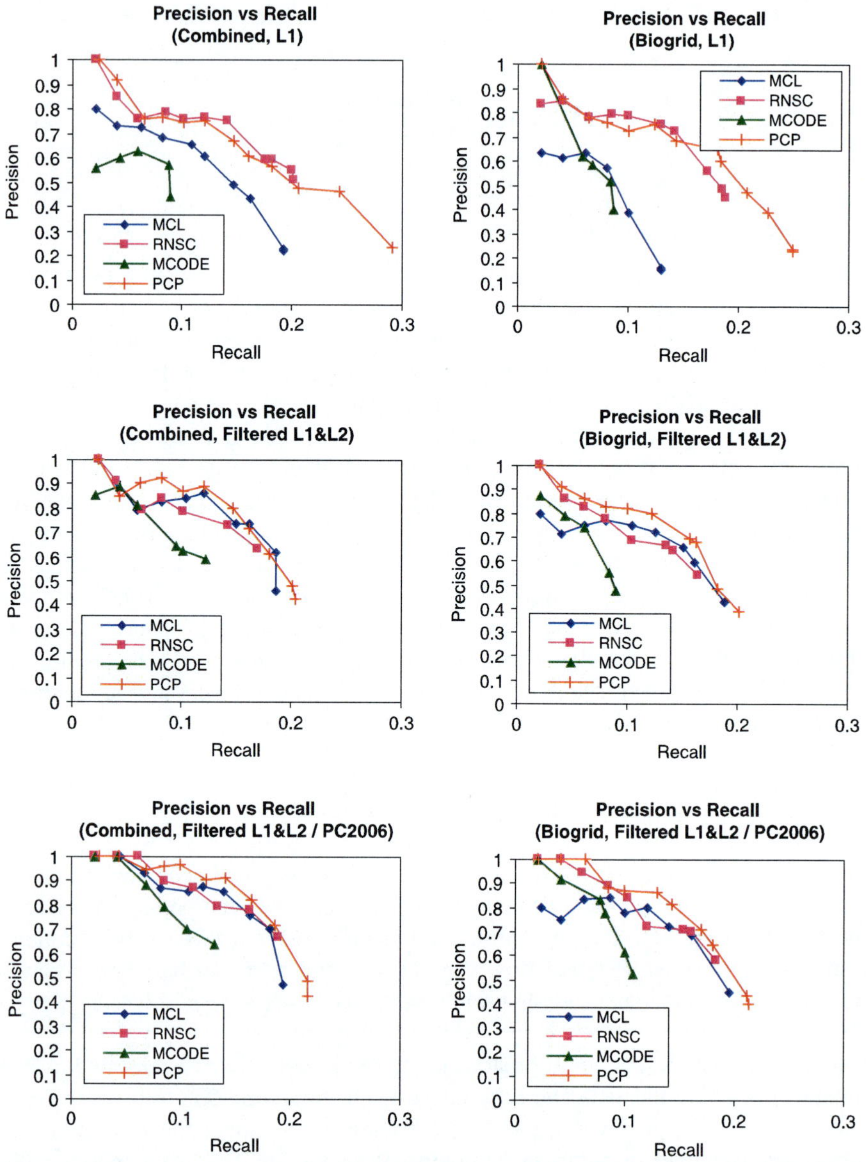

Fig. 7 Precision–recall curves for complex predictions using MCL, RNSC, MCODE and PCP for the combined (*left column*) and BioGRID (*right column*) datasets. Predictions are made using the original networks (*top row*) and the modified networks (*middle row*) and evaluated against complexes from the 2004 MIPS dataset. Predictions made using the modified networks are also evaluated against complexes from the 2006 MIPS dataset (*bottom row*). Figure from [30]

two recent studies that this concept can be used to improve upon local topological measures such as the CD-Distance or FS-Weight in identifying functionally-related interactions and improve complex prediction performance [52, 53].

Iterative Scoring

We define the iterative scoring function from a base topological score function. In the study we used a variant of the CD-Distance as the base measure:

$$AdjustCD\,(u,v) = \frac{2\,|N_u \cap N_v|}{|N_u| + \lambda_u + |N_v| + \lambda_v} \tag{12}$$

λ_u and λ_v are pseudo counts used to penalize proteins with few neighbors, and are defined similarily as $\lambda_{u,v}$ used in FS-weight. The iterative version of AdjustCD is defined as:

$$w^k(u,v) = \frac{\sum_{x \in N_u \cap N_v} \left(w^{k-1}(x,u) + w^{k-1}(x,v)\right)}{\sum_{x \in N_u} w^{k-1}(x,u) + \lambda_u^k + \sum_{x \in N_v} w^{k-1}(x,v) + \lambda_v^k} \tag{13}$$

where $w^{k-1}(u,v)$ is the weight of the edge (u,v) at the $(k\text{–}1)$-th iteration. At the initial stage $(k = 0), w^0(u,v) = 1$ if the edge (u,v) exists and $w^0(u,v) = 0$ otherwise.

$$\lambda_u^k = \max\left\{0, \frac{\sum_{x \in V}\sum_{y \in N_x} w^{k-1}(x,y)}{|V|} - \sum_{x \in N_u} w^{k-1}(x,u)\right\}$$
$$\lambda_v^k = \max\left\{0, \frac{\sum_{x \in V}\sum_{y \in N_x} w^{k-1}(x,y)}{|V|} - \sum_{x \in N_v} w^{k-1}(x,v)\right\} \tag{14}$$

are the weighted variants of λ_u and λ_v at the k-th iteration and V is the set of all nodes in the network. At iteration $k = 1, w^k(u,v) = \text{AdjustCD}(u,v)$. We refer to the k-iteration version of this scoring function as AdjustCD^k.

We showed in [52], that the use of this iterative scoring function reaches best performance at $k = 2$. The weights assigned to interactions using the score function were significantly more predictive of functional similarity and co-localization than FS-Weight and CD-Distance. The weights assigned to indirect level-2 interactions with the iterative function are also more relevant to functional homogenity and localization coherence. These observations suggested that the iterative weighting function may be used to improve the protein complex prediction approach we visited in the previous section.

Complex Discovery Using AdjustCD^k Weighted Interactions

In [53] we conducted a detailed analysis on protein complex finding using interactions that are weighted using AdjustCD^k. Two reference sets of protein complexes are used. The first set is the set of hand-curated complexes from MIPS [39]. The other set of complexes are modeled from three-dimensional structures that were screened using electron microscopy by Aloy et al. [54]. Using the 6 physical protein-protein interaction datasets used in [30, 31], we study how the performance of MCL, MCODE, CFinder [55] and a new clustering algorithm, which we called CMC (Clustering Based on Maximal Cliques), is affected when the input interaction is weighted using AdjustCD^k.

The CMC Algorithm

Like the PCP algorithm, the CMC algorithm starts by finding all maximal cliques in the network using the algorithm described in [40]. However, unlike PCP, CMC do not iteratively merge cliques through building higher-level abstract networks. Instead, a heuristic procedure is used to quickly merge well overlapping cliques into larger clusters. Each clique C is first scored based on its weighted network density:

$$score(C) = \frac{\sum_{u \in C, v \in C} w(u, v)}{|C| \cdot (|C| - 1)} \tag{15}$$

where $w(u,v)$ is the weight of edge (u,v) scored using AdjustCD^k. The cliques are then sorted into a list based on their score in a decreasing order. Each clique C_i is in turn examined beginning from the top of the sorted list. For every other clique C_j in the list which overlaps with C_i above a predefined threshold (i.e. $|C_i \cap C_j| / |C_j| \geq$ *overlap_thres*) and $score(C_j) < score(C_i)$, C_j is removed from the list. A weighted inter-connectivity score is then computed between C_i and C_j to decide if C_j should be merged with C_i:

$$inter{-}score\,(C_1, C_2) = \sqrt{\frac{\sum_{u \in (C_1 - C_2)} \sum_{v \in C_2} w(u, v)}{|C_1 - C_2| \cdot |C_2|} \cdot \frac{\sum_{u \in (C_2 - C_1)} \sum_{v \in C_1} w(u, v)}{|C_2 - C_1| \cdot |C_1|}} \tag{16}$$

If $inter{-}score(C_i, C_j) \geq merge\_thres$, then C_j will be merged with C_i, otherwise it is discarded. *merge_thres* is a pre-defined parameter. The parameters *overlap_thres* and *merge_thres* are empirically determined.

Performance Evaluation

In this study we considered a predicted cluster to match a protein complex if the Jaccard index between them is at least 0.5. To ensure that random matches are unlikely, we randomly swapped complex members to see if the resulting random complexes match with any predicted clusters from the CMC algorithm. We found no matches over 1000 such runs. Precision and recall are defined similarly as described in the previous section of this chapter. We found that all 4 clustering methods achieved significant improvement in precision when using weighted networks compared to unweighted networks. Using k=2 in the AdjustCD^k weighting function result in the best performance among most of the clustering algorithms that are evaluated, and further increase in k to 20 showed little change in performance for CMC and Cfinder.

Robustness Against Noise in the Interaction Network

Perhaps the most interesting observation we made from this study is the robustness of the weighted network to random additive noise. By randomly adding edges to the original network, we examine the impact of additive noise on the prediction performance of CMC using k=1, k=2 and k= 20 for AdjustCD^k weighted versions of the interaction network. Evaluating against the complex dataset from [54], we find that when k=1, the performance of the CMC algorithm degrades significantly when random interactions amounting to 50% of the original network are added, and continues to degrade quickly with higher levels of noise (Fig. 8, top). When k=2, however, the performance of CMC shows only a slight decrease when 50% random interactions are added, and only exhibited significant degradation when the added random interactions exceed 300% of the original network. At k=20, the performance of CMC only shows signs of degradation when the number of added random interactions is 5 times that of the original network. These observations suggest that the iterative scoring approach can potentially be used to benefit downstream analyses that make use of protein-protein interaction data by accentuating the biologically relevant subset of interactions within noisy datasets.

Conclusions

In this chapter, we briefly review some of the works we have done on using protein-protein interactions for computational approaches related to protein function discovery. The key concepts introduced here include indirect functional association between proteins that do not interact directly, the use of topological weights such as FS-weight to identify functionally relevant interactions so that such indirect interactions can be feasible for practical use, and the impact of using topological weighting techniques (such as FS-weight and the iterative AdjustCD^k) to improve interaction data quality on protein complex prediction. It is noteworthy that while protein-protein interaction data is highly relevant to understanding and inferring protein functions, it captures a limited aspect of protein functionality. Greater success in computational function prediction is likely to be achievable through the use of a multitude of biological data such as expression profiles, sequence homology and more. Such holistic approaches are actively being researched on [56–59], and

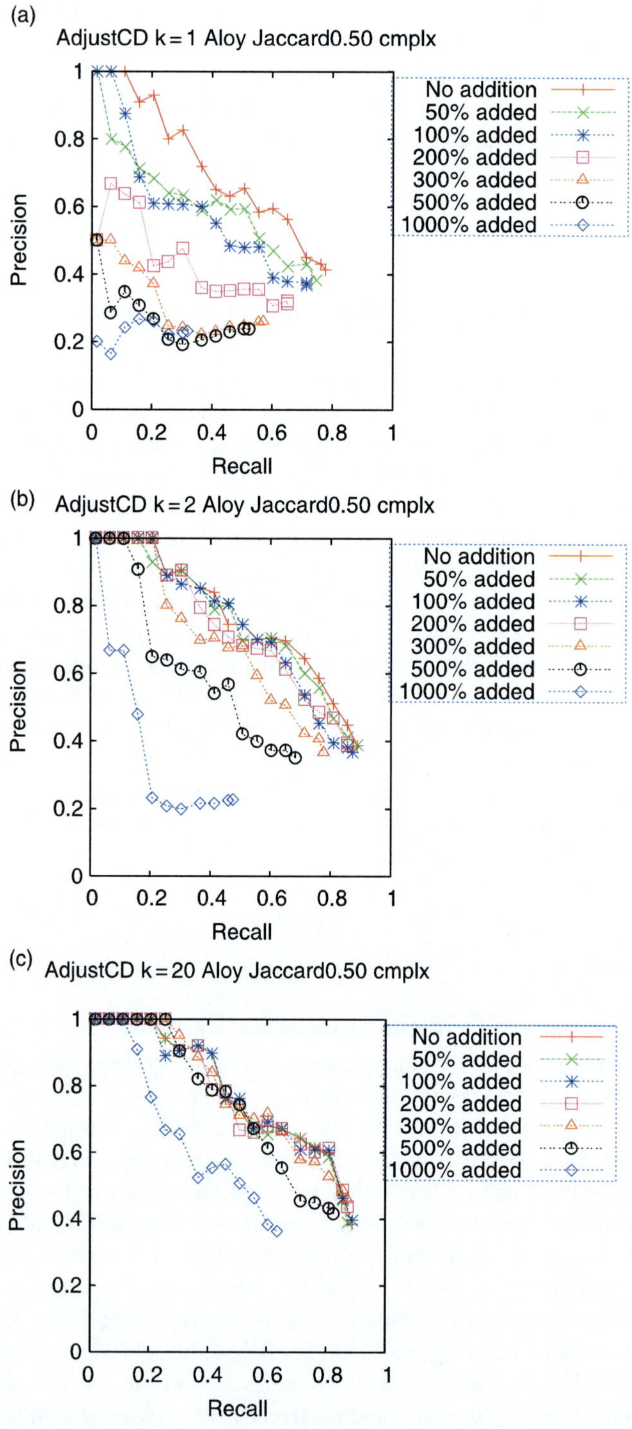

Fig. 8 Precision–recall curves for Aloy reference set when different amount of interactions are randomly added. Overlap thres=0.5, match thres=0.5. Figure from [53]

hold promise for the eventual goal of reliable characterization of protein functionality in a high-throughput fashion. Protein-protein interaction data is an important source of data for these approaches, and research on the analysis and processing of protein–protein interactions will continue be a key area of research in protein function prediction.

References

1. Puig, O., Caspary, F., Rigaut, G., Rutz, B., Bouveret, E., Bragado-Nilsson, E., Wilm, M., Séraphin, B. The Tandem Affinity Purification (TAP) Method: a general procedure of protein complex purification. Methods **24**: 218–229 (2001).
2. Gavin, A., Bosche, M., Krause, R., Grandi, P., Marzioch, M., Bauer, A., Schultz, J., Rick, J.M., Michon, A., Cruciat, C., Remor, M., Hofert, C., Schelder, M., Brajenovic, M., Ruffner, H., Merino, A., Klein, K., Hudak, M., Dickson, D., Rudi, T., Gnau, V., Bauch, A., Bastuck, S., Huhse, B., Leutwein, C., Heurtier, M., Copley, R.R., Edelmann, A., Querfurth, E., Rybin, V., Drewes, G., Raida, M., Bouwmeester, T., Bork, P., Seraphin, B., Kuster, B., Neubauer, G., Superti-Furga, G. Functional organization of the yeast proteome by systematic analysis of protein complexes. Nature **415**: 141–147 (2001).
3. Fromont-Racine, M., Rain, J., Legrain, P. Toward a functional analysis of the yeast genome through exhaustive two-hybrid screens. Nat. Genet. **16**: 277–282 (2001).
4. Ito, T., Tashiro, K., Muta, S., Ozawa, R., Chiba, T., Nishizawa, M., Yamamoto, K., Kuhara, S., Sakaki, Y. Toward a protein–protein interaction map of the budding yeast: a comprehensive system to examine two-hybrid interactions in all possible combinations between the yeast proteins. Proc. Natl. Acad. Sci. USA **97**: 1143–1147 (2001).
5. Uetz, P., Giot, L., Cagney, G., Mansfield, T.A., Judson, R.S., Knight, J.R., Lockshon, D., Narayan, V., Srinivasan, M., Pochart, P., Qureshi-Emili, A., Li, Y., Godwin, B., Conover, D., Kalbfleisch, T., Vijayadamodar, G., Yang, M., Johnston, M., Fields, S., Rothberg, J.M. A comprehensive analysis of protein–protein interactions in Saccharomyces cerevisiae. Nature **403**: 623–627 (2001).
6. Krogan, N.J., Cagney, G., Yu, H., Zhong, G., Guo, X., Ignatchenko, A., Li, J., Pu, S., Datta, N., Tikuisis, A.P., Punna, T., J.M. Peregrín-Alvarez, Shales, M., Zhang, X., Davey, M., Robinson, M.D., Paccanaro, A., Bray, J.E., Sheung, A., Beattie, B., Richards, D.P., Canadien, V., Lalev, A., Mena, F., Wong, P., Starostine, A., Canete, M.M., Vlasblom, J., Wu, S., Orsi, C., Collins, S.R., Chandran, S., Haw, R., Rilstone, J.J., Gandi, K., Thompson, N.J., Musso, G., St Onge, P., Ghanny, S., Lam, M.H.Y., Butland, G., Altaf-Ul, A.M., Kanaya, S., Shilatifard, A., O'Shea, E., Weissman, J.S., Ingles, C.J., Hughes, T.R., Parkinson, J., Gerstein, M., Wodak, S.J., Emili, A., Greenblatt, J.F. Global landscape of protein complexes in the yeast Saccharomyces cerevisiae. Nature **440**: 637–643 (2001).
7. Dixon, S.J., Costanzo, M., Baryshnikova, A., Andrews, B., Boone, C. Systematic mapping of genetic interaction networks. Annu. Rev. Genet. **43**: 601–625 (2001).
8. Tong, A.H.Y., Evangelista, M., Parsons, A.B., Xu, H., Bader, G.D., Page, N., Robinson, M., Raghibizadeh, S., Hogue, C.W.V., Bussey, H., Andrews, B., Tyers, M., Boone, C. Systematic genetic analysis with ordered arrays of yeast deletion mutants. Science **294**: 2364–2368 (2001).
9. Tong, A.H.Y., Lesage, G., Bader, G.D., Ding, H., Xu, H., Xin, X., Young, J., Berriz, G.F., Brost, R.L., Chang, M., Chen, Y., Cheng, X., Chua, G., Friesen, H., Goldberg, D.S., Haynes, J., Humphries, C., He, G., Hussein, S., Ke, L., Krogan, N., Li, Z., Levinson, J.N., Lu, H., Menard, P., Munyana, C., Parsons, A.B., Ryan, O., Tonikian, R., Roberts, T., Sdicu, A., Shapiro, J., Sheikh, B., Suter, B., Wong, S.L., Zhang, L.V., Zhu, H., Burd, C.G., Munro, S., Sander, C., Rine, J., Greenblatt, J., Peter, M., Bretscher, A., Bell, G., Roth, F.P., Brown, G.W., Andrews, B., Bussey, H., Boone, C. Global mapping of the yeast genetic interaction network. Science **303**: 808–813 (2001).

10. Davierwala, A.P., Haynes, J., Li, Z., Brost, R.L., Robinson, M.D., Yu, L., Mnaimneh, S., Ding, H., Zhu, H., Chen, Y., Cheng, X., Brown, G.W., Boone, C., Andrews, B.J., Hughes, T.R. The synthetic genetic interaction spectrum of essential genes. Nat. Genet. **37**: 1147–1152 (2001).
11. Stark, C., Breitkreutz, B.J., Reguly, T., Boucher, L., Breitkreutz, A., Tyers, M. BioGRID: a general repository for interaction datasets. Nucleic Acids Res. **34**: D535 (2001).
12. Schwikowski, B., Uetz, P., Fields, S., others. A network of protein–protein interactions in yeast. Nat. Biotechnol. **18**: 1257–1261 (2001).
13. Hishigaki, H., Nakai, K., Ono, T., Tanigami, A., Takagi, T. Assessment of prediction accuracy of protein function from protein–protein interaction data. Yeast **18**: 523–531 (2001).
14. Deng, M., Zhang, K., Mehta, S., Chen, T., Sun, F. Prediction of protein function using protein–protein interaction data. J. Comput. Biol. **10**: 947–960 (2001).
15. Letovsky, S. Kasif, S. Predicting protein function from protein/protein interaction data: a probabilistic approach. Bioinformatics **19**: i197–204 (2001).
16. Vazquez, A., Flammini, A., Maritan, A., Vespignani, A. Global protein function prediction from protein–protein interaction networks. Nat. Biotechnol. **21**: 697–700 (2001).
17. Chua, H.N., Sung, W.K., Wong, L. Exploiting indirect neighbours and topological weight to predict protein function from protein–protein interactions. Bioinformatics **22**: 1623–1630 (2001).
18. Ruepp, A., Zollner, A., Maier, D., Albermann, K., Hani, J., Mokrejs, M., Tetko, I., Guldener, U., Mannhaupt, G., Munsterkotter, M., Mewes, H.W. The FunCat, a functional annotation scheme for systematic classification of proteins from whole genomes. Nucleic Acids Res. **32**: 5539–5545 (2001).
19. Samanta, M.S, Liang, P. Predicting protein functions from redundancies in large-scale protein interaction networks, Proc. Natl. Acad. Sci. USA **100**: 12579–12583 (2001).
20. Brun, C., Chevenet, F., Martin, D., Wojcik, J., A. Guénoche, Jacq, B. Functional classification of proteins for the prediction of cellular function from a protein–protein interaction network. Genome Biol. **5**: 6–6 (2001).
21. Serebriiskii, I.G., Golemis, E.A. Two-hybrid system and false positives. Approaches to detection and elimination. Methods Mol. Biol. **177**: 123–134 (2001).
22. Friedel, C.C., Zimmer, R. Identifying the topology of protein complexes from affinity purification assays. Bioinformatics **25**: 2140–2146 (2009).
23. Zhou, X., Kao, M.J., Wong, W.H. Transitive functional annotation by shortest-path analysis of gene expression data. Proc. Natl. Acad. Sci. USA **99**: 12783–12788 (2009).
24. Chua, H., Sung, W.K., Wong, L. Using indirect protein interactions for the prediction of gene ontology functions. BMC Bioinformatics **8**: S8 (2009).
25. Spirin, V., Mirny, L.A. Protein complexes and functional modules in molecular networks. Proc. Natl. Acad. Sci. USA **100**: 12123–12128 (2009).
26. Bu, D., Zhao, Y., Cai, L., Xue, H., Zhu, X., Lu, H., Zhang, J., Sun, S., Ling, L., Zhang, N., Li, G., Chen, R. Topological structure analysis of the protein–protein interaction network in budding yeast. Nucleic Acids Res. **31**: 2443–2450 (2009).
27. Bader, G.D., Hogue, C.W. An automated method for finding molecular complexes in large protein interaction networks. BMC Bioinformatics **4**: 2 (2009).
28. Shannon, P., Markiel, A., Ozier, O., Baliga, N.S., Wang, J.T., Ramage, D., Amin, N., Schwikowski, B., Ideker, T. Cytoscape: a software environment for integrated models of biomolecular interaction networks. Genome Res. **13**: 2498–2504 (2009).
29. Chua, H.N., Ning, K., Sung, W.K., Leong, H.W., Wong, L. *Using indirect protein–protein interactions for protein complex prediction*. Computational systems bioinformatics: proceedings of the CSB 2007 Conference. Markstein, P., Xu, Y. London: Imperial College Press, pp. 97–110 (2009).
30. Chua, H.N., Ning, K., Sung Wing-Kin, Leong, H.W., Wong, L. Using indirect protein–protein interactions for protein complex prediction. J. Bioinform. Comput. Biol. **6**: 435–466 (2009).

31. Brohee, S., van Helden, J. Evaluation of clustering algorithms for protein–protein interaction networks. BMC Bioinformatics **7**: 488 (2009).
32. King, A.D., Przulj, N., Jurisica, I. Protein complex prediction via cost-based clustering. Bioinformatics **20**: 3013–20 (2009).
33. Van Dongen, S.M. Graph clustering by flow simulation. PhD thesis, Universiteit Utrecht (2000).
34. Blatt, M., Wiseman, S., Domany, E. Superparamagnetic clustering of data. Phys. Rev. Lett. **76**: 3251–3254 (1996).
35. Ito, T., Chiba, T., Ozawa, R., Yoshida, M., Hattori, M., Sakaki, Y. A comprehensive two-hybrid analysis to explore the yeast protein interactome. Proc. Natl. Acad. Sci. USA **98**: 4569–74 (1996).
36. Gavin, A.C., Aloy, P., Grandi, P., Krause, R., Boesche, M., Marzioch, M., Rau, C., Jensen, L.J., Bastuck, S., Dmpelfeld, B., Edelmann, A., Heurtier, M.A., Hoffman, V., Hoefert, C., Klein, K., Hudak, M., Michon, A.M., Schelder, M., Schirle, M., Remor, M., Rudi, T., Hooper, S., Bauer, A., Bouwmeester, T., Casari, G., Drewes, G., Neubauer, G., Rick, J.M., Kuster, B., Bork, P., Russell, R.B., Superti-Furga, G. Proteome survey reveals modularity of the yeast cell machinery. Nature **440**: 631–636 (1996).
37. Ho, Y., Gruhler, A., Heilbut, A., Bader, G.D., Moore, L., Adams, S.L., Millar, A., Taylor, P., Bennett, K., Boutilier, K., Yang, L., Wolting, C., Donaldson, I., Schandorff, S., Shewnarane, J., Vo, M., Taggart, J., Goudreault, M., Muskat, B., Alfarano, C., Dewar, D., Lin, Z., Michalickova, K., Willems, A.R., Sassi, H., Nielsen, P.A., Rasmussen, K.J., Andersen, J.R., Johansen, L.E., Hansen, L.H., Jespersen, H., Podtelejnikov, A., Nielsen, E., Crawford, J., Poulsen, V., Sorensen, B.D., Matthiesen, J., Hendrickson, R.C., Gleeson, F., Pawson, T., Moran, M.F., Durocher, D., Mann, M., Hogue, C.W.V., Figeys, D., Tyers, M. Systematic identification of protein complexes in Saccharomyces cerevisiae by mass spectrometry. Nature **415**: 180–183 (1996).
38. Krogan, N.J., Cagney, G., Yu, H., Zhong, G., Guo, X., Ignatchenko, A., Li, J., Pu, S., Datta, N., Tikuisis, A.P., Punna, T., Peregrn-Alvarez, J.M., Shales, M., Zhang, X., Davey, M., Robinson, M.D., Paccanaro, A., Bray, J.E., Sheung, A., Beattie, B., Richards, D.P., Canadien, V., Lalev, A., Mena, F., Wong, P., Starostine, A., Canete, M.M., Vlasblom, J., Wu, S., Orsi, C., Collins, S.R., Chandran, S., Haw, R., Rilstone, J.J., Gandi, K., Thompson, N.J., Musso, G., Onge, P.S., Ghanny, S., Lam, M.H.Y., Butland, G., Altaf-Ul, A.M., Kanaya, S., Shilatifard, A., O'Shea, E., Weissman, J.S., Ingles, C.J., Hughes, T.R., Parkinson, J., Gerstein, M., Wodak, S.J., Emili, A., Greenblatt, J.F. Global landscape of protein complexes in the yeast Saccharomyces cerevisiae. Nature **440**: 637–643 (2006).
39. Mewes, H.W., Amid, C., Arnold, R., Frishman, D., Guldener, U., Mannhaupt, G., Munsterkotter, M., Pagel, P., Strack, N., Stumpflen, V., Warfsmann, J., Ruepp, A. MIPS: analysis and annotation of proteins from whole genomes. Nucleic Acids Res **32**: D41–44 (2004).
40. Tomita, E., Tanaka, A., Takahashi, H. The worst-case time complexity for generating all maximal cliques and computational experiments. Theor. Comput. Sci. **363**: 28–42 (2006).
41. Deane, C.M., Salwinski, L., Xenarios, I., Eisenberg, D. Protein interactions: two methods for assessment of the reliability of high throughput observations. Mol. Cell Proteomic. **1**: 349–356 (2002).
42. Deng, M., Sun, F., Chen, T. Assessment of the reliability of protein–protein interactions and protein function prediction. Biocomputing 2003: Proceedings of the Pacific Symposium Hawaii, USA, 3–7 January 2002, p. 140 (2003).
43. von Mering, C., Krause, R., Snel, B., Cornell, M., Oliver, S.G., Fields, S., Bork, P. Comparative assessment of large-scale data sets of protein–protein interactions. Nature **417**: 399–403 (2002).
44. Bader, J.S., Chaudhuri, A., Rothberg, J.M., Chant, J. Gaining confidence in high-throughput protein interaction networks. Nat. Biotechnol. **22**: 78–85 (2004).

45. Huang, H., Jedynak, B.M., Bader, J.S. Where have all the interactions gone? Estimating the coverage of two-hybrid protein interaction maps. PLoS Comput. Biol. **3**: e214 (2007).
46. Huang, H., Bader, J.S. Precision and recall estimates for two-hybrid screens. Bioinformatics **25**: 372–378 (2009).
47. Gilchrist, M.A., Salter, L.A., Wagner, A. A statistical framework for combining and interpreting proteomic datasets. Bioinformatics **20**: 689–700 (2004).
48. Saito, R., Suzuki, H., Hayashizaki, Y. Construction of reliable protein–protein interaction networks with a new interaction generality measure. Bioinformatics **19**: 756–763 (2003).
49. Goldberg, D.S., Roth, F.P. Assessing experimentally derived interactions in a small world. Proc. Natl. Acad. Sci. USA **100**: 4372–4376 (2003).
50. Chen, J., Chua, H.N., Hsu, W., Lee, M.L., Ng, S.K., Saito, R., Sung, W.K., Wong, L. Increasing confidence of protein–protein interactomes. Genome Inform. Ser. **17**: 284–297 (2006).
51. Chen, J., Hsu, W., Lee, M.L., Ng, S.-K. Discovering reliable protein interactions from high-throughput experimental data using network topology. Artif. Intell. Med. **35**: 37–47 (2005).
52. Liu, G., Li, J., Wong, L. Assessing and predicting protein interactions using both local and global network topological metrics. Proceedings of 19th International Conference on Genome Informatics, pp. 138–149 (2008).
53. Liu, G., Wong, L., Chua, H.N. Complex discovery from weighted PPI networks. Bioinformatics **25**: 1891–1897 (2009).
54. Aloy, P., Bottcher, B., Ceulemans, H., Leutwein, C., Mellwig, C., Fischer, S., Gavin, A., Bork, P., Superti-Furga, G., Serrano, L., Russell, R.B. Structure-based assembly of protein complexes in yeast. Science **303**: 2026–2029 (2004).
55. Adamcsek, B., Palla, G., Farkas, I.J., Derenyi, I., Vicsek, T. CFinder: locating cliques and overlapping modules in biological networks. Bioinformatics **22**: 1021–1023 (2006).
56. Chen, Y., Xu, D. Global protein function annotation through mining genome-scale data in yeast Saccharomyces cerevisiae. Nucleic Acids Res. **32**: 6414–6424 (2004).
57. Troyanskaya, O.G., Dolinski, K., Owen, A.B., Altman, R.B., Botstein, D. A Bayesian framework for combining heterogeneous data sources for gene function prediction (in Saccharomyces cerevisiae). Proc. Natl. Acad. Sci. USA **100**: 8348–8353 (2003).
58. Chua, H.N., Sung, W.K., Wong, L. An efficient strategy for extensive integration of diverse biological data for protein function prediction. Bioinformatics **23**: 3364 (2007).
59. Tian, W., Zhang, L., Tasan, M., Gibbons, F., King, O., Park, J., Wunderlich, Z., Cherry, J.M., Roth, F. Combining guilt-by-association and guilt-by-profiling to predict Saccharomyces cerevisiae gene function. Genome Biol. **9**: S7 (2008).

KEGG and GenomeNet Resources for Predicting Protein Function from Omics Data Including KEGG PLANT Resource

Toshiaki Tokimatsu, Masaaki Kotera, Susumu Goto, and Minoru Kanehisa

Abstract With the rise of experimental technologies for omics research in recent years, considerable quantitative data related to transcription, protein and metabolism are available for predicting protein functions. To predict protein functions from large omics data, reference knowledge databases and bioinformatics tools play considerable roles. KEGG (http://www.genome.jp/kegg/) database we have been establishing is an integrated database of biological systems including genomic, chemical and systemic functional information. Our group has also been developing the tools for genome or chemical analysis as GenomeNet Bioinformatics Tools (http://www.genome.jp/en/gn_tools.html). In this chapter, we introduce the KEGG database resources and the GenomeNet Bioinformatics Tools for predicting protein functions from the viewpoint of omics research, as well as some recent topics (KEGG PLANT Resource and PathPred). KEGG PLANT Resource is one of the contents in the KEGG EDRUG database, and contains links for plant secondary metabolite biosynthesis pathways, plant genomes and EST sequences, chemical information of plant natural products and the prediction tool for plant secondary metabolism pathway. PathPred is a recently developed pathway prediction tool based on the chemical structure transformation patterns of enzyme reactions found in metabolic pathways.

Introduction

In recent years, high-throughput omics data such as transcriptome and metabolome data is continuously increasing. Genomics, transcriptomics and proteomics provide the data of genes and proteins in individual organisms. On the other hand, metabolomics, glycomics, and lipidomics provide information for endogenous molecules, and chemical genomics provides information for exogenous molecules.

T. Tokimatsu (✉)
Bioinformatics Center, Institute for Chemical Research, Kyoto University, Uji 611-0011, Japan
e-mail: tokimatu@kuicr.kyoto-u.ac.jp

D. Kihara (ed.), *Protein Function Prediction for Omics Era*,
DOI 10.1007/978-94-007-0881-5_14,

For environmental studies, metagenomics and meta-metabolomics data are becoming available as genomic and chemical information, respectively. One of the main objectives of these high-throughput experiment projects is to uncover molecular building blocks of life. Integration and of high throughput genomics and chemical spaces data and interpretation of high-order function is a powerful technique for understanding molecular building block of life such as protein function. Bioinformatics approaches are required to predict protein function by analyzing exclusively increasing omics data.

KEGG (http://www.kegg.jp/) [1] is a computer representation of biological systems, consisting of a number of sub-databases, such as those including genomic and chemical information. Among these, systems information database is the most unique feature in KEGG. They have been manually collected from review and original articles, other publications, specialists' website, and other resources. In KEGG project, several useful bioinformatics tools have also been developed for genome analysis and chemical analysis. These tools are released as GenomeNet bioinformatics tools at GenomeNet website (http://www.genome.jp/).

Plants are known to produce vast and diverse secondary metabolites, and the total number of plant metabolites are estimated to be over 200,000 [2]. Plant secondary metabolites support our life either directly or indirectly as foods, medicines, and industrial materials. Notably, physiologically active natural products mainly from plants are used as crude drugs and traditional medicine in our lives since ancient times. Physiological active plant natural products have been main resources for drug seed compounds. Thus, elucidating the biosynthetic pathways of plant secondary metabolites is a valuable research area for plant biotechnology, agricultural sciences and pharmaceutical sciences. In just the past decade, transcriptome [3, 4] and metabolome [5, 6] analysis of model plant species such as Arabidopsis (*Arabidopsis thaliana*) and rice (*Oryza sativa*) has become an active area of research. Accordingly, the necessity of high-quality database resource of plants gains the importance for predicting plant secondary metabolite biosynthesis pathway and protein functions involved in the pathway. Therefore, we are currently accumulating crude drugs and other plant natural product information as KEGG EDRUG database.

In this chapter, we introduce the overview of the KEGG database and the recent topics on the KEGG and GenomeNet resources from the viewpoint of protein function prediction, including KEGG EGENES, KEGG PLANT Resource and PathPred: an enzyme-catalyzed metabolic pathway prediction server.

Outline of KEGG Resource

Overview of KEGG Database

Table 1 shows the list of KEGG main databases [1], and their contents. As of July 2010, KEGG comprises 19 main databases, categorized into systems information, genomic information and chemical information as shown in Table 1. Genomic and

Table 1 KEGG databases

Category	Database	Contents
Systems information	KEGG PATHWAY	Pathway maps
	KEGG BRITE	Functional hierarchies
	KEGG MODULE	Pathway modules
	KEGG DISEASE	Human diseases
	KEGG DRUG	Drugs
	KEGG EDRUG	Crude drugs and other natural products
Genomic information	KEGG ORTHOLOGY	KEGG Orthology (KO) groups
	KEGG GENOME	KEGG organism
	KEGG GENES	Genes in completely sequenced genomes
	KEGG SSDB	Best hit relation within GENES
	KEGG DGENES	Genes in draft genomes
	KEGG EGENES	Genes as EST contigs
	KEGG MGENES	Genes in metagenomes
Chemical information (KEGG LIGAND)	KEGG COMPOUND	Metabolite and other small molecules
	KEGG GLYCAN	Glycans
	KEGG REACTION	Biochemical reactions
	KEGG RPAIR	Reactant pair chemical transformations
	KEGG RCLASS	Reaction classification
	KEGG ENZYME	Enzyme nomenclature

chemical information databases are collection of molecular building blocks of life in the genomic and chemical spaces, respectively, and systems information represent the molecular systems that are built from the molecular building blocks.

KEGG is a computer representation of biological systems. Systems information is the most characteristic feature in KEGG database, and is manually collected from review articles, other publications, specialists' website, and other resources. Six databases in KEGG describe systems information. They are classified into two types. The former three databases (PATHWAY, BRITE, and MODULE) are the databases for pathway and functional classification. The latter three databases (DISEASE, DRUG, and EDRUG) are the databases for analysis of the molecular network-disease association. DISEASE, DRUG, and EDRUG contain data of disease, drug and bioactive natural products, respectively. The following seven databases (ORTHOLOGY, GENOME, GENES, SSDB, DGENES, EGENES, and MGENES) are categorized as genomic information. They are gene catalogs in the completely sequenced genomes, manually defined ortholog groups, computationally calculated sequence similarity information, and supplementary gene catalog data (for draft genomes, EST contigs, and metagenomes). The six databases in chemical information category (COMPOUND, GLYCAN, REACTION, RPAIR, RCLASS, and ENZYME) are collectively called as KEGG LIGAND. They contain the information of small molecules, glycans, biochemical reaction of these molecules, chemical structure transformation patterns derived from reaction data, reaction classification according to the chemical structure transformation pattern, and supplemental information of enzyme nomenclatures. SSDB, DGENES, EGENES, and

MGENES in the genomic information category are computationally generated, but all other 15 databases are manually curated.

KEGG Orthology (KO): Basis of Genome Annotation in KEGG

KEGG Orthology (KO) is the basis for the protein function annotation in KEGG. KEGG ortholog annotation procedure is described as follows. Protein sequences with experimental evidences in specific organisms are used as seeds, and the homologous sequences from other organisms are automatically collected. Consequently, these sequence groups are manually curated and defined as the KO groups in the context of molecular networks; i.e., as the nodes in the KEGG PATHWAY and BRITE. KO groups are given K numbers for identification. Next, cross-species annotation is added as follows. Gene catalogs of all complete genomes are generated from RefSeq database and other public resources. They are computationally processed to generate what we refer to as the GFIT tables, containing the list of genes in a genome with the data of the best-hit genes (i.e., the most homologous genes) against the all other genomes. The automatic cross-species annotation is performed for a set of the "safe" K numbers, representing clearly defined ortholog groups. Manual curation of this automatic annotation is performed using the KOALA and GFIT tools. As of July 2010, genes data taken from 1135 prokaryotes and 131 eukaryotes species are stored in the GENES database. We developed KEGG Automatic Annotation Server (KAAS) as functional annotation tool of genes. This system automatically assigned KO for query genes. Detailed information about KAAS is described in section "KAAS – KEGG Automatic Annotation Server".

PATHWAY and BRITE: Systems Representation in KEGG

KEGG PATHWAY maps describe the dual aspects of metabolic network. The first aspect is genomic information network, i.e., the network of enzyme genes or enzymes. In KEGG PATHWAY, genes and proteins are identified by the K numbers as mentioned in the previous section. EC numbers are shown as the node names in the pathway maps, but they are not used as the identifiers in KEGG. The second aspect is chemical information network, the network of small molecules (chemical compounds) and chemical structure transformations. Chemical compounds are identified by the C numbers and reactions are identified by the R numbers.

The KEGG reference pathway maps and BRITE reference hierarchies are created as to be applicable to all organisms; the exceptions are those describing human diseases. The organism-specific pathways and hierarchies can be generated by using the K numbers as the gene identifiers in particular organisms. Genes in an organism, take *Arabidopsis thaliana* as an example, are annotated with the K numbers, representing manually defined ortholog groups corresponding to the nodes in the KEGG pathway maps.

PATHWAY also provides the global metabolism maps, which are created by manually combining about 120 existing traditional metabolic pathway maps. Circular nodes represent chemical compounds, and the lines connecting two nodes are series of reactions. These global pathway maps allow users to view and compare the entire metabolism, by such means as mapping transcriptome data and/or metabolome data.

Color Objects in KEGG Pathways and BRITE Hierarchies

Integrating large-scale data of genomic (e.g., transcriptome) and/or chemical (e.g., metabolome) spaces onto the systems space (e.g., KEGG PATHWAY, BRITE) helps our understanding for protein function prediction. This section explains the methods for mapping molecular datasets to the KEGG pathway and BRITE hierarchies.

The first method is to use the options "Pathway Mapping" and "Brite Mapping" available on the web pages (http://www.genome.jp/kegg/tool/color_pathway.html and http://www.genome.jp/kegg/tool/color_brite.html), respectively. From the Search Object page, the user can find the objects (genes, metabolites, etc.) of interest in the PATHWAY maps or the BRITE hierarchies by coloring them. Consequently, the user can obtain PATHWAY maps or BRITE hierarchies with these objects favorably colored through the Color Object page. The objects of interest have to be specified by the KEGG identifiers. The user can input the list of objects either directly from the input box or by uploading the file including the list.

Another method for mapping dataset on pathway is accessing KEGG through KEGG API (http://www.genome.jp/kegg/soap/). KEGG API is a web service to use the KEGG system from your program via SOAP/WSDL. The service enables users to develop software that accesses and manipulates vast amount of KEGG data that are constantly updated. KEGG API provides function for coloring pathways. For the general information on KEGG API, please refer to the KEGG API page at GenomeNet (http://www.genome.jp/kegg/soap/).

KEGG REACTION: Chemical Structure Transformation Information in KEGG

KEGG REACTION database contains enzyme reactions taken from KEGG ENZYME database and from the metabolic pathway maps in KEGG PATHWAY database. Each reaction is identified by the R number. KEGG RPAIR database is a collection of reactant pair defined for each reaction in KEGG reaction, together with the chemical structure transformation patterns characterized by the RDM patterns. Each reaction pair is identified by the RP number. In general, a reaction consists of multiple reactant pairs, and the one that appears on the KEGG metabolic pathway maps is called as the main pair. The RDM pattern is defined as KEGG atom type change at the reaction center "R", the difference atom next to reaction center "D", and the matched atom next to reaction center "M", respectively. KEGG RCLASS database represents classification of reaction based on the RDM patterns of main reactant pairs. The transformation pattern may consist of multiple RDM patterns. Each reaction class is identified as RC number. We developed PathPred and E-zyme for predicting pathway and enzymatic functions. The RDM patterns are the basis of these prediction tools. Detailed information about PathPred and E-zyme are described in sections "PathPred: Pathway Prediction Server" and "E-zyme for Prediction of Enzymatic Reactions", respectively.

KEGG Resources and GenomeNet Bioinformatics Tools for Predicting Protein Function

KEGG EDRUG and KEGG PLANT Resource

Overview of KEGG EDRUG and KEGG PLANT Resource

Natural resources including bioactive natural products, such as crude drugs and foods, have been used usefully since ancient times. These natural resources are mostly taken from plants. For this reason, we developed new database, KEGG EDRUG (http://www.genome.jp/kegg/drug/edrug.html), which is a database of crude drugs, essential oils and other useful natural product resources including plant information resources.

Plants are known to produce diverse chemical compounds including those with medicinal, nutritional and industrial values. These plant secondary metabolites can be divided into groups that share the same core substructure, originated from the same biosynthetic pathways and chemical building blocks. In this context, KEGG EDRUG is also considered as a part of KEGG PLANT Resource, which is an interface to the KEGG resource for plant research, especially for understanding relationships between genomic and chemical information of plant natural products. KEGG PLANT Resource links to the biosynthetic pathway of plant secondary metabolites, sequences of plant genomes and ESTs, structural classification of plant secondary metabolites, and pathway prediction tools for plant secondary metabolites.

Pathway Maps of Plant Secondary Metabolite Biosynthesis

Plants produce vast and diverse secondary metabolites, but core structures of these metabolites are synthesized from several important precursors. Biosynthesis pathways of plant secondary metabolites are classifiable by precursors and their biosynthesis pathways. Grouping secondary metabolite biosynthesis pathways by their biosynthetic origins and mapping the core structures on overview pathway maps will help our understanding about plant secondary metabolisms.

Therefore, the KEGG PLANT Resource provides the links to the pathway maps for secondary metabolism. There are three types of pathway maps in KEGG PLANT Resource, e.g., KEGG traditional pathway maps, a global map, and overview maps.

In recent years, the repertoire of the KEGG pathway maps for plant secondary metabolism is expanded, and some of the maps are renewed incorporating recent information. As a result, the secondary metabolite biosynthesis subclass in KEGG PATHWAY is divided to "Metabolism of Terpenoids and Polyketides" and "Biosynthesis of Other Secondary Metabolites".

Recently, we developed new global map of secondary metabolism pathways. Figure 1a is a screenshot of the reference global maps of secondary metabolite

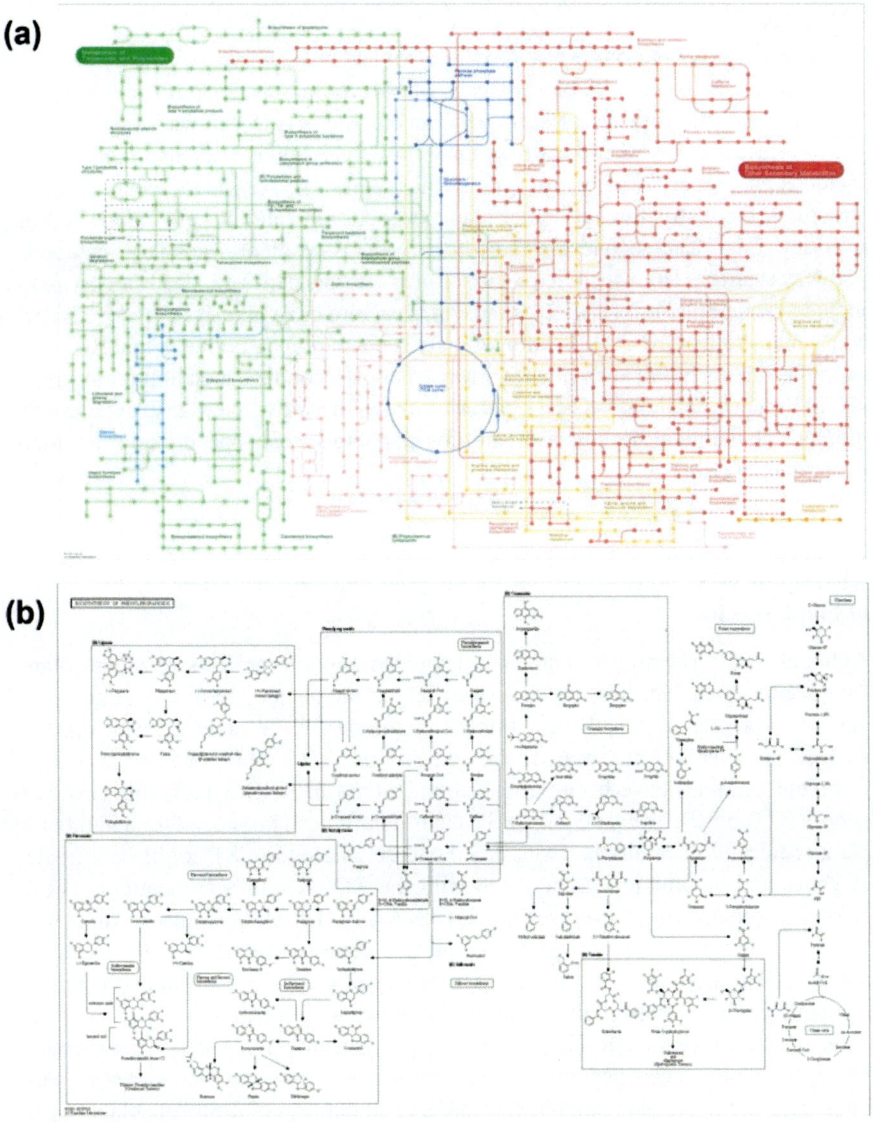

Fig. 1 Example PATHWAY maps of global pathway and overview pathway. (**a**) Reference global pathway map of secondary metabolites. (**b**) Overview pathway map of Biosynthesis of phenylpropanoids

biosynthesis, which include secondary metabolite biosynthesis pathways and related pathway maps. This map allows users to view and compare the omics data or mapping species-specific pathways on the secondary metabolite specific pathway. The major part of this pathway is related to plant metabolism, intended for the usefulness for plant scientists. This new global map of secondary metabolism also allows users to map the species-specific pathways by using K numbers, e.g. *Arabidopsis thaliana*. Users can easily figure out the secondary metabolism of specific species at a glance.

KEGG PLANT Resource also provides two levels of overview pathway maps for plant secondary metabolite biosynthesis. First level map is general overview pathway map for biosynthesis of secondary metabolites. This pathway map is based on overview of biosynthetic pathways map in KEGG PATHWAY, and modification includes plant-specific pathways such as biosynthesis of secondary metabolite core structures. Core structures of major plant secondary metabolites are mapped on the overview map. The second are the category maps of plant secondary metabolite pathways starting from specific precursor biosynthesis pathway. The category maps include the detailed information of the core structure biosynthesis pathway. The category maps reflect the classification of plant secondary metabolites. Figure 1b is the example screenshot of overview pathway map of secondary metabolite biosynthesis. Different to standard KEGG pathway maps, these overview maps contain the graphics of chemical structures and not for mapping species-specific information or experimental data.

KEGG GENES and EGENES of Plants: Sequence Information of Plant Species

Although some research groups have launched genome projects of model plants for specific plant families and industrial important plants, the available complete genomes for plants are still very limited in comparison to other organism groups such as animals and bacteria. At the end of June 2010, only 13 plant species of complete genomes have been published and stored in KEGG, including two draft genomes (Table 2). Thus, massive EST dataset have been processed for a number of plant species to generate the EGENES database where the EST contigs are treated as genes and automatically annotated with the KO (K number) identifiers [7] by KAAS automatic annotation (see section "KAAS – KEGG Automatic Annotation Server"). Currently, 77 plant species of the EST datasets are stored in the EGENES database. The EST dataset covers wider variety of plant families and species than complete genomes, especially in asterids as shown in Table 2.

We also provide the BRITE hierarchical lists of plant phylogenetic classification. The phylogenetic classification of angiosperm is based on the second Angiosperm Phylogeny Group classification for the orders and families of flowering plants (APG II 2003) [8]. APG II classification is based on molecular systematic of flowering plants.

Table 2 Number of plant families and species in complete genomes and EST datasets in KEGG

Classification	Complete genomes (GENES, DGENES)	EST datasets (EGENES)
Eudicots: asterids	0(0)	7(18)
Eudicots: resides	4(4)	8(31)
Eudicots: others	1(1)	4(4)
Monocots	1(3)	3(11)
Basal angiosperms	0(0)	1(1)
Gymnosperms	0(0)	2(5)
Ferns	0(0)	1(2)
Mosses	1(1)	2(2)
Green algae	2(3)	2(2)
Red algae	1(1)	0(0)
Glaucophytes	0(0)	1(1)
Total	10(13)	31(77)

Figures before parentheses are the number of families and figures in parenthesis are number of species

Classification of Plant Secondary Metabolites

Based on the biosynthetic origin, major plant secondary metabolites are classified to polyketides (from acetate-malonate pathway), phenylprpoanoids and related compounds (from shikimate pathway), terpenoids and steroids (from mevalonate pathway or deoxyxylulose-phosphate pathway), and nitorogen-containing alkaloids and sulfur-containing compounds (from amino-acids and related compounds). This classification also reflects the core chemical structures of plant secondary metabolites. We classified the plant secondary metabolites in the KEGG COMPOUND database by their biosynthetic origins and the core chemical structures. Currently, about 2600 plant secondary metabolites are collected in the BRITE classification of phytochemical compounds (Table 3). The top seven metabolite classes contain over 150 metabolites. All these metabolite classes are well studied, and contain many bioactive metabolites such as components of crude drugs and essential oils. This classification system is also used for categorizing drugs and other bioactive compounds derived from plant metabolites. In a future, phytochemical compound classification will be more refined and categorized into detail core structures, which will help to link compound classification to biosynthetic pathways.

KEGG EDRUG Database

KEGG EDRUG database is a collection of bioactive natural products, which are ingested in our bodies. These natural products, such as crude drugs, essential oils, etc., are mostly supplied from plant. The E number is used for identifier of each KEGG EDRUG entry, and is associated with the chemical components, efficacy information, and source species information. At present, crude drugs and essential oils are the two main components of the EDRUG entries.

Table 3 Phytochemical compounds classification in KEGG BRITE as of July 2010

Classification	# of compounds
Total phytochemical compounds	2617
Phenylpropanoids and related compounds	744
Monolignols	37
Lignans	71
Coumarins	61
Flavonoids	507
Stilbenoids	39
Hydrolysable tannins	24
Misc. Phenylpropanoids	5
Polyketides	97
Quinones	41
gamma-Pyrones	56
Terpenoids	965
Hemiterpenoids (C5)	3
Monoterpenoids (C10)	164
Sesquiterpenoids (C15)	298
Diterpenoids (C20)	170
Triterpenoids (C30), sterols and steroids	286
Tetraterpenoids (C40) (Carotenods)	36
Polyterpenoids	8
Alkaloids	720
Alkaloids derived from ornithine	103
Alkaloids derived from lysine	76
Alkaloids derived from nicotinic acid	18
Alkaloids derived from tyrosine	201
Alkaloids derived from tryptophan and anthranillic acid	199
Alkaloids derived from histidine	3
Alkaloids derived by amination reactions	103
Misc. alkaloids	17
Amino acid derivatives other than alkaloids	91
Betalains	30
Cyanogenic glucosides	25
Glucosinolates	36

This table shows the first classes and second classes of the Phytochemical compounds classification in KEGG BRITE

Pathway Prediction for Plant Secondary Metabolism

Predicting biosynthetic pathways of plant secondary metabolites and linking them to the plant genomes are challenging problems. We recently developed a web-based server named PathPred, which is designed to predict secondary metabolite biosynthesis pathways for a given compound using the information of known enzyme reactions (i.e., RDM patterns and chemical structure alignments of substrate-product pairs). Detailed information about PathPred is described in the next section.

PathPred: Pathway Prediction Server

PathPred (http://www.genome.jp/tools/pathpred/) [9] is a recently developed web-based server for predicting metabolic pathway of a given compound. Current version of PathPred provides a multi-step reaction prediction of xenobiotics biodegradation pathways and secondary metabolite biosynthesis pathways, and we aim to improve this toward more sophisticated prediction for metabolic pathway reconstruction.

Prediction procedure consists of the following three steps. The first step is a global similarity search of a query compound against the KEGG COMPOUND database by the SIMCOMP program [10, 11]. The second step is a local pattern match against the RDM pattern library to select the matched patterns that are applicable to the query compound. Specific category of the KEGG pathways, such as xenobiotics biodegradation pathways or secondary metabolite biosynthesis pathways, have their specific subsets of the RDM patterns [9, 12, 13]. Thus, we extracted and use the specific RDM patterns library for xenobiotics biodegradation and secondary metabolite biosynthesis, respectively. The third step is to apply the structure transformation to the query compound based on the selected matched patterns. PathPred has a function to assign plausible EC numbers to the suggested reaction steps. This function is based on the E-zyme program. Further information about E-zyme is described in section "E-zyme for Prediction of Enzymatic Reactions ".

We have to mention that the meaning of the query compound is different depending on whether users would like to predict biodegradation pathways or biosynthesis pathways. In the case of biodegradation, the query compound is the molecule that will break down. In other words, it is located at the beginning of the pathway. On the contrary, in the case of biosynthesis, the query compound is the molecule that was synthesized. In other words, it is located at the end of the pathway. This makes sense when we consider what we would like to do, but this is sometimes confusing when we actually use this application. Users may optionally input the end product in biodegradation or the start compound in the biosynthesis. If the users have an idea what the origin of the query secondary metabolite might be, specifying them might help better prediction. Users can use input query compound in the MDL mol format, the SMILES representation, or KEGG COMPOUND/DRUG identifier (C/D numbers). We provide KegDraw for drawing chemical compound structures and glycan structures. Compound structures drawn by KegDraw are also used as queries for PathPred. KegDraw is java application and software for MacOSX, Microsoft Windows, and Linux provided from KegTools download page (http://www.genome.jp/download/).

Figure 2 shows an example of the prediction result for plant secondary metabolite pathway (shown as a tree-like structure) by PathPred. This example includes the biosynthetic pathway from umbelliferon (7-Hydroxycoumarin) to fraxidin (8-Hydroxy-6,7-dimethoxycoumarin). Fraxidin is a major component of a crude drug Saposhnikovia root [14] and chemically classified to coumarins. Figure 2a shows a PathPred prediction pathway tree from umbelliferon to fraxidin. Red, blue, and gray compound numbers indicate the query or final compounds, compounds in the KEGG database, and hypothetical molecules that are also generated elsewhere in

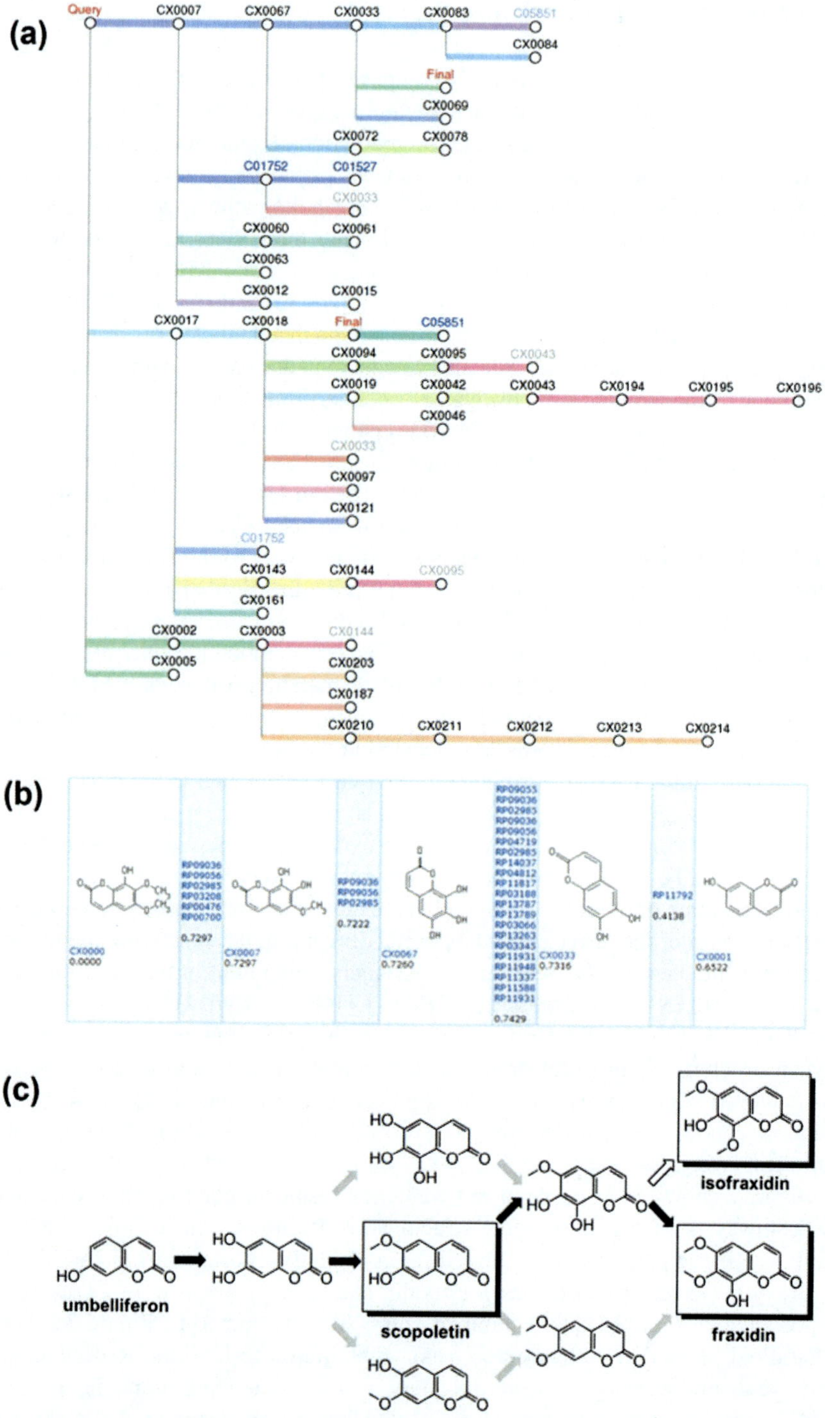

Fig. 2 (continued)

the tree, respectively. In the case of secondary metabolite biosynthesis pathway prediction, "query" and "final" mean "end product" and "starting substrate" of the synthetic pathway, respectively. Figure 2b shows one of the predicted paths from query (fraxidin) to final (umbelliferon) compounds in the prediction pathway tree. The identification numbers (which we refer to as the RP numbers) between two chemical structures indicate the links to the template reaction pairs for predicting the transformation. Figure 2c shows the total predicted pathway network from umbelliferon to fraxidin and 7-hydroxycoumarin related components of crude drug Saposhinicovia root. Black and gray arrows are the paths predicted by PathPred. Three Saposhinicovia components, fraxidin, isofraxidin, and scopoletin [14], are located on or linked to the predicted pathway. According to the components of Saposhinicovia root, the predicted path indicated by the black arrows pathway is highly likely to exist (black arrows pathway is same as Fig. 2b).

As PathPred is knowledge-based prediction system, the quality of the knowledge-base is crucial for the prediction accuracy. We are continuously updating the KEGG RPAIR, REACTION and PATHWAY databases. We also categorized the plant secondary metabolite biosynthesis pathway into subclasses, such as phenylpropanoids, polyketides, terpenoids and alkaloids, to use only the frequent RDM patterns depending on the compound subclasses and to improve the efficiency in terms of specificity and computational time.

E-zyme for Prediction of Enzymatic Reactions

Enzyme Commission (EC) number [15, 16] is a hierarchical classification system for enzyme reactions established by International Union of Biochemistry and Molecular Biology (IUBMB). This EC number system is widely accepted as the standard classification system in the field of biochemical and enzymatic studies. The EC numbers also play key roles in linking the enzyme genes or proteins to reactions and in the computational representations of enzymatic reactions in metabolic pathways. E-zyme (http://www.genome.jp/tools/e-zyme/) [12, 17] is the GenomeNet bioinformatics tool for prediction of enzyme reactions, i.e., to automatically assign the EC numbers up to the sub-sub classes for a given enzyme reaction. The prediction process is based on the relationships between the EC numbers and the corresponding RDM patterns (See section "Color Objects in KEGG Pathways and BRITE Hierarchies").

Fig. 2 Prediction result of biosynthetic pathway from umbelliferon (7-Hydroxycoumarin) to fraxidin (8-Hydroxy-6,7-dimethoxycoumarin) by PathPred. (**a**) Predicted pathway tree consist compounds (*node*) and reactions (*edge*). (**b**) One of successfully predicted pathway from umbelliferon to fraxidin. (**c**) Possible predicted pathway from umbelliferon to fraxidin by PathPred (*black* and *gray arrows*) and Saposinicovia coumarin components (in *boxes*). *Brack arrowed* pathway is same pathway as shown in figure (**b**)

Prediction of EC classes by E-zyme consists of the following steps. First, the chemical structures of a substrate and a product are compared by the SIMCOMP chemical structure alignment program [10, 11], and outputs the changes occurred during the reaction in a form of the RDM patterns. Consequently, the possible EC numbers are suggested based on the pre-computed correlations between the RDM patterns and the EC numbers. Users can input query compounds in the MDL mol format, or KEGG COMPOUND identifier (C number).

Figure 3 shows an example screenshot of the output by E-zyme. It includes the chemical alignment of the two compound structures (i.e., a substrate and a product

E-zyme result

Pair 1 [KCF] Edit alignment

C00077 C00327

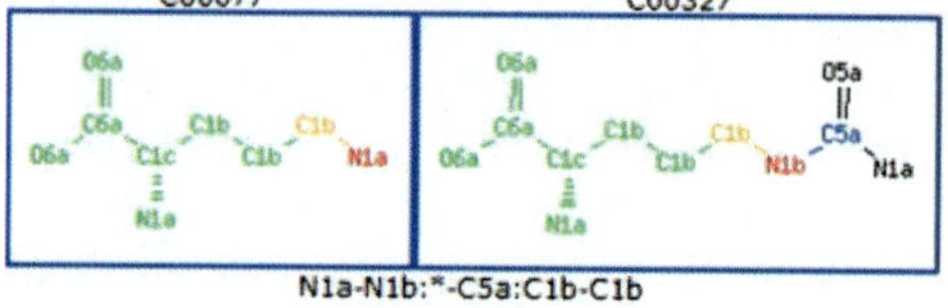

N1a-N1b:*-C5a:C1b-C1b

EC number assignments

EC number	Weighted score	Observed freq.	Reactions having the same RDM	
3.5.1	160.2	36	R00059 R00458 R00545 R00665 R00905 R00909 R01077 R01152 R01156 R01424 R01425 R01918 R01987 R02041 R02276 R02474 R02797 R02930 R02973 R03579 R03975 R03977 R04056 R04174 R04486 R04487 R04666 R05226 R05835 R07300 R07301 R07419 R08228 R08350 R08351	Get_all
2.3.1	61.9	24	R00395 R00454 R01154 R01617 R01620 R01889 R01944 R02152 R02385 R02452 R02911 R03552 R03718 R03720 R03760 R03910 R03992 R04192 R04211 R04777 R05841 R06978 R08744 R08745	Get_all
6.3.2	40.0	5	R00497 R02473 R02741 R03095 R07643	Get_all
2.3.2	30.3	9	R01687 R03956 R03971 R04613 R05055 R08776 R08777 R08778 R08779	Get_all
6.3.1	21.5	11	R01917 R02929 R03822 R06529 R07302 R07414 R08352 R08353 R08354 R08355	Get_all
6.3.4	16.7	8	R04562 R04563 R04582 R04604 R05145	Get_all
3.4.13	12.7	2	R00899 R04951	Get_all
3.4.11	6.0	6	R00899 R04951	Get_all
2.1.3	5.3	4	R01396 R01398 R01399 R07245	Get_all
3.1.1	3.6			
2.1.2	3.0			
6.3.5	2.7			
3.5.4	1.4			
3.4.17	1.2			
3.5.3	1.2			
6.2.1	1.0			
2.7.7	0.9			
3.1.2	0.9			
3.4.19	0.8			
1.2.1	0.4			
2.3.3	0.3			
6.1.1	0.3			
4.1.3	0.2			
2.6.1	0.2			

Confidence level of prediction results: RDM

Fig. 3 Example screenshot of E-zyme out put page of the prediction result of a possible reaction) and the list of the predicted EC numbers. The alignment of the compounds and assigned RDM patterns are shown in the upper section of the result page. In the lower section of the result page, EC number prediction results are displayed.

For further information, we recommend to refer Yamanishi et al. [17] and documents in E-zyme page (http://www.genome.jp/tools/e-zyme/).

KAAS – KEGG Automatic Annotation Server

In recent years, the number of complete and draft genomes, EST and metagenome sequences are rapidly increasing. This makes it increasingly important to automatically annotate functional properties and biological roles to genes. We provide KAAS (http://www.genome.jp/tools/kaas/) [18] for this purpose as a GenomeNet bioinformatics tool. KAAS serves functional annotation of genes in genomes (or large number of genes) by BLAST comparisons against the manually curated KEGG GENES database. Genes in KEGG DGENES (draft genomes), KEGG EGENES (EST contigs) and KEGG MGENES (metagenomes) are automatically annotated by KAAS.

We also provide a web service for the general public users. Overall procedure of the KAAS annotation is as follows. KAAS accepts three types of query sets, i.e., complete or draft genome, partial genome, or EST sequences. Query sequence data should be in multi-FASTA format of amino-acid or nucleotide sequences with unique ids. The user can choose one or more species from the latest KEGG GENES entries as the reference data set. We recommend to choose more closely related species to the species of interest as possible, in order to obtain better result.

KAAS provides three types of outputs as the results. "KO list" is the flat list of the correspondence table with query genes and K numbers assigned by the KAAS program. "BRITE hierarchies" is the hierarchical list of the annotated genes, which is incorporated into the classification of the BRITE database. "Pathway map" is the list of pathways that include the annotated query genes. The list is linked to the graphical pathway maps, and the annotated query genes are highlighted.

For further information, refer to Moriya et al. [18] and the documents in the KAAS webpage (http://www.genome.jp/tools/kaas/).

KegArray

To predict protein function from omics data, one of the effective ways is the integrated analysis of omics data by using systems information such as pathway network diagram. KegArray is a standalone desktop application for analyzing both transcriptome data (gene expression profiles) and metabolome data (compound profiles) in conjunction with the KEGG databases (http://www.genome.jp/download/kegtools.html). KegArray software is a Java application, and users can download the software

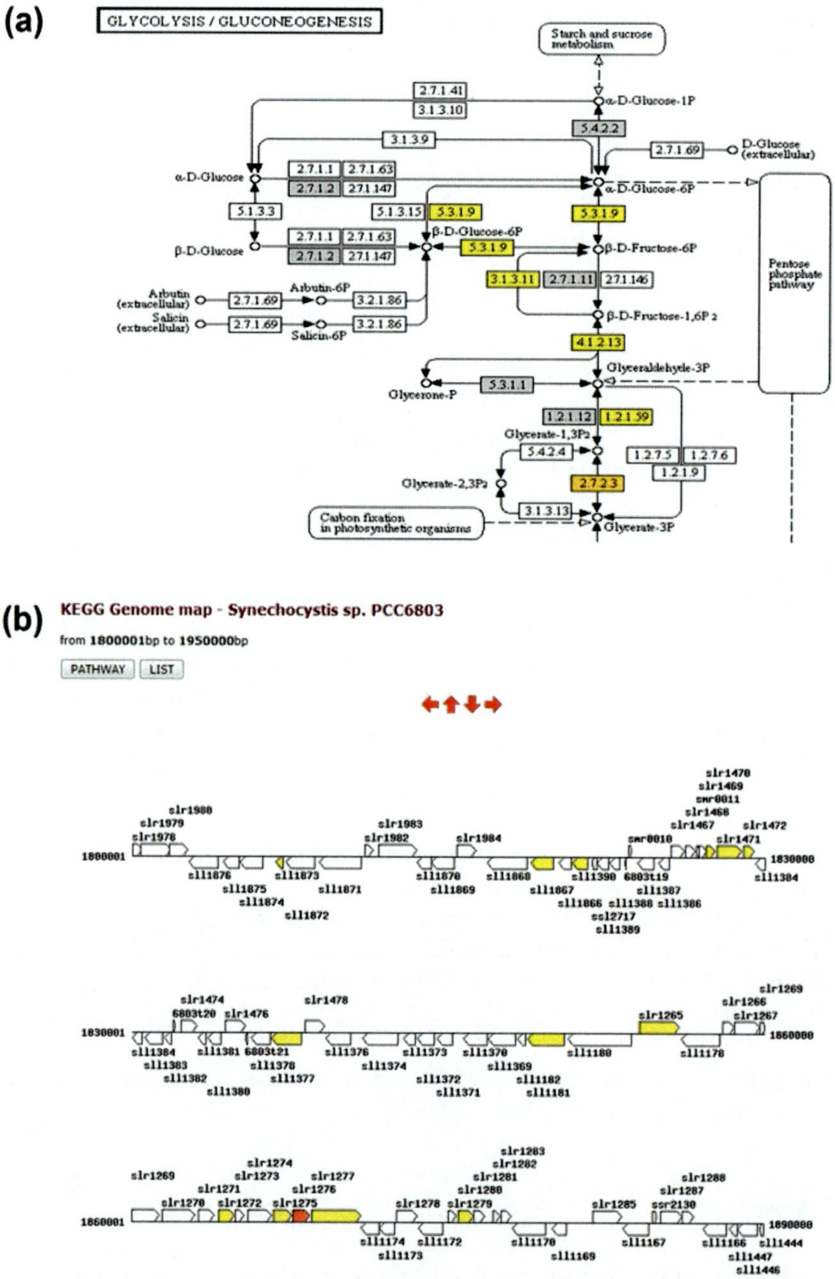

Fig. 4 Example screenshot of (**a**) pathway mapping and (**b**) genome mapping by KegArray

(for Mac OS X, Microsoft Windows, and Linux) from the download page (http://www.genome.jp/download/).

Transcriptome data format for KegArray is KEGG EXPRESSION format or tab-deliminated text similar to the KEGG EXPRESSION format. KEGG EXPRESSION format is original data format for KEGG EXPRESSION database (http://www.genome.jp/kegg/expression/). KEGG EXPRESSION database is a repository of microarray gene expression profile data for *Synechosystis*, *Bacillus subtilis* and other species. KegArray can convert external database IDs (e.g. NCBI GI) to the KEGG GENES IDs. Only ratio values can be used for metabolome data. Main function of KegArray is to map the transcriptome and metabolome data to the KEGG resources including PATHWAY, BRITE and genome maps. Figure 4 shows example screen shots of pathway mapping and genome mapping by KegArray tools. As shown in the figure, users can visualize the up- or down-regulated genes on various KEGG systems information. Users can also visualize increasing or decreasing metabolites on various KEGG objects.

Detailed usage information for KegArray, refer to the ReadMe file provided in the KegTools download page (http://www.genome.jp/download/).

Summary

In the post-genomic era, bioinformatics approach is necessary to analyze increasing omics data. Also, high quality database and bioinformatics approach will play important roles to predict protein function. KEGG is a manually curated integrated database for computer representation of biological systems. KEGG and GenomeNet also provide several useful tools to support protein function prediction from omics data.

In this chapter, we briefly outlined a perspective of the KEGG database and several tools for predicting protein functions from omics data, with introducing some recent topics. KEGG EDRUG is the database for crude drugs, essential oils, other natural products and related plant resources. KEGG EDRUG provides useful information for protein function related to plant secondary metabolism pathway. PathPred and E-zyme are tools for predicting enzyme reaction pathway from metabolites. These tools help users to predict unknown pathway. KAAS is automatic annotation and pathway prediction server for large set of sequences. KegArray is desktop application for transcriptome and metabolome analysis. KegArray will help mapping those data to the KEGG systems information such as KEGG PATHWAY, KEGG BRITE etc.

Acknowledgements The computational resources were provided by the Bioinformatics Center, Institute for Chemical Research, Kyoto University. The KEGG project is supported by the Institute for Bioinformatics Research and Development of the Japan Science and Technology Agency, and a grant-in-aid for scientific research on the priority area "Comprehensive Genomics" from the Ministry of Education, Culture, Sports, Science and Technology of Japan.

References

1. Kanehisa, M., Goto, S., Furumichi, M., et al. KEGG for representation and analysis of molecular networks involving diseases and drugs. Nucleic Acids Res. **38**: D355–D360 (2010).
2. Dixon, R.A., Strack, D. Phytochemistry meets genome analysis, and beyond. Phytochemistry **62**: 815–816 (2003).
3. Donson, J., Fang, Y., Espiritu-Santo, G., et al. Comprehensive gene expression analysis by transcript profiling. Plant Mol. Biol. **48**: 75–97 (2002).
4. Aharoni, A., Vorst, O. DNA microarrays for functional plant genomics. Plant Mol. Biol. **48**: 99–118 (2002).
5. Sumner, L.W., Mendes, P., Dixon, R.A. Plant metabolomics: large-scale phytochemistry in the functional genomics era. Phytochemistry **62**: 817–836 (2003).
6. Sato, S., Soga, T., Nishioka, T., et al. Simultaneous determination of the main metabolites in rice leaves using capillary electrophoresis mass spectrometry and capillary electrophoresis diode array detection. Plant J **40**: 151–163 (2004).
7. Masoudi-Nejad, A., Goto, S., Jauregui, R., et al. EGENES: transcriptome-based plant database of genes with metabolic pathway information and expressed sequence tag indices in KEGG. Plant Physiol. **144**: 857–866 (2007).
8. The Angiosperm Phylogeny Group. An update of the Angiosperm Phylogeny Group classification for the orders and families of flowering plants: APG II. Bot. J. Linnean Soc. **141**: 399–436 (2003).
9. Moriya, Y., Shigemizu, D., Hattori, M., et al. PathPred: an enzyme-catalyzed metabolic pathway prediction server. Nucleic Acid Res. **38**: W138–W143 (2010).
10. Hattori, M., Okuno, Y., Goto, S., et al. Development of a chemical structure comparison method for integrated analysis of chemical and genomic information in the metabolic pathways. J. Am. Chem. Soc. **125**: 11853–11865 (2003).
11. Hattori, M., Tanaka, N., Kanehisa, M., et al. SIMCOMP/SUBCOMP: chemical structure search servers for network analyses. Nucleic Acid Res. **38**: W652–W656 (2010).
12. Kotera, M., Okuno, Y., Hattori, M., et al. Computational assignment of the EC numbers for genomic-scale analysis of enzymatic reactions. J. Am. Chem. Soc. **126**: 16487–16498 (2004).
13. Oh, M., Yamada, T., Hattori, M., et al. Systematic analysis of enzyme-catalyzed reaction patterns and prediction of microbial biodegradation pathways. J. Chem. Inf. Model. **47**: 1702–1712 (2007).
14. Okuyama, E., Hasegawa, T., Matsushita, T., et al. Analgestic Components of Saposhinikovia Root (*Saposhinikovia divaricata*). Chem. Pharm. Bull. **49**: 154–160 (2001).
15. Barrett, A.J., Cantor, C.R., Liebecq, C., et al. *Enzyme nomenclature*. San Diego, CA: Academic (1992).
16. Tipton, K.F., Boyce, S. History of the enzyme nomenclature system. Bioinformatics **16**: 34–40 (2000).
17. Yamanishi, Y., Hattori, M., Kotera, M., et al. E-zyme: prediction potential EC numbers from the chemical transformation pattern of substrate-product pairs. Bioinformatics **25**: i79–i86 (2009).
18. Moriya, Y., Itoh, M., Okuda, S., et al. KAAS: an automatic genome annotation and pathway reconstruction server. Nucleic Acid Res. **35**: W182–W185 (2007).

Towards Elucidation of the *Escherichia coli* K-12 Unknowneome

Yukako Tohsato, Natsuko Yamamoto, Toru Nakayashiki, Rikiya Takeuchi, Barry L. Wanner, and Hirotada Mori

Abstract Advances in genome sequencing have revolutionized biology by providing the molecular blueprints for thousands of living organisms. Yet, the functions of a large fraction, as much as one-half, of the component parts remain unknown even for the best understood organisms, including *Escherichia coli*, *Bacillus subtilis*, and *Saccharomyces cerevisiae*. Here, we describe our development of comprehensive genomic resources (ORFeome clone sets and mutant libraries) for systematic functional analysis of *E. coli*, summaries on our use of these resources, the GenoBase information resource for handling high-throughput experimental data obtained with them, and our creation of user workspaces at our *P*rotein *F*unction *E*lucidation *T*eam (www.PrFEcT.org) website.

Defining the Unknowneome

Understanding the functionality of the protein components of living cells demands application of next-generation biology approaches. Not only are current annotations of genes encoding proteins incomplete and often times inaccurate, but up to 30% of the genes of newly sequenced bacteria cannot be annotated (or even recognized) using our current knowledge about proteins [1].

The development of comprehensive genomic resources for *Escherichia coli* K-12 [2, 3] have not only made systematic functional analyses feasible [4] but also have opened up new avenues for protein function elucidation, e.g., [5, 6], which otherwise cannot even be considered. The ability to carry out systematic analyses has even led biologists to re-consider the definition of biological function. *E. coli* K-12, which is one the best studied model organisms, is not exceptional in this regard. Similar genome-wide functional genomics approaches are underway in yeast ([7]; http://www.yeastgenome.org/), *Acinetobacter* species [8], *Bacillus subtilis* ([9]; http://www.genoscope.cns.fr/agc/microscope/home/index.php), *Pseudomonas*

Y. Tohsato (✉)
Department of Bioscience and Bioinformatics, Ritsumeikan University, Kusatsu, Shiga 525-8577, Japan
e-mail: yukako@sk.ritsumei.ac.jp

D. Kihara (ed.), *Protein Function Prediction for Omics Era*,
DOI 10.1007/978-94-007-0881-5_15,

aeruginosa ([10]; http://www.pseudomonas.com/), and many others (e.g., http://pfgrc.jcvi.org/index.php/gateway_clones/about_libraries.html). As a prototype of what can now be achieved through the use of genomic resources, we describe approaches that have already been proven to be useful in systematic functional analyses of *E. coli* K-12 with the Keio single-gene deletion library [2, 4, 11].

The Need for Gene Ontologies

Gene functions have traditionally been defined with natural language terms much like how humans speak. However, just like colloquialisms are specific to a region, gene terminologies often relate only to specific species, which creates difficulties when making comparisons between species. Biochemical, physiological, and phenotypic functions of proteins can differ dramatically even for seemingly similar proteins. Clear examples exist for particular enzyme and crystalline protein families, e.g., argininosuccinate lyase and δ-crystallin, enolase and τ-crystallin, glutathione S-transferase and SIII-crystallin, and lactate dehydrogenase and ε-crystallin [12]. On the basis of DNA and amino acid sequence similarities, these enzyme and crystalline families are closely related, however their physiological functions are quite different.

The rapid generation of new genome sequences has led to recognizing an urgent need for consistent descriptions of proteins across different organisms. The Gene Ontology (GO) Consortium (http://www.geneontology.org/) was launched to consolidate efforts for development of systematic and standardized terminologies [13]. The GO Consortium was initiated as a collaboration among three model organism databases [14]: the *Saccharomyces* Genome Database (SGD; http://www.yeastgenome.org/); a Database of Drosophila Genes and Genomes (FlyBase; http://flybase.org/); and Mouse Genome Database/Informatics (MGD/MGI; http://www.informatics.jax.org). GO uses a controlled ontology in which proteins are described in a species-independent manner in terms of a biological process (es), cellular component(s) and molecular function(s) (Table 1). The GO consortium has since been expanded to include a wide range of eucaryotic organisms, as well as many bacteria and Archaea [15] and *E. coli* K-12 among its twelve "reference genomes" [16].

Systematic Screening for Gene Functions Using the Keio Collection Single-Gene Deletion Library

Our development of the Keio collection single-gene deletion library [2], in which each of nearly 4000 of the ca. 4300 *E. coli* genes is individually deleted, has allowed testing for mutant phenotypes on a genome-wide scale. Because all mutants are in the same genetic background, genome-wide screening is expected to show effects resulting from loss of the respective gene. Questions are how to detect the effect(s) of a gene deletion and how to elucidate the function based on phenotypic changes. Two issues require mention. First, most single-gene deletions showed no observable

Table 1 Gene ontology terms[a]

Cellular component	
	The cellular component ontology describes locations, at the levels of subcellular structures and macromolecular complexes. As is true for the other ontologies, not all terms are applicable to all organisms; the set of terms is meant to be inclusive. Cellular component includes such terms as ATP synthase, cell envelope, inner membrane, periplasm, and ribosome, where many gene products are found.
Biological process	
	A biological process is series of events accomplished by one or more ordered assemblies of molecular functions. Examples are metabolic processes, regulation, response to stimulus, signaling, and transport and transporters.
Molecular function	
	The biochemical activity (including specific binding to ligands or structures) of a gene product. This definition also applies to the capability that a gene product (or gene product complex) carries as a potential. It describes only what is done without specifying where or when the event actually occurs.

[a]Gene Ontology (GO) Consortium, http://www.geneontology.org/

phenotype during growth on a rich (LB) medium. Second, many that did show an effect on rich medium grew poorly, thus making it difficult to distinguish primary and secondary mutational effects.

Many research groups have now used the Keio collection to study different biological processes, like osmolarity, salt stress, heat stress, DNA repair, antibiotic sensitivity, etc. In general, these groups have used the Keio collection to perform genome-wide screens in two ways: (i) by screening mutants for ones that display a novel phenotype under a particular growth condition or (ii) by testing mutants for ones that display an altered cellular behavior. In these ways, systematic and comprehensive screening of the Keio collection [4] have led to finding genes whose loss affects antibiotic hypersensitivity [17, 18], swarming motility [19], biofilm formation [20], growth in human blood [21], recipient ability in conjugation [22], cysteine tolerance and production [23], colicin import and cytotoxicity [24], deethylation of 7-ethoxycoumarin [25], and glycogen metabolism [26]. In most cases, several single-gene deletion mutants were identified that affected the biological process of interest. For example, Samant et al. [21] discovered that purine and pyrimidine biosynthesis is critical for growth of *E. coli* in human serum; they uncovered 17 of the 22 Keio mutants that have deletions of pyrimidine or purine biosynthetic genes among mutants to grow in human serum.

In many cases, mutants were recovered which had deletions of genes of unknown function, which made interpretations difficult. In other cases, mutants were found which had deletions of genes that appeared to be unrelated to the process being investigated. For instance, a genome-wide screening for mutants altered in swarming motility revealed mutants with deletions of unrelated functions, such as translation or DNA replication as well as ones of unknown functions, which caused strong repression of swarming [19]. These results show how strongly biological processes are interconnected within the cell, as well as how complicated the link can be between cause and effect. That is, in some cases, a deletion that directly affects one process can in turn indirectly affect a second process that is coordinately regulated with the first process. Such examples are well known in transcriptional regulatory networks where a global regulator(s) can affect the expression of genes beyond those that are directly controlled by the said regulator. Complex biochemical pathways are arranged in hubs with many connectivities. Hence, mutants lacking a hub protein are likely to display many more phenotypes than mutants lacking a protein with fewer connectivities. Understanding how various cellular networks interact requires much further studies in next-generation biology.

We recently screened the Keio collection for hydroxyurea (HU)-sensitive mutants. HU is believed to interfere with DNA replication by inhibiting ribonucleotide reductase (RNR, encoded by *nrd*), which is required for conversion of NTPs to dNTPs [27]. High-throughput screening was done using robotics as illustrated in

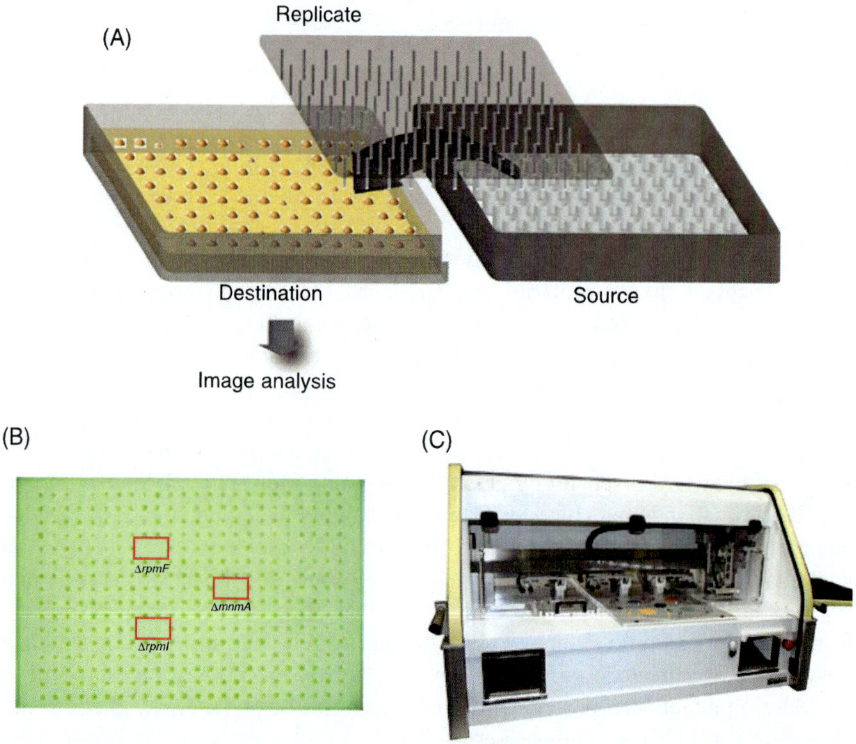

Fig. 1 High-throughput screening of the Keio collection. A schematic view of our screening protocol is shown. (**a**) The Keio collection is maintained as frozen glycerol stocks in 96-well microplates (Source). Portions are replicated in 384-spot format onto the surfaces of LB agar plates without or with 10 m*M* HU (Destination) such that duplicate replicas are juxtaposed horizontally. (**b**) Growth is measured by imaging plates at various times with a CCD camera. *Red rectangles* show candidate single-gene deletion mutants unable to grow on HU-containing agar for which both replicas grew only in the absence of HU. (**c**) Stamping is done with a Singer RoToR (Singer Instruments, UK) colony pinning robot (shown) or until recently with a Biomek FX robot (Beckman Coulter, Brea, CA)

Table 2 Hydroxyurea-sensitive mutants in the Keio collection[a]

Gene	Encoded protein	GO biological process
ahpC	alkyl hydroperoxide reductase, C22 subunit	6805 xenobiotic metabolic process
dnaT	DNA biosynthesis protein	6261 DNA-dependent DNA replication
fis	global DNA-binding transcriptional dual regulator	6310 DNA recombination
hda[b]	ATPase regulatory factor involved in DnaA inactivation	6261 DNA-dependent DNA replication
holC	DNA polymerase III, X subunit	6261 DNA-dependent DNA replication
iscS	cysteine desulfurase	9451 RNA modification
mnmA[b]	tRNA-methyltransferase	9451 RNA modification
nusB	transcription antitermination protein	transcription
priA[b]	primosome factor n′ (replication factor Y)	6261 DNA-dependent DNA replication
rplA	50S ribosomal subunit protein L1	6412 translation
rpmF	50S ribosomal subunit protein L32	6412 translation
rpmJ	50S ribosomal subunit protein L36	6412 translation
rrmJ	23S rRNA methyltransferase	6364 rRNA processing
sirA (*tusA*)	2-thiolation step of mnm^{5}s^{2}U34-tRNA synthesis	6400 tRNA modification
yheL (*tusB*)	2-thiolation step of mnm^{5}s^{2}U34-tRNA synthesis	6400 tRNA modification
yheM (*tusC*)	2-thiolation step of mnm^{5}s^{2}U34-tRNA synthesis	6400 tRNA modification
yheN (*tusD*)	2-thiolation step of mnm^{5}s^{2}U34-tRNA synthesis	6400 tRNA modification
yfaE	ferredoxin involved with ribonucleotide reductase cofactor	6124 ferredoxin metabolic process

[a]Annotations were found by using the PrFEcT WebSearch (www.prfect.org) and include results from the EcoGene (www.EcoGene.org), EcoliWiki (www.EcoliWiki.org) and PEC (http://www.shigen.nig.ac.jp/ecoli/pec/index.jsp) databases
[b]*hda*, *mnmA*, and *priA* are annotated as essential in the PEC database, yet we isolated single-gene *hda*, *mnmA*, and *priA* deletion mutants [2, 3]

Fig. 1. HU-sensitive mutants were identified as ones unable to form colonies on rich (LB) agar containing 10 m*M* HU. We uncovered 18 different HU-sensitive mutants (Table 2). Unexpectedly, ten HU-sensitive mutants (*iscS*, *mnmA*, *rplA*, *rpmF*, *rpmJ*, *rrmJ*, *sirA* (*tusA*), *yheL* (*tusB*), *yheM* (*tusC*), and *yheN* (*tusD*) are deleted of genes connected to translation, including six (*iscS*, *mnmA*, *sirA* (*tusA*), *yheL* (*tusB*), *yheM* (*tusC*), and *yheN* (*tusD*)) in a specific tRNA modification, the thiolation step of mnm^{5}s^{2}U34-tRNA synthesis. Further analysis revealed that these mutants showed strongly reduced synthesis of Nrd (Fig. 2).

From the point of view of cellular function, tRNA modification and dNTP supply would seem to be quite distinct biological processes. Although it is well known that DNA replication and protein synthesis are coordinated in bacteria [28], the precise mechanism is unknown. Conducting global analyses with the Keio collection may shed new light on many undiscovered or poorly understood cellular networks.

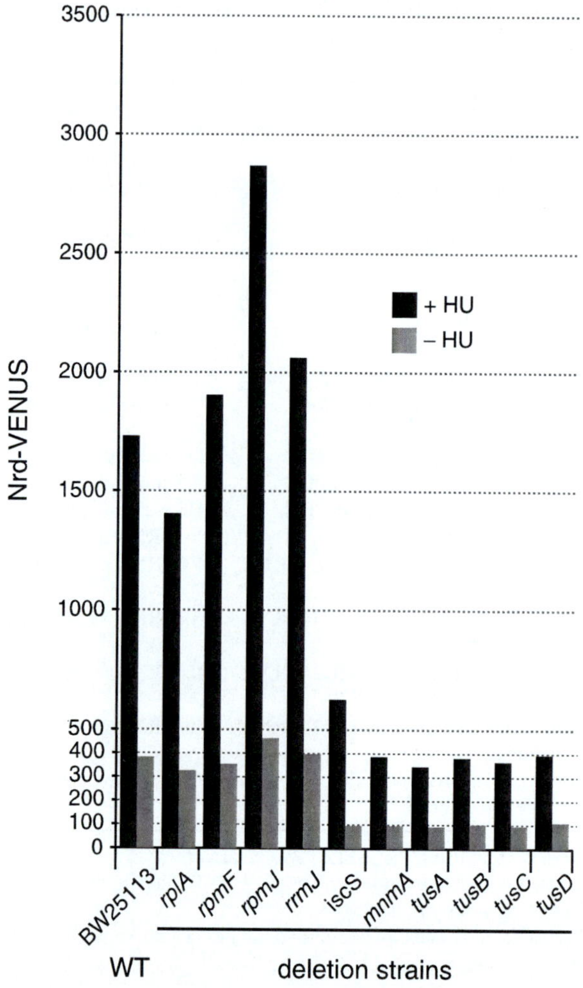

Fig. 2 Effect of HU on Nrd synthesis. Nrd synthesis was monitored by measuring fluorescence in cells carrying a low-copy plasmid with an *nrd*-Venus fusion during growth in LB without (–HU) or with 5 m*M* HU (+HU). Fluorescence was measured by flow cytometry (FACScan; BD, Franklin Lakes NJ). *E. coli* K-12 BW25113, the parent of the Keio collection [2], was used as the "wild-type (WT) control"

Screening the Keio collection for effects on many different biological processes is expected to provide new insights into the elucidation of functions of proteins belonging to uncharacterized or poorly characterized protein families – a critical challenge in the post-genomics era.

In the case of HU-sensitive screening, mutants were categorized into several groups. First, they are divided groups based on whether HU inhibited growth or killed the cells. Second, the latter were tested for whether cell death resulted from a membrane stress response or non-membrane stress response [29]. Classifying the mutants by suitable methods should yield new clues regarding function.

In addition to the Keio collection, the construction of a second single-gene deletion library, the ASKA deletion collection, is underway. Among other new features of the ASKA collection, mutants belonging to this library contain a 20-nt molecular barcode. This feature allows identification of individual mutants within a mixed population of all single-gene deletion mutants. Studying mixed populations has several benefits over studying individual cultures or stamping protocols. The selection process is closer to natural environmental conditions. Because competition occurs among the mutants, one can quantitatively assess growth advantage(s) and disadvantage(s) of all mutants simultaneously under various culture conditions. It is also beneficial for examining effects of drugs and inhibitors because much smaller amounts are required for mixed cultures, which is especially important when the drug or inhibitor of interest is expensive or hard to purify. Depending upon the purpose, one needs to choose the most appropriate procedure (simple screening, stamping, or competition). Regardless of purpose, the availability of genome-wide single-gene deletion mutants provides many advantages over traditional methods: (1) by conducting systematic and comprehensive genome-wide screens, one can quickly determine whether all (non-essential) genes for a biological process have been identified; (2) genome-wide screening can provide clues of value for identification of unknown gene functions; and (3) Genome-wide screening can help elucidate complex intracellular networks. Due to the vast amount of detailed knowledge already available for *E. coli*, developing deeper understanding of "intracellular networks" will be especially informative towards construction of a whole-cell model of unicellular organisms.

Systematic Screening of Single-Gene Deletion Mutants for Phenotypes Using Phenotype MicroArray^{TM} Technology

We employed Phenotype MicroArray^{TM} (PM) technology to perform systematic phenotype screening of selected single-gene deletion mutants [30, 31] (Fig. 3). PM technology was originally developed as a method for finding unique traits of individual organisms and for recognizing traits common to groups of organisms, such as species, and has been expanded as a high-throughput tool for global analysis of cellular phenotypes in post-genomic era [32]. This system monitors cellular respiration during growth in 96-well microplates under 1536 different chemical environments over a period of 24 h. Growth in each well is detected colorimetrically by quantifying the generation of purple colored formazan from tetrazolium which corresponds to the intracellular reducing state by NADH simultaneously. The effect of single-gene deletions on this screen provides information on the importance of the corresponding protein in response to diverse chemical stresses, as well as its contribution to a wide variety of different metabolic pathways. This high-throughput assay provides direct information on the contribution of the protein to the environmental fitness of the organism.

We performed PM analysis on ca. 300 single-gene mutants from the Keio collection and clustered them as reported in part previously deletion strains of *E. coli*

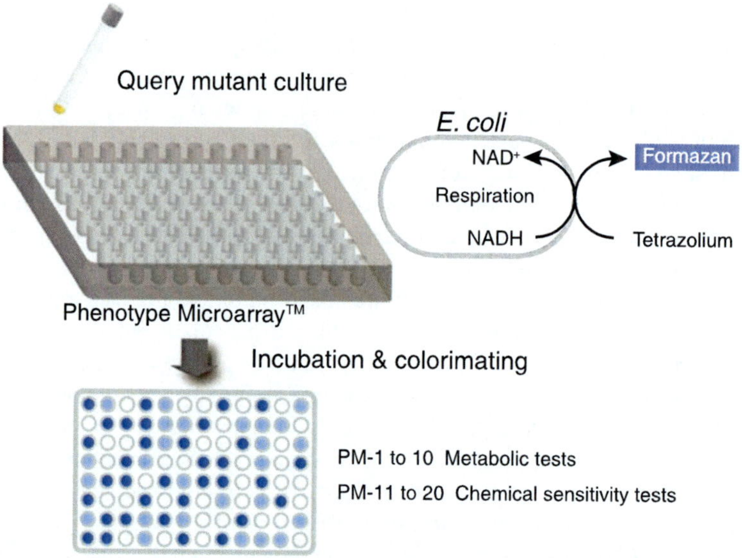

Fig. 3 Schematic view of Phenotype MicroArray™ analysis. Respiration is measured by quantifying the generation of NAD^+ by formation of purple-colored formazan from tetrazolium. A culture of the query mutant is dispensed into BIOLOG Phenotype MicroArray™ plate. Color development is automatically quantified during incubation in an *OmniLog*® instrument. PM-1 to PM-10 include 1 blank culture well and PM-11-20 have different chemical concentrations, resulting in a total of 1536 different chemical environments (1920 conditions). More precise information is available at http://www.biolog.com

and the clustering analysis using the part of the results was reported previously [30, 31]. Based on the entire PM results, we show here statistically the effectiveness of systematic phenotype screening using PM technology. The precise method of statistical measurement of PM will be reported elsewhere [31]. As shown in Fig. 4, 709 (36.9%) of 1920 conditions showed no respiration in our control strain and 12 (0.6%) of the conditions are negative (water) and positive (LB medium) controls. Only 8 of the remaining 1199 conditions had no significant phenotype change in any of the 300 mutants tested.

Figure 5 shows the medium condition effects and environmental dependencies of the ca. 300 mutants tested. On average, 45 mutants showed significant phenotypic changes under each condition (Fig. 5a). Further, each mutant showed differences under 183 conditions (Fig. 5b). It is worth mentioning that our PM tests were done in duplicate and showed a high degree of reproducibility (94.5%).

Although PM technology is a powerful tool for functional screening, it alone provides insufficient sensitivity to identify functions for proteins of unknown function. The most likely causes are robustness and the existence of unknown alternative metabolic pathways. *E. coli* K-12 has 98 genes encoding isozymes, including 80 encoding pairs of isozymes based on annotations in EcoCyc version 13.5 [22]. In some cases, expression patterns of these isozymes differ; integrating results from

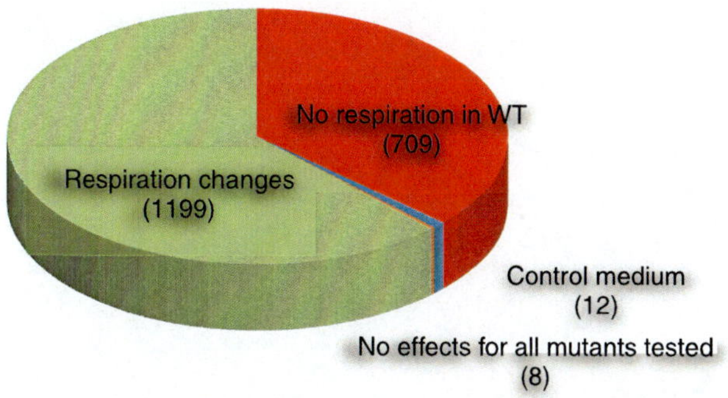

Fig. 4 Classification of 1920 medium conditions by color change. The number of conditions showing no respiration in *E. coli* K-12 BW25113 (WT), respiration changes, and no effects for all mutants tested are given in *parentheses*. Conditions included negative (water) and positive (LB) control medium

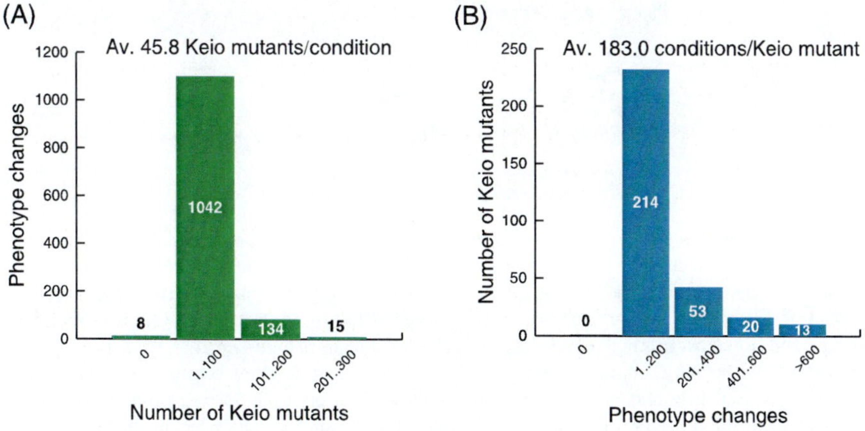

Fig. 5 Medium condition effects and environmental dependencies. (**a**) *Abscissa* and ordinate axes show number of single-gene deletions resulting in significant phenotype changes using 100 gene deletions as the bin size. (**b**) *Abscissa* and ordinate axes show the number of phenotype changes observed in the single-gene deletion mutants using 200 medium conditions as the bin size

transcriptome analysis from DNA microarrays, or preferably RNA-seq, with PM analysis may provide deeper insight into physiological function.

Systematic Screening for Genetic Interactions

Synthetic lethal screens are an effective experimental approach for revealing mechanisms of cellular robustness [33]. We have developed a strategy to screen comprehensively for effects of double gene deletions in *E. coli* [34, 35] (Fig. 6). Preliminary results using such strategies have been reported [5, 6]. Even though

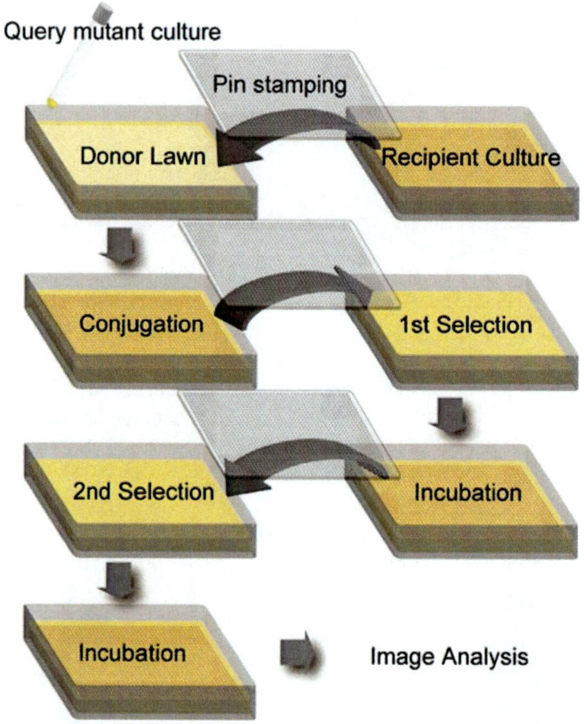

Fig. 6 Schematic view of genetic interaction analysis by double gene deletion. A schematic view of the protocol for creating and examining double-gene deletion mutants in a high-throughput manner by conjugation is shown. The query gene mutant, which serves as an Hfr donor, is evenly spread on LB agar to form a donor lawn. Single-gene deletion library, which serves as recipient culture and stored as frozen glycerol stocks or colonies on agar in high-density format, is replicated onto the donor lawn by robotic pin stamping. Following growth to allow conjugation to occur, pin stamping is used to replicate from the conjugation surface onto the 1st selection. These plates are incubated for 6 h and then replicated by pin stamping on the 2nd selection plate, which is necessary to eliminate background growth. The 2nd selection plate is imaged with a CCD camera over time, and image analysis is done to identify double mutants growing poorer or better than control matings. Details will be reported elsewhere

the genome sequencing and large-scale genetic analyses have revealed the enormous amount of genetic information of the target organisms, our knowledge of cellular system is still very limited. A major challenge is to understand physiological networks of genes in a living cell. As described above, single-gene deletion mutants generally show limited phenotype changes because of the redundancy or compensatory pathways. This phenomenon is called robustness; there can be many mechanisms that can lead to re-construction of physiological steps or gene product networks. The structure of the cellular network may not be rigid but rather be dynamically changing according to the environment, which can result not only from the extracellular environment but also by genetic alterations (mutation or deletion).

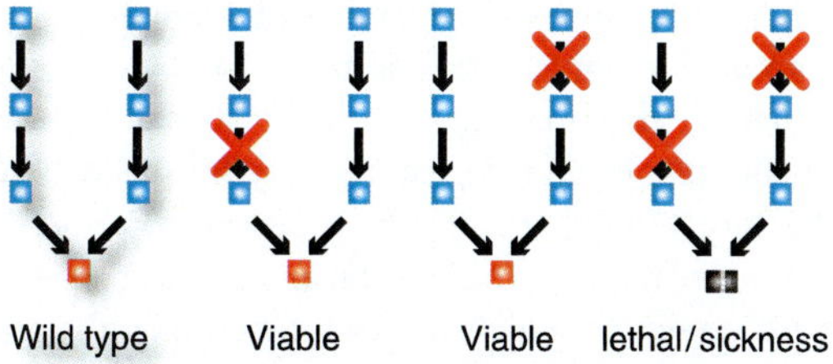

Fig. 7 Concept of synthetic lethality analysis by a double-gene deletion strategy. Normally organisms may have multiple pathways to generate essential substrates. In such cases, elimination of one pathway step by mutation is without effect because the other pathway can provide the missing step. The cell is unable to survive only when both pathways are disrupted simultaneously. The consequence of the corresponding double mutation leads to synthetic lethality (or sickness), which reveals a genetic interaction

Robustness in a cellular network is similar to the operation of a transportation network. Although a shortest path exists, an alternative longer detour pathway(s) is often available if the shortest one is blocked. The concept of the synthetic lethality analysis by a double-gene deletion strategy is shown in Fig. 7. When two genes are found to interact such that loss of one is without (major) effect but loss of both results in a new mutant phenotype, the genetic interaction is called epistasis. Epistasis effects can cause many kinds of effects; only a subset cause cell lethality or sickness. However, those causing severe growth effects are easiest to score. Large-scale genetic interaction studies have provided the basis for defining gene function and gene networks. Recent results from comprehensive genetic interaction analyses have greatly accelerated deeper insight into physiological gene functions and networks from bacteria to humans [33].

Information Resources

GenoBase was originally developed for the *E. coli* genome project, which was launched in Japan in 1989, to select phage clones from the ordered Kohara library [36] for sequencing [37]. Because *E. coli* is one of the best studied organisms, many gene sequences had already been accumulated at the time. To facilitate selection of the target phage clone, we constructed the GenoBase database to help distinguish sequenced and non-sequenced regions of the chromosome. First, we collected all of the *E. coli* sequence information from publicly available databases, such as GenBank, EMBL and DDBJ, and made consensus sequence by assembling these followed by mapping them onto the chromosome to identify sequenced regions. Upon completion of genome sequencing [38–42], GenoBase was further developed for supporting systematic functional genomics and systems analyses for *E. coli*. The development of biological resources for systematic studies of *E. coli* K-12, like the single-gene deletion Keio collection [2] and the ASKA ORFeome clone library [43], has proven especially valuable worldwide. The advent of technologies for acquisition of high-throughput data types (series of comprehensive network analyses, e.g., transcriptome, proteome, interactome and genetic interaction) has created need to preserve and share information.

GenoBase originally displayed information only for the W3110 strain of *E. coli* K-12, which was the target in the Japanese *E. coli* genome project [27, 28], whereas the MG1655 strain was the target in the Wisconsin genome sequencing project in the USA [29]. GenoBase version 7, which was developed in collaboration with Purdue University (www.PrFEcT.org/GenoBase), has been enhanced to permit the user to choose displaying information for *E. coli* K-12 MG1655 or W3110. GenoBase ver. 7 has also been enhanced to support image data and other high-throughput data.

GenoBase (Fig. 8) is especially rich in experimental resources (mutants and plasmids) and experimental data from a large *E. coli* functional genomics project in Japan, which far exceeds all other resources combined. Information in GenoBase is public or private (password accessible), depending upon whether the data have been published. Current resources include (1) two types of ASKA ORFeome libraries [30], including one with a C-terminal GFP tag and one without; and (2) the single-gene deletion library known as the Keio collection [2]. Comprehensive experimental

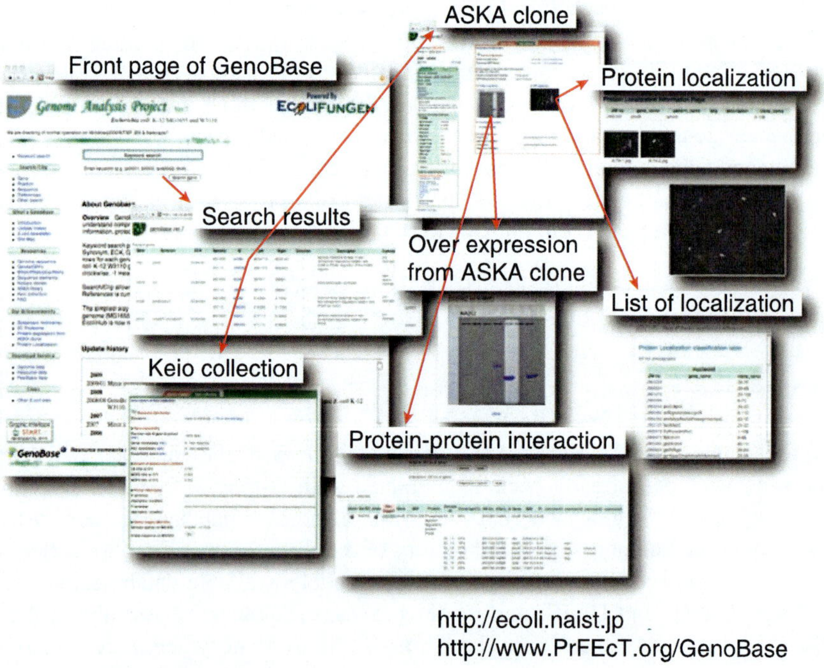

Fig. 8 The GenoBase Information Resource. GenoBase version 6 is fully operational at http://ecoli.naist.jp, while GenoBase version 7 is at www.PrFEcT.org/Genobase. Version 8 is now under construction. Once development is completed, mirroring will be deployed to maintain synchrony between these sites. Querying from home page gives search results in a table with links to pages for resources and experimental results based on the use of these resources

resources and data generated by systematic analysis using those resources are continuously growing. To facilitate both systems and individual research approaches using *E. coli* K-12 as a model system, integrative databases provide essential information.

GenoBase activities have not only involved the collection of high-throughput experimental data but also improvement in the quality of K-12 genome annotation. One example was the re-confirmation of the *E. coli* genome sequence, which resulted in correction of sequencing errors of previously published W3110 and MG1655 sequences [28]. Correction of the K-12 genome sequence provided major stimulus for cooperative re-annotation of the K-12 genome at international annotation workshops held in Woods Hole in 2003 and 2005 [31].

GenoBase is a searchable database devoted to systems biology of *E. coli* K-12. Querying GenoBase is done from the home page (Fig. 8). Any term, such as an id, gene name, product name, is accepted. Searching results are displayed in a tabular format with links to gene pages, which show additional information about the target gene. Contents show genome annotation information together with the biological resources and systematic analysis data using those resources. GenoBase is based on the predicted genes and all data are stored associated with the genes.

High-throughput systematic experimental data currently includes three large data sets:

- Protein–protein interaction data are based on using His-tagged ASKA ORF clone library without GFP [32]. All of the interaction data including data produced from TOF-MAS analysis is stored and specific partner candidates as prey proteins are available from each target protein as bait.
- DNA microarray data from analysis of single-gene deletion mutants were generated using full length cDNA type arrays, which were made with PCR-amplified fragments from ASKA clone library. Quantitative data were generated by ImaGene for about 150 deletion mutants, mostly for ones lacking transcription factors.
- Protein localization data are displayed for transformants carrying the GFP-tagged ASKA ORFeome clones. Transformants were analyzed by confocal microscopy of transformants expressing each protein at a low basal level in absence of an inducer to avoid misfolding of the target protein that can accompany protein overproduction. Images captured with a CCD camera are stored in our database.

Future Perspectives

(A) Quality control of resources

Our group continues to improve the quality of the biological resources created in Japan. For example, one quality control issue for the Keio collection has been the occasional discovery of partial duplications. Accordingly, the entire collection has now been validated. Upon publication, these will be open to the public through GenoBase.

(B) New experimental resources
Continuous efforts are underway to improve current resources and to construct new resources to expand systematic studying of *E. coli*. A new single gene deletion library with a different antibiotic resistance marker has been created for construction of double mutants to test for genetic interactions. The same library is bar coded to permit population studies. A second new resource near completion is a Gateway-fitted ASKA clone entry. All precise information on these new resources will also be stored in GenoBase.

(C) Systematic approaches using resources
As described above, systematic analyses, such as protein–protein interaction and protein localization, were performed and the data stored in GenoBase. Recently, we reported high-throughput systems for studying genetic interactions [5, 6] and the results from these analyses will be stored in GenoBase or partner databases.

Acknowledgements Y. T. is supported by the Asahi Glass Foundation; B. L. W. is supported by GM092047 from the N.I.H.; and H. M. is supported by a Grant-in-Aid for Scientific Research (A) and KAKENHI (Grant-in-Aid for Scientific Research) on Priority Areas "System Genomics" from the Ministry of Education, Culture, Sports, Science and Technology of Japan and the Bio Innovation funding program in Okinawa, Japan.

References

1. Roberts, R.J. Identifying protein function-a call for community action. PLoS. Biol. **2**: E42 (2004).
2. Baba, T., Ara, T., Hasegawa, M., Takai, Y., Okumura, Y., Baba, M., Datsenko, K.A., Tomita, M., Wanner, B.L., Mori, H. Construction of *Escherichia coli* K-12 in-frame, single-gene knockout mutants: the Keio collection. Mol. Syst. Biol. **2**: 2006 (2006).
3. Yamamoto, N., Nakahigashi, K., Nakamichi, T., Yoshino, M., Takai, Y., Touda, Y., Furubayashi, A., Kinjyo, S., Dose, H., Hasegawa, M., Datsenko, K.A., Nakayashiki, T., Tomita, M., Wanner, B.L., Mori, H. Update on the Keio collection of *Escherichia coli* single-gene deletion mutants. Mol. Syst. Biol. **5**: 335 (2009).
4. Baba, T., Huan, H.C., Datsenko, K., Wanner, B.L., Mori, H. The applications of systematic in-frame, single-gene knockout mutant collection of *Escherichia coli* K-12. Methods Mol. Biol. **416**: 183–194 (2008).
5. Butland, G., Babu, M., Diaz-Mejia, J.J., Bohdana, F., Phanse, S., Gold, B., Yang, W., Li, J., Gagarinova, A.G., Pogoutse, O., Mori, H., Wanner, B.L., Lo, H., Wasniewski, J., Christopolous, C., Ali, M., Venn, P., Safavi-Naini, A., Sourour, N., Caron, S., Choi, J.Y., Laigle, L., Nazarians-Armavil, A., Deshpande, A., Joe, S., Datsenko, K.A., Yamamoto, N., Andrews, B.J., Boone, C., Ding, H., Sheikh, B., Moreno-Hagelseib, G., Greenblatt, J.F., Emili, A. eSGA: *E. coli* synthetic genetic array analysis. Nat. Methods **5**: 789–795 (2008).
6. Typas, A., Nichols, R.J., Siegele, D.A., Shales, M., Collins, S.R., Lim, B., Braberg, H., Yamamoto, N., Takeuchi, R., Wanner, B.L., Mori, H., Weissman, J.S., Krogan, N.J., Gross, C.A. High-throughput, quantitative analyses of genetic interactions in *E. coli*. Nat. Methods **5**: 781–787 (2008).
7. Giaever, G., Chu, A.M., Ni, L., Connelly, C., Riles, L., Veronneau, S., Dow, S., Lucau-Danila, A., Anderson, K., Andre, B., Arkin, A.P., Astromoff, A., El-Bakkoury, M., Bangham, R., Benito, R., Brachat, S., Campanaro, S., Curtiss, M., Davis, K., Deutschbauer, A., Entian, K.D., Flaherty, P., Foury, F., Garfinkel, D.J., Gerstein, M., Gotte, D., Guldener, U., Hegemann, J.H., Hempel, S., Herman, Z., Jaramillo, D.F., Kelly, D.E., Kelly, S.L., Kotter, P., LaBonte, D., Lamb, D.C., Lan, N., Liang, H., Liao, H., Liu, L., Luo, C., Lussier, M., Mao, R., Menard,

P., Ooi, S.L., Revuelta, J.L., Roberts, C.J., Rose, M., Ross-Macdonald, P., Scherens, B., Schimmack, G., Shafer, B., Shoemaker, D.D., Sookhai-Mahadeo, S., Storms, R.K., Strathern, J.N., Valle, G., Voet, M., Volckaert, G., Wang, C.Y., Ward, T.R., Wilhelmy, J., Winzeler, E.A., Yang, Y., Yen, G., Youngman, E., Yu, K., Bussey, H., Boeke, J.D., Snyder, M., Philippsen, P., Davis, R.W., Johnston, M. Functional profiling of the Saccharomyces cerevisiae genome. Nature **418**: 387–391 (2002).

8. de Berardinis, V., Vallenet, D., Castelli, V., Besnard, M., Pinet, A., Cruaud, C., Samair, S., Lechaplais, C., Gyapay, G., Richez, C., Durot, M., Kreimeyer, A., Le, F.F., Schachter, V., Pezo, V., Doring, V., Scarpelli, C., Medigue, C., Cohen, G.N., Marliere, P., Salanoubat, M., Weissenbach, J. A complete collection of single-gene deletion mutants of *Acinetobacter baylyi* ADP1. Mol. Syst. Biol. **4**: 174 (2008).
9. Kobayashi, K., Ehrlich, S.D., Albertini, A., Amati, G., Andersen, K.K., Arnaud, M., Asai, K., Ashikaga, S., Aymerich, S., Bessieres, P., Boland, F., Brignell, S.C., Bron, S., Bunai, K., Chapuis, J., Christiansen, L.C., Danchin, A., Debarbouille, M., Dervyn, E., Deuerling, E., Devine, K., Devine, S.K., Dreesen, O., Errington, J., Fillinger, S., Foster, S.J., Fujita, Y., Galizzi, A., Gardan, R., Eschevins, C., Fukushima, T., Haga, K., Harwood, C.R., Hecker, M., Hosoya, D., Hullo, M.F., Kakeshita, H., Karamata, D., Kasahara, Y., Kawamura, F., Koga, K., Koski, P., Kuwana, R., Imamura, D., Ishimaru, M., Ishikawa, S., Ishio, I., Le, C.D., Masson, A., Mauel, C., Meima, R., Mellado, R.P., Moir, A., Moriya, S., Nagakawa, E., Nanamiya, H., Nakai, S., Nygaard, P., Ogura, M., Ohanan, T., O'Reilly, M., O'Rourke M, Pragai, Z., Pooley, H.M., Rapoport, G., Rawlins, J.P., Rivas, L.A., Rivolta, C., Sadaie, A., Sadaie, Y., Sarvas, M., Sato, T., Saxild, H.H., Scanlan, E., Schumann, W., Seegers, J.F., Sekiguchi, J., Sekowska, A., Seror, S.J., Simon, M., Stragier, P., Studer, R., Takamatsu, H., Tanaka, T., Takeuchi, M., Thomaides, H.B., Vagner, V., van Dijl, J.M., Watabe, K., Wipat, A., Yamamoto, H., Yamamoto, M., Yamamoto, Y., Yamane, K., Yata, K., Yoshida, K., Yoshikawa, H., Zuber, U., Ogasawara, N. Essential *Bacillus subtilis* genes. Proc. Natl. Acad. Sci. USA **100**: 4678–4683 (2003).
10. Jacobs, M.A., Alwood, A., Thaipisuttikul, I., Spencer, D., Haugen, E., Ernst, S., Will, O., Kaul, R., Raymond, C., Levy, R., Chun-Rong, L., Guenthner, D., Bovee, D., Olson, M.V., Manoil, C. Comprehensive transposon mutant library of *Pseudomonas aeruginosa.* Proc. Natl. Acad. Sci. USA **100**: 14339–14344 (2003).
11. Baba, T., Mori, H. The construction of systematic in-frame, single-gene knockout mutant collection in *Escherichia coli* K-12. Methods Mol. Biol. **416**: 171–181 (2008).
12. Wistow, G., Piatigorsky, J. Recruitment of enzymes as lens structural proteins. Science **236**: 1554–1556 (1987).
13. Ashburner, M., Ball, C.A., Blake, J.A., Botstein, D., Butler, H., Cherry, J.M., Davis, A.P., Dolinski, K., Dwight, S.S., Eppig, J.T., Harris, M.A., Hill, D.P., Issel-Tarver, L., Kasarskis, A., Lewis, S., Matese, J.C., Richardson, J.E., Ringwald, M., Rubin, G.M., Sherlock, G. Gene ontology: tool for the unification of biology. The Gene Ontology Consortium. Nat. Genet. **25**: 25–29 (2000).
14. The Gene Ontology Consortium: Creating the gene ontology resource: design and implementation. Genome Res. **11**: 1425–1433 (2001).
15. Carbon, S., Ireland, A., Mungall, C.J., Shu, S., Marshall, B., Lewis, S. AmiGO: online access to ontology and annotation data. Bioinformatics **25**: 288–289 (2009).
16. Reference Genome Group of the Gene Ontology Consortium. The Gene Ontology's Reference Genome Project: a unified framework for functional annotation across species. PLoS Comput. Biol. **5**: e1000431 (2009).
17. Tamae, C., Liu, A., Kim, K., Sitz, D., Hong, J., Becket, E., Bui, A., Solaimani, P., Tran, K.P., Yang, H., Miller, J.H. Determination of antibiotic hypersensitivity among 4,000 single-gene-knockout mutants of *Escherichia coli.* J. Bacteriol. **190**: 5981–5988 (2008).
18. Liu, A., Tran, L., Becket, E., Lee, K., Chinn, L., Park, E., Tran, K., Miller, J.H. Antibiotic sensitivity profiles determined with an *Escherichia coli* gene knockout collection: generating an antibiotic bar code. Antimicrob. Agents Chemother. **54**: 1393–1403 (2010).

19. Inoue, T., Shingaki, R., Hirose, S., Waki, K., Mori, H., Fukui, K. Genome-wide screening of genes required for swarming motility in *Escherichia coli* K-12. J. Bacteriol. **189**: 950–957 (2007).
20. Niba, E.T., Naka, Y., Nagase, M., Mori, H., Kitakawa, M. A genome-wide approach to identify the genes involved in biofilm formation in *E. coli*. DNA Res. **14**: 237–246 (2007).
21. Samant, S., Lee, H., Ghassemi, M., Chen, J., Cook, J.L., Mankin, A.S., Neyfakh, A.A. Nucleotide biosynthesis is critical for growth of bacteria in human blood. PLoS Pathog. **4**: e37 (2008).
22. Perez-Mendoza, D., de la, C.F. *Escherichia coli* genes affecting recipient ability in plasmid conjugation: are there any? BMC Genomics **10**: 71 (2009).
23. Wiriyathanawudhiwong, N., Ohtsu, I., Li, Z.D., Mori, H., Takagi, H. The outer membrane TolC is involved in cysteine tolerance and overproduction in *Escherichia coli*. Appl. Microbiol. Biotechnol. **81**: 903–913 (2009).
24. Sharma, O., Datsenko, K.A., Ess, S.C., Zhalnina, M.V., Wanner, B.L., Cramer, W.A. Genome-wide screens: novel mechanisms in colicin import and cytotoxicity. Mol. Microbiol. **73**: 571–585 (2009).
25. Zhou, Y., Minami, T., Honda, K., Omasa, T., Ohtake, H. Systematic screening of *Escherichia coli* single-gene knockout mutants for improving recombinant whole-cell biocatalysts. Appl. Microbiol. Biotechnol. **87**: 647–655 (2010).
26. Montero, M., Eydallin, G., Viale, A.M., Almagro, G., Munoz, F.J., Rahimpour, M., Sesma, M.T., Baroja-Fernandez, E., Pozueta-Romero, J. *Escherichia coli* glycogen metabolism is controlled by the PhoP-PhoQ regulatory system at submillimolar environmental Mg2+ concentrations, and is highly interconnected with a wide variety of cellular processes. Biochem. J. **424**: 129–141 (2009).
27. Krakoff, I.H., Brown, N.C., Reichard, P. Inhibition of ribonucleoside diphosphate reductase by hydroxyurea. Cancer Res. **28**: 1559–1565 (1968).
28. Maaloe, O. The control of normal DNA replication in bacteria. Cold Spring Harb. Symp. Quant. Biol. **26**: 45–52 (1961).
29. Davies, B.W., Kohanski, M.A., Simmons, L.A., Winkler, J.A., Collins, J.J., Walker, G.C. Hydroxyurea induces hydroxyl radical-mediated cell death in *Escherichia coli*. Mol. Cell **36**: 845–860 (2009).
30. Tohsato, Y., Mori, H. Phenotype profiling of single gene deletion mutants of *E. coli* using Biolog technology. Genome Inform. **21**: 42–52 (2008).
31. Tohsato, Y., Baba, T., Mazaki, Y., Ito, M., Wanner, B.L., Mori, H. Environmental dependency of gene knockouts on phenotype microarray analysis in *Escherichia coli*. J. Bioinform. Comput. Biol. **8**(Suppl 1): 83–99.
32. Bochner, B.R. Global phenotypic characterization of bacteria. FEMS Microbiol. Rev. **33**: 191–205 (2009).
33. Dixon, S.J., Costanzo, M., Baryshnikova, A., Andrews, B., Boone, C. Systematic mapping of genetic interaction networks. Annu. Rev. Genet. **43**: 601–625 (2009).
34. Typas, A., Nichols, R.J., Siegele, D.A., Shales, M., Collins, S.R., Lim, B., Braberg, H., Yamamoto, N., Takeuchi, R., Wanner, B.L., Mori, H., Weissman, J.S., Krogan, N.J., Gross, C.A. High-throughput, quantitative analyses of genetic interactions in E. coli. Nat. Methods **5**(9): 781–787 (2008).
35. Butland, G., Babu, M., Díaz-Mejía, J.J., Bohdana, F., Phanse, S., Gold, B., Yang, W., Li, J., Gagarinova, A.G., Pogoutse, O., Mori, H., Wanner, B.L., Lo, H., Wasniewski, J., Christopolous, C., Ali, M., Venn, P., Safavi-Naini, A., Sourour, N., Caron, S., Choi, J.Y., Laigle, L., Nazarians-Armavil, A., Deshpande, A., Joe, S., Datsenko, K.A., Yamamoto, N., Andrews, B.J., Boone, C., Ding, H., Sheikh, B., Moreno-Hagelseib, G., Greenblatt, J.F., Emili, A. eSGA: E. coli synthetic genetic array analysis. Nat. Methods **5**(9): 789–795 (2008).
36. Kohara, Y., Akiyama, K., Isono, K. The physical map of the whole *E. coli* chromosome: application of a new strategy for rapid analysis and sorting of a large genomic library. Cell **50**: 495–508 (1987).

37. Yura, T., Mori, H., Nagai, H., Nagata, T., Ishihama, A., Fujita, N., Isono, K., Mizobuchi, K., Nakata, A. Systematic sequencing of the *Escherichia coli* genome: analysis of the 0–2.4 min region. Nucleic Acids Res. **20**: 3305–3308 (1992).
38. Blattner, F.R., Plunkett, G., III, Bloch, C.A., Perna, N.T., Burland, V., Riley, M., Collado-Vides, J., Glasner, J.D., Rode, C.K., Mayhew, G.F., Gregor, J., Davis, N.W., Kirkpatrick, H.A., Goeden, M.A., Rose, D.J., Mau, B., Shao, Y. The complete genome sequence of *Escherichia coli* K-12. Science **277**: 1453–1474 (1997).
39. Aiba, H., Baba, T., Hayashi, K., Inada, T., Isono, K., Itoh, T., Kasai, H., Kashimoto, K., Kimura, S., Kitakawa, M., Kitagawa, M., Makino, K., Miki, T., Mizobuchi, K., Mori, H., Mori, T., Motomura, K., Nakade, S., Nakamura, Y., Nashimoto, H., Nishio, Y., Oshima, T., Saito, N., Sampei, G., Horiuchi, T. A 570-kb DNA sequence of the *Escherichia coli* K-12 genome corresponding to the 28.0–40.1 min region on the linkage map. DNA Res. **3**: 363–377 (1996).
40. Itoh, T., Aiba, H., Baba, T., Hayashi, K., Inada, T., Isono, K., Kasai, H., Kimura, S., Kitakawa, M., Kitagawa, M., Makino, K., Miki, T., Mizobuchi, K., Mori, H., Mori, T., Motomura, K., Nakade, S., Nakamura, Y., Nashimoto, H., Nishio, Y., Oshima, T., Saito, N., Sampei, G., Seki, Y., Horiuchi, T. A 460-kb DNA sequence of the *Escherichia coli* K-12 genome corresponding to the 40.1–50.0 min region on the linkage map. DNA Res. **3**: 379–392 (1996).
41. Oshima, T., Aiba, H., Baba, T., Fujita, K., Hayashi, K., Honjo, A., Ikemoto, K., Inada, T., Itoh, T., Kajihara, M., Kanai, K., Kashimoto, K., Kimura, S., Kitagawa, M., Makino, K., Masuda, S., Miki, T., Mizobuchi, K., Mori, H., Motomura, K., Nakamura, Y., Nashimoto, H., Nishio, Y., Saito, N., Horiuchi, T. A 718-kb DNA sequence of the *Escherichia coli* K-12 genome corresponding to the 12.7–28.0 min region on the linkage map. DNA Res. **3**: 137–155 (1996).
42. Hayashi, K., Morooka, N., Yamamoto, Y., Fujita, K., Isono, K., Choi, S., Ohtsubo, E., Baba, T., Wanner, B.L., Mori, H., Horiuchi, T. Highly accurate genome sequences of *Escherichia coli* K-12 strains MG1655 and W3110. Mol. Syst. Biol. **2**: 2006 (2006).
43. Kitagawa, M., Ara, T., Arifuzzaman, M., Ioka-Nakamichi, T., Inamoto, E., Toyonaga, H., Mori, H. Complete set of ORF clones of *Escherichia coli* ASKA library (a complete set of *E. coli* K-12 ORF archive): unique resources for biological research. DNA Res. **12**: 291–299 (2005).

Index

DOI 10.1007/978-94-007-0881-5, © Springer Science+Business Media B.V. 2011